D1444387

METHODS OF BIOCHEMICAL

Suppl

Ad

METHODS OF
BIOCHEMICAL ANALYSIS

ANALYSIS OF BIOGENIC AMINES AND THEIR RELATED ENZYMES

Edited by **DAVID GLICK**

Stanford University Medical School
Stanford, California

SUPPLEMENTAL VOLUME

INTERSCIENCE PUBLISHERS a division of
John Wiley & Sons, Inc., New York • London • Sydney • Toronto

PREFACE

The interest in quantitative methods of measurement of biogenic amines has followed from the remarkable advance in understanding the role of these substances in many aspects of biology and medicine, but in some cases this interest has led the advance by making possible means of investigation. However, whether leading or following, the influence of the methodology, particularly on development of the neurosciences, has been profound. In this highly active field advances in methods have been so rapid that it is essential to provide critical evaluations from time to time and to present selected procedures of special value.

The aim of this volume was to help meet the current need. The individual authors have had considerable direct experience with the methods they treated, and they have provided practical analytical information on biogenic amines of particular interest and also on the enzymes involved in their biosynthesis and degradation. In bringing this material together into a single convenient source, I felt that the gain would be greater if the authors were allowed more freedom to deal with their subjects as they considered best, in spite of the lack of uniformity and other advantages that might accrue from a more rigidly set structure for the chapters. Although this has indeed resulted in noticeable unevenness between individual treatments, the minimization of restrictions has enabled the authors to exploit their resources more effectively. In a few instances some minor duplication of material among the chapters is evident, but it is minimal and it was considered too unimportant to justify reorganization of the writing for its elimination.

Ideas, concepts, and vision usually run far ahead of actually attainable reality at a given time, so that the limiting factor in the advance of science most often is the instrumentation and technique necessary to obtain information—the methodology on which the experimental approach depends. Thus it is hoped that this volume will contribute to the further advance of the understanding of the biological and medical role of biogenic amines.

DAVID GLICK

JUN 7- 1971

CONTENTS

Assay of Serotonin, Related Metabolites, and Enzymes

WALTER LOVENBERG AND KARL ENGELMAN, *Experimental Therapeutics Branch, National Heart and Lung Institute, National Institutes of Health, Bethesda, Maryland 20014*

I. INTRODUCTION

It is now more than a decade since the previous review (81) was published in this series. During this time several thousand scientific articles have been published which have considered serotonin and have reported measurements of either serotonin or its metabolities. Much of

1

this work utilized the methodology presented in the original review. Several new or modified procedures have been developed and knowledge of the enzymatic transformation of hydroxyindoles has increased. This chapter attempts to integrate the new methods and point out situations in which it is advantageous to use a particular procedure.

The huge literature which has appeared concerning serotonin indicates that the biological scientist has been extremely fascinated with this compound; yet, although it appears to be a neurohumoral agent, no unique or specific physiological function can be assigned. Its pharmacological effects have been studies in detail. Serotonin was first isolated from beef serum in 1948 (57) as a vasoconstrictor substance, and identified as 5-hydroxytryptamine in 1949 (56). The material has since been found widely distributed in nature. Many animal tissues contain significant amounts of serotonin (25). It is notably present in brain and innervated tissues, in blood platelets, and in intestinal mucosa. Many plants also contain large amounts of 5-hydroxyindoles (79). Bananas contain as much as 150 μg of serotonin per gram in peel and 30 μg/g in pulp. Although the role of serotonin in plant physiology is unknown, its presence in foods makes dietary restrictions essential in attempting to examine man or animals for alterations in hydroxyindole metabolism.

The synthetic route for serotonin in plants is largely unexplored. In animals however it is clear that serotonin is derived from the amino acid tryptophan as shown in Fig. 1. The enzymes catalyzing the two reactions, tryptophan hydroxylase and aromatic L-amino acid decarboxylase, have been observed in numerous animal tissues. It is evident that tryptophan hydroxylase is the rate-limiting factor in the conversion of tryptophan to serotonin. The major serotonin metabolic pathway is oxidation by monoamine oxidase and aldehyde dehydrogenase to yield 5-hydroxyindoleacetic acid (5-HIAA). In certain species however appreciable amounts of conjugation occurs and large amounts of the O-glucuronide or O-sulfate are excreted in the urine after a loading dose of serotonin. Another important pathway is the formation of melatonin in the pineal gland. Melatonin content and the enzymes catalyzing its formation undergo dramatic diurnal variations in pineal (4) and considerable attention has been devoted to its physiological role. We will attempt to define the most suitable means for measuring the enzymes and the reaction products in the 5-hydroxyindole pathway in the ensuing sections.

II. METHODS OF DETECTION AND QUANTITATION

Serotonin was first isolated on the basis of its biological activity, and subsequently assayed in several bioassay systems (18,20,52,77).

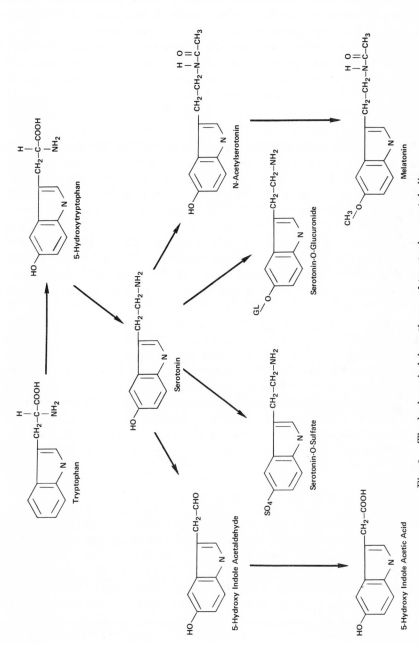

Fig. 1. The hydroxyindole pathway of tryptophan metabolism.

Although the bioassay techniques for serotonin are very sensitive, they are not applicable to many of the 5-hydroxyindole compounds and not completely specific for individual compounds. In the case of melatonin, however, the bioassay is still the only practical approach. The optical properties of 5-hydroxyindoles and specific derivatives of the 5-hydroxyindole nucleus coupled with specific isolation procedures provide the most adequate means of analyzing tissue for these coupounds.

1. Spectrophotometric Analysis

A. ULTRAVIOLET ABSORPTION

The ultraviolet absorption spectra of serotonin and 5-hydroxytryptophan and their nonhydroxyindole analogues are shown in Fig. 2. In acid solutions the 5-hydroxyindole nucleus has an absorption maximum at 275 mμ ($\epsilon = 5.3 \times 10^3 M^{-1}$ cm^{-1}) with a subsidiary peak at 298 mμ ($\epsilon = 4.3 \times 10^3 M^{-1}$ cm^{-1}). The indole nucleus has maxima at 287.5 mμ ($\epsilon = 4.55 \times 10^3 M^{-1}$ cm^{-1}), 278 mμ ($\epsilon = 5.55 \times 10^3 M^{-1}$ cm^{-1}), and a shoulder at 272 mμ ($\epsilon = 5.35 \times 10^3 M^{-1}$ cm^{-1}). Although changing the solvent to $0.1M$ NaOH has minimal effects on the indole nucleus, the ionization of the phenolic hydroxyl of the 5-hydroxyindoles has a dramatic effect on the spectra. The 5-hydroxyindoles now have two distinct maxima at 275 mμ ($\epsilon = 4.9 \times 10^3 M^{-1}$ cm^{-1}) and 322 mμ ($\epsilon = 3.9 \times 10^3 M^{-1}$ cm^{-1}). Although the ultraviolet absorption spectrum of 5-hydroxyindoles is unique the lack of sensitivity and specificity preclude the use of this technique for measuring hydroxyindoles in tissues. The large difference in absorbancy of tryptophan and 5-hydroxytryptophan (5-HTP) at 305 mμ would suggest that a spectrophotometric assay might be useful in following the enzymatic conversion of tryptophan to 5-HTP. The extremely low rate of tryptophan hydroxylation in mammalian tissues and the rather large absorbancy of the reaction mixture ingredients at this wavelength make this type of assay impractical at the present time.

B. COLOR REACTIONS OF THE 5-HYDROXYINDOLE NUCLEUS

Serotonin and related compounds undergo two major types of chemical reactions which are useful in analysis. These compounds react with many of the typical phenol reagents to yield compounds with intense chromophores. In terms of assay perhaps the most useful reagent has been 1 nitroso-2-napthol. This compound which reacts with phenolic compounds in the presence of nitric acid and sodium nitrite reacts rather specifically with 5-hydroxyindoles in the presence of sodium nitrite and dilute sulfuric acid (82) to yield a violet chromophore. Of the compounds

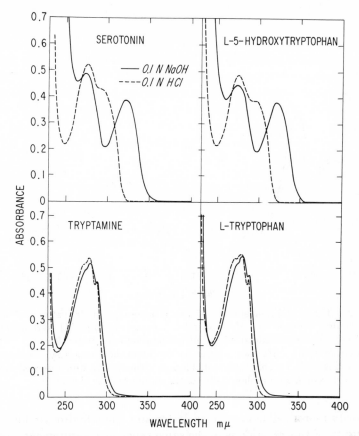

Fig. 2.   The ultraviolet absorption spectra of tryptophan and related indoles.   The spectra were observed in a Cary model 15 recording spectrophotometer using a 0.1 mM solution.

tested in this early work only *p*-nitroacetamide in addition to hydroxyindoles gave a positive response.   The same coupling procedure (82) can be used for all hydroxyindole compounds and specificity of assay is introduced by the appropriate isolation procedure before reaction. The absorption spectra of the chromophores obtained by reacting 1 nitroso-2-napthol with serotonin, 5-HTP or 5-HIAA are shown in Fig. 3. Although similar chromophores are obtained with each compound, the color yield at 540 mμ varies somewhat with each of the 5-hydroxyindole compounds.

The nature of the chromophore is not known and probably represents a mixture of several reaction products (60).   This procedure is however

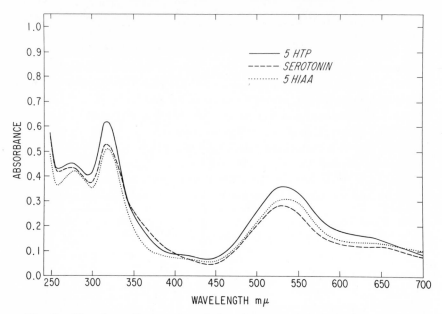

Fig. 3.   The absorption spectra of the nitroso-napthol derivatives of some 5-hydroxyindole compounds.   Hydroxyindole compounds (0.1 μM) were dissolved in 1 ml H_2O to which was added 0.5 ml nitroso-napthol reagent and 0.5 ml nitrous acid reagent (*vida infra*).   After warming at 55° for 5 min the excess reagent was removed by extracting with 5 ml of ethylene dichloride.   The spectra were recorded in a Cary model 15 recording spectrophotometer using an appropriate blank.

extremely reproducible in color yield for any one 5-hydroxyindole.   The sensitivity of this reaction is in the same range as measurement of native absorption; however, a great increase of specificity is obtained by using the 540-mμ absorption band of the chromophore.   Other colorimetric reactions appear to be much less specific and are useful mainly in qualitative detection on paper or thin layer chromatography.   Xanthydrol reacts with tryptophan and related indoles to give a colored compound (28) and can be used for 5-hydroxyindole assay.   A variety of phenol and indole specific reagents react with hydroxyindoles and their use for chromatographic "detection has been well summarized" (32).   Perhaps the most useful of these is *p*-dimethyl-amino-benzaldehyde (Ehrlichs reagent) which reacts with many indole compounds in strong acid.

2. Fluorometric Analysis

The concentrations of serotonin and related compounds in most tissues of biological interest are too low to use the colorimetric assay described above.   Workers therefore have turned almost exclusively to fluorometric techniques for measuring 5-hydroxyindoles in tissues.   The oldest and most widely used technique is that of Bogdanski et al. (8).   5-Hydroxy-indoles fluoresce in neutral solutions, the excitation and fluorescence maxima being of approximately 295 and 330 mμ, respectively.*   These wavelength maxima are similar to those of indoles not substituted on the ring (287 and 348 mμ) and similar to other substances found in biological materials.   A great deal of specificity was added to the fluorometric determination of 5-hydroxyindoles by the observation of Udenfriend et al. (78) that the fluorescent maximum is shifted to 550 mμ when the solutions are made 3N with respect to hydrochloric acid.   Figure 4 shows the excitation and fluorescent spectra of serotonin (1 mμM/ml) at pH 7.0 and

* The wavelength given for the various fluorometric analyses throughout this chapter are reported as uncorrected instrument values.

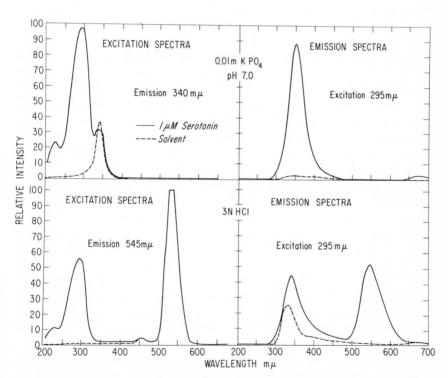

Fig. 4.   Fluorescent spectra of serotonin in neutral and strong acid solution.   The concentration of serotonin is 1 μM.   The spectra were recorded using an Aminco-Bowman recording spectrophotofluorometer.

in $3N$ HCl.   The excitation spectra are similar at either pH, but the fluorescence maximum is shifted in strong acid.   This is true for all 5-hydroxyindoles and is thought to result from the protonation of the excited hydroxyindole nucleus (13).   The following structure has been proposed (13):

The quantum yield of the strong acid fluorescence at 550 mμ is considerably less than the neutral fluorescence at 340 mμ.   The presence of interfering substances in biological materials that fluoresce at the shorter wavelengths make the strong acid fluorescence the method of choice.

In an effort to obtain greater sensitivity for measuring 5-hydroxyindoles two new techniques have been developed based on the formation of highly fluorescent derivatives.   The first of these, which involves a ninhydrin reaction, was reported by Vanable (83) and developed by Snyder et al. (71).   This analytical procedure increases the sensitivity of serotonin assay about 10-fold and introduces specificity in that other 5-hydroxyindoles do not yield appreciable fluorescence.   Figure 5 shows the excitation and emission spectra of 0.1 mμM of serotonin in a reaction mixture of 1.3 ml.   The structure of the fluorophor is unknown but the conditions for its formation have been carefully standardized (71).   A second fluorescent derivative assay has been developed by Maickel et al. (44–46), based on the reaction of 5-hydroxy- or 5-alkoxy-indoles with o-pthaldialdehyde in the presence of mineral acids.   The reaction product which is as yet uncharacterized yields intense fluorescence at 470 mμ when excited with light of 360 mμ.   Figure 6 shows the excitation and fluorescent spectra of 0.2 mμM of serotonin in 0.7 ml.   This assay has a similar sensitivity to the ninhydrin assay but is much less specific in that all 5-hydroxy or 5-alkoxy-indoles having a side chain of at least 1 carbon at the 3-position yield a fluorescent product.   Maickel and Miller (46) have described techniques for fractionation of 5-hydroxyindole derivatives so that serotonin, N-acetylserotonin, 5-methoxyserotonin, melatonin, and 5-methoxyindoleacetic acid can all be assayed from an individual tissue sample.

Use of the native fluorescence in strong acid appears to be the procedure of choice for samples containing sufficient serotonin, however, the fluorescent derivative methods are extremely valuable when the samples to be analyzed contain very small amounts of 5-hydroxyindoles.

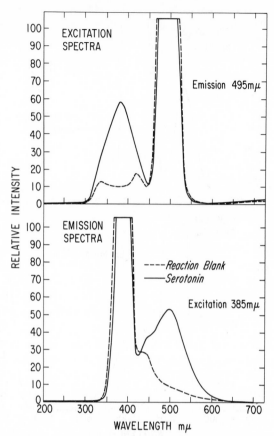

Fig. 5.   Fluorescent spectra of serotonin which has been reacted with ninhydrin. Serotonin (0.1) mμM) in 1.2 ml of 0.05M KPO buffer was mixed with 0.1 ml of 0.1M ninhydrin and heated at 75° for 1 hr.   After cooling spectra were recorded in an Aminco-Bowman recording spectrophotofluorometer.

3. Chromatography

Gas-liquid chromatography is potentially a very useful tool in the analysis of 5-hydroxyindole compounds.   Although this approach is not widely used in experimental situations, several procedures for analyzing these compounds have been reported.   Fales and Pisano (22) described conditions for the separation of many aromatic amines using SE-30 on "GAS Chrom P."   The following compounds can be completely resolved and partially quantitated using a column of 4% SE-30 at 205°: tryptamine, N,N-dimethyl-tryptamine, 5-methoxyserotonin, N-acetylsero-

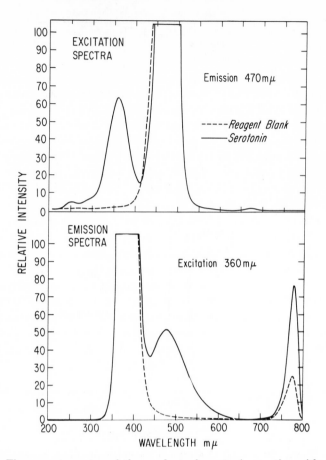

Fig. 6.   Fluorescent spectra of the product of serotonin reaction with o-pthaldi-aldehyde (OPT).   Serotonin (0.2 mμM) in 0.1 ml is mixed with 0.6 ml OPT reagent (4 mg/100 ml 10N HCl) and heated in a boiling water bath for 15 min.   After cooling the spectra were recorded as described in Fig. 5.

tonin, serotonin, and melatonin.   Gas-liquid chromatography has also been used for many phenolic acids including the serotonin metabolite 5-HIAA (30).   It has been found that optimal results are obtained if one prepares the methyl ester-trimethylsilyl ether of 5-HIAA for chromatography.   The 5-HIAA derivative appears to be well separated from other phenolic acids found in biological tissues when chromatographed on a 10% F-60 column* using a temperature program from 100 to 240°.   This

* The type of column used in these experiments was 10% F-60 (DC-560, a methyl p-chlorophenyl siloxane polymer) on silanized "Gas Chrom P."

methodology is not commonly used largely because of the specialized equipment required and difficulty in quantitative analysis, however, it is a useful approach to be considered for specialized situations.

Paper and thin layer chromatography are still used extensively for qualitative identification of the 5-hydroxyindole compounds. Two solvent systems that are widely used with paper chromatography are n-butanol:acetic acid:water and n-propanol:ammonia:water. Each of these solvents can be used with either ascending or descending techniques to give adequate separation of all common 5-hydroxyindole compounds, but require approximately 8 to 16 hours running time. Thin layer systems are also adequate and have the advantage of a much shorter running time (15–120 min) (73). Using Silica Gel G coated plates one of the most useful systems is methyl acetate-isopropanol-25% ammonia. Table I gives the R_f values of the common 5-hydroxyindole compounds in several solvent systems. The 14% NaCl system has been used successfully (29) in the assay of pineal tryptophan hydroxylase. This system is useful because the hydroxyindoles run ahead of tryptophan and the conversion of radioactive tryptophan to labeled 5-HTP can be determined by counting the Silica Gel in the 5-HTP area. In all instances, however, it is essential that the compounds be desalted and removed from protein-containing extracts before chromatography. Treatment of the tissues extracts with cold acetone (10 volumes) and ethanol (3 volumes) results in the precipitation of most of the proteins and salts. The recovery of

TABLE I
Rf Values for Indole Compounds Using
Thin Layer and Paper Chromatography

	Solvent[a]				
	1[b]	2[b]	3[c]	4[c]	5[c]
Serotonin	0.65	0.85	0.65	0.42	0.39
5-Hydroxytryptophan	0.14	0.85	0.10	0.17	0.44
5-Hydroxyindole acetic acid	0.19		0.30	0.76	0.49
Tryptophan	0.23	0.65	0.55	0.41	0.59
Tryptamine	0.77	0.65			

[a] The solvents were: 1. Methyl acetate–isopropanol–25% ammonia (45:35:20); 2. 14% NaCl; 3. n-propanol–ammonia–water (200:10:20); 4. n-butanol–acetic acid–water (12:3:5); and 5. 20% KCl.

[b] Thin layer chromatographic system. The values are from Refs. 73 and 29.

[c] Paper chromatographic system. The values are from Ref. 17.

5-hydroxyindole compounds from the supernatant fraction is variable and this procedure should only be used for qualitative identification of a compound. Samples may also be desalted electrolytically (70), or with the appropriate use of ion exchange resins which are described in detail elsewhere (70).

III. ISOLATION AND DETERMINATION OF 5-HYDROXYINDOLES

1. Serotonin

This hormone is present in significant quantities in a variety of normal mammalian tissues including brain, spleen, gastrointestinal mucosa, lung and blood (platelets) as well as in human carcinoid tumors (Table II). The most satisfactory method for extracting and assaying serotonin from tissue is still the method proposed by Bogdanski et al. (8). This procedure is only useful for tissues containing at least 0.2 μg serotonin per specimen and involves the extraction of serotonin from a salt-saturated alkaline homogenate into n-butanol and then a return from the organic solvent into an acid aqueous phase. The serotonin content of the specimen is then determined by its native fluorescence in strong mineral acid. This method has also been found useful for assay of separated platelet preparations, but whole blood and urine do not give satisfactory assays.

TABLE II
Tissue Content of Serotonin[a] (μg/g)

	Mouse	Rat	Guinea pig	Rabbit	Dog	Cat	Human
Whole blood (μg/ml)	0.2–1.5	0.2–0.8	0.1–0.2	3.5–5.0	0.4–0.5	0.68	0.1–0.3
Brain (whole)	0.6–0.9	0.3–0.6	0.3–1.2	0.3–0.6	1.0	0.2–0.3	—
Hypothalamus	—	2.75	—	0.4	1.0–1.8	1.8–2.0	0.3–0.8
Small intestine	5.2–5.4	2–5	4–5	4–20	4–4.5	0.7	3.7
Stomach	1–8	1.5–6.0	1.4	1–9	0.4–5.2	0.5	0.6–0.9
Spleen	1.6–2.7	2.5–5.0	1.1	12–24	1.4–4.6	8.5	2.37
Heart	—	0.1	—	<0.1	0.2	—	—
Lung	1.0–3.0	1–3.5	<0.2	1.6–2.1	0.2–0.3	0.6	—
Liver	0.7–0.9	<0.1	<0.1	0.5–0.9	0.1–0.5	0.6	—
Malignant carcinoid tumor							
Serotonin—producing type							500–2000[b]
5-HTP—producing type							<5[b]

[a] These values were obtained from published data as summarized in Refs. 25 and 81.
[b] Ref. 19.

A. GENERAL TECHNIQUES FOR TISSUE ASSAY

Reagents. *n-Butanol* and *heptane.* Reagent grade *n*-butanol and practical grade heptane are purified by shaking with an equal volume of $0.1N$ NaOH, then twice with equal volumes of distilled water followed by shaking with an equal volume of $0.1N$ HCl and finally twice with equal volumes of distilled water.

Borate buffer. To 3 L of water are added 94.2 g boric acid and 165 ml $10N$ NaOH. The buffer solution is then saturated with *n*-butanol and NaCl. The final pH of this solution as measured with a pH electrode should be 11.2–11.4 [rather than pH 10.0 as originally reported (81)].

Procedure. One part tissue is homogenized in a glass homogenizer with 2 parts (W/V) $0.1N$ HCl. An aliquot containing at least 0.2 μg serotonin is transferred to a glass-stoppered bottle or centrifuge tube (45–60 ml capacity) and adjusted to approximately pH 10 by addition of anhydrous sodium carbonate. Borate buffer (5 ml) and water are added to a final volume of 15 ml. Excess NaCl (5–10 g) is added for salt saturation and the aqueous phase is then shaken for 10 min in the presence of 15 ml *n*-butanol. After centrifugation* 10 ml butanol are removed and placed in a second shaking bottle or tube containing 20 ml heptane and 1.5 ml $0.1N$ HCl. After 10 min of shaking and centrifugation the organic solvents are removed and 1.0 ml of the aqueous layer is placed in a quartz cuvette with 0.3 ml conc. HCl. The fluorescence of the serotonin is measured in a spectrophotofluorometer with an excitation wavelength of 295 mμ and fluorescence at 550 mμ.

Recovery of serotonin through this assay usually ranges from 85–100%.† However, at least several sample recoveries should be determined by the addition of serotonin to an aliquot of the sample (internal standard) in an amount approximately equal to the sample content. As with all other fluorescence assays described in this chapter, a reagent (external) standard curve for fluorescence equivalence should be determined with each assay. This method does not provide a tissue fluorescence blank so the blank value is derived from the reagents themselves

---

* Under normal circumstances washing of the butanol phase is not necessary since other 5-hydroxyindoles are not present in tissue in appreciable quantities. In studies where 5-HTP is administered a significant portion may extract into the organic phase yielding a spurious high value for serotonin, therefore it is necessary to wash the butanol layer three times with equal volumes of the borate buffer to remove residual 5-HTP.

† In this chapter sample recoveries refer to the ratio of color or fluorescence yield of authentic standard material added to a duplicate tissue sample and the yield of standard material added only to reagent solution and carried through the procedure.

as carried through the procedure. Therefore tissue samples with fluorescence values approaching that of the reagent blank cannot be assayed with certainty. Attempts to produce accurate tissue blanks have been made by ultraviolet irradiation (1) or addition of H_2O_2 (15) with resultant oxidative destruction of the serotonin content of the sample. The lower limit of sensitivity for this assay is 0.2 μg. Increased sensitivity can be achieved with micro methods (35,58) or with more sensitive fluorometric techniques which now permit the assay of serotonin in smaller tissue samples (such as parts of brain) or in tissues where serotonin content is low.

Two other assays with approximately 10-fold increased sensitivity depend on the formation of highly fluorescent indole derivatives.

Ninhydrin Method. Following the observation by Jepson and Stevens (33) that serotonin developed strong fluorescent properties after exposure to ninhydrin and heat, Vanable developed a quantitative fluorescence assay for serotonin using standard solutions (83). This technique was later applied to tissue samples and specificity of the reaction was determined by Snyder et al. (71).

The tissue samples may be handled as with the Bogdanski method with the exception that the serotonin is extracted from the n-butanol-heptane organic mixture into 1.5 ml of $0.05M$ phosphate buffer, pH 7.0. Since peak fluorescence of the serotonin ninhydrin product is achieved at pH 6.5 (33) phosphate buffer rather than $0.1N$ HCl is used for convenience so that the ninhydrin reaction may be performed directly. A 1.2-ml aliquot of the aqueous phase is mixed with 0.1 ml $0.1M$ ninhydrin in distilled water, heated for 30 min at 75°C and one hour later the fluorescence of the product is measured with an excitation wavelength of 385 $m\mu$ and fluorescence at 490 $m\mu$. The fluorescence of the product increases slightly during the first 20 minutes following the incubation and then remains stable for six hours. The assay is linear over a range of serotonin content from 0.005–0.5 μg/ml. When a wide variety of related hydroxyindole and catechol compounds were reacted with ninhydrin (83), only bufotenin (N,N'-dimethylserotonin) and 5-HTP were reported to yield an appreciable amount of fluorescence (2 and 1% of serotonin, respectively). However, recently Barchas (5) found that spurious high values for serotonin were obtained using the ninhydrin method on tissues of animals treated with the tryptophan hydroxylase inhibitor, p-chlorophenylalanine (PCPA) (5). He has shown that phenylethylamine and presumably the PCPA metabolite, p-chlorophenylethylamine, yield a fluorescent product with the ninhydrin reagent which spuriously increases the apparent serotonin content in the sample. The possibility of similar effects should be excluded in studies using the ninhydrin method where structurally related drugs or metabolic products are administered.

Ortho-pthaldialdehyde (OPT) Method.   Aromatic aldehydes have been shown to form colored or fluorescent derivatives with a variety of amino acids and amines (16).   In 1966 Maickel and Miller (45) systematically studied the fluorescence properties of indole and hydroxyindole substances which were reacted with aldehyde reagents.   They later proposed a new sensitive assay for serotonin and related compounds based on the fluorophores developed after reaction with OPT (44,46).   This procedure may be employed using the final acid aqueous phase of the Bogdanski method.   A 0.1-ml aliquot of the tissue extract is mixed with 0.6 ml of freshly prepared OPT reagent (4 mg/100 ml 10N HCl) and the mixture is incubated in boiling water for 15 min.   After cooling to room temperature the fluorescence of the product is determined in a spectrophotofluorometer with excitation wavelength 360 mμ and emission wavelength 470 mμ. Fluorescence is linear with serotonin content over a range of 0.025–3.5 μg and is stable for an hour following incubation.   Since exposure to strong ultraviolet light (as in the well of the fluorometer) rapidly results in diminished fluorescence (65) the values for each sample must be determined quickly.

Using this assay a number of N-acetyl or 5-O-methyl substituted 5-hydroxyindoles yield equal or greater fluorescence than serotonin. Those substances which would potentially interfere with the specificity of the OPT assay for serotonin in biological studies using the Bogdanski extraction procedure would be melatonin (N-acetyl-5-methoxytryptamine), 5-methoxytrpytamine, and N-acetylserotonin.   Though these compounds yield 2–6 times the fluorescence of serotonin after reaction with OPT, they would not normally interfere significantly except in assays of pineal gland where concentrations of these other substances are high.   Maickel and Miller (45) have developed a complicated extraction procedure so that these compounds may be measured independently in tissues.

B. BLOOD ASSAY

Whole blood analysis performed directly following the Bogdanski extraction method does not give satisfactory results unless either the protein content is first removed or the platelets containing at least 95% of the circulating serotonin are first separated and concentrated for assay. Since platelets aggregate, lyse and release their serotonin on exposure to glass surfaces it is necessary to take special precautions in the handling of blood samples prior to assay.   Siliconized or plastic disposable syringes, tubes and pipets are used during the collection and transfer of the blood; hemostasis and rapid drawing or ejection of blood should be avoided in obtaining the sample.

a. Whole Blood Assay.   The only acceptable method of protein precipitation providing quantitative recovery of serotonin requires the use of zinc hydroxide.   The serotonin in the clear supernatant may be assayed directly or it may then be extracted and concentrated as in the Bogdanski method.

After the blood is drawn into a heparinized syringe the needle should be removed before the blood is transferred to an iced tube to avoid hemolysis in expelling the blood.   Five ml distilled water are then mixed vigorously with 1–2 ml blood to effect hemolysis; 1.0 ml 10% $ZnSO_4$ is added, mixed and then 0.5 ml $1.0N$ NaOH is added with mixing.   The precipitated proteins are sedimented by centrifugation in the cold (2500–3000 rpm for 10 min) and the clear supernatant containing the serotonin is then removed.   The serotonin in the supernatant may be assayed quantitatively by fluorescence after acidification (1.0 ml supernatant plus 0.3 ml conc. HCl) in those species with high blood serotonin concentrations (chicken, rabbit and perhaps dog).   Serotonin content in normal human whole blood is too low to measure accurately by this method, but levels in blood from patients with the serotonin-producing carcinoid tumor (68) may be approximated by this method with the understanding that the values are spuriously increased by elevated plasma levels of 5-HIAA. This problem may be obviated by an extraction procedure for serotonin or by platelet assays (see below).   Furthermore, this method is not applicable in patients receiving 5-hydroxyindoles (such as 5-HTP) or in those rare patients with carcinoid syndrome whose tumors (gastric, pancreatic, biliary system and sometimes lung) produce 5-HTP rather than serotonin. In this latter group whole blood and platelet serotonin are normal (50) and elevated plasma levels of both 5-HTP and 5-HIAA will yield spurious high values for serotonin unless an extraction technique is used.

Where blood levels of serotonin are less than 0.5 $\mu g/ml$ or where interfering substances may be present it is necessary to extract the serotonin from the supernatant solution as described above for tissues homogenates.   Quantitation may be achieved by the fluorescence techniques in strong acid or by the formation of fluorescent derivatives with ninhydrin or OPT.

Another method for estimation of whole blood serotonin which may find use is that introduced by Contractor in 1964 (15).   This method introduces three important contributions: first, the chromatographic separation of serotonin from blood proteins without the need for precipitation; second, the recognition that serotonin may be destroyed in the presence of free oxyhemoglobin, and finally, the use of an oxidized tissue blank. This procedure (15) is summarized as follows: Venous blood (5 ml) is obtained, transferred to a plastic tube containing 10 mg EDTA and 15 mg ascorbic acid and mixed by gentle inversion. Two milliliters of blood are then transferred dropwise with a plastic pipet into a centrifuge tube containing 30 ml 0.05M tris buffer, pH 8.0, and 0.1 ml of octyl alcohol. Carbon monoxide is bubbled slowly through the tube for 2 min to convert the free oxyhemoglobin from the lysed cells to carbon-monoxy hemoglobin and the tubes are then centrifuged for 15 min at 3000 rpm.

The supernatant is removed and passed through a column of carboxymethyl cellulose (CMC) (1.0 × 10.0 cm). (The carboxymethyl cellulose is prepared by mixing 100 g resin in 1.5 L of 0.1N HCl for 30 min followed by repeated washing with distilled water to remove fines and excess acid; the resin is then suspended in 0.05M tris buffer, pH 8.0, and the pH is titrated to 8.0 with 2M tris solution.) The column is washed with 20 ml buffer followed by 7 ml 0.2N HCl, and the serotonin is then eluted with 6 ml 0.2N HCl. Quantitation may be achieved with any of the methods already described. Since individual batches of CMC may vary in their preparation, the optimum size of the column and its elution characteristics must be determined for each batch. It was found that the carbon monoxide increased whole blood values by 30–40% if the gas was bubbled through the lysed cells but reduced the values if whole cells were exposed to the gas. Recovery of serotonin through the procedure ranged from 90–100%. Non-amine 5-hydroxyindoles should not interfere with this assay.

b. Platelet Assay. Since virtually all blood serotonin is bound to the platelets it is possible to increase significantly the quantity of serotonin for measurement by concentrating the platelet fraction of a larger sample of blood. This assay is especially desirable in human studies where serotonin values are normally low (0.1–0.3 μg/ml blood). Since the recovery of platelets during the isolation procedure is variable it is necessary to relate the serotonin content either to platelet protein or to the number of platelets present.

The blood sample (10–20 ml) is collected, mixed with 0.1 volume of 0.7% saline containing 1% EDTA as anticoagulant, and centrifuged at 30 × g for 30 min at 4°C. If results are to be determined in terms of platelet protein, the supernatant platelet-rich plasma is removed with care not to disturb the buffy coat layer containing sedimented white cells and some trapped platelets, but if serotonin values are to be related to the number of platelets then the platelet rich plasma as well as the buffy coat may be removed to give a greater platelet yield.

If platelets are to be isolated, the platelet rich plasma is centrifuged again (300–500 × g, 30 min, 4°C) and the supernatant platelet free plasma is removed from the platelet button. The platelets are resuspended by gently disrupting the button with 10–20 ml cold isotonic (0.9%) saline and the centrifugation is repeated. Finally, the platelets are resuspended in 1–3 ml H_2O and are frozen and thawed twice to release the serotonin. After a small aliquot of the mixture is taken for protein analysis (75) aliquots of the remainder are assayed for serotonin after protein precipitation with zinc sulfate and sodium hydroxide. Usually the clear supernatant solution contains sufficient serotonin so that it may be acidified with conc. HCl and the serotonin fluorescence determined directly. Since the 5-hydroxyindole content of platelets is thought to be solely serotonin (82) and the removal of the plasma eliminates the possible presence of other 5-hydroxyindoles this method is quite specific and serotonin content is expressed in terms of μg serotonin/mg platelet protein.

If serotonin is to be expressed in terms of platelet content then the platelet rich plasma including the buffy coat may be used. An aliquot of the suspension is taken for determination of the platelet count either by phase contrast light microscopy (10) or by automated electronic counting devices (12). Aliquots of the remainder containing 0.5–1.0 μg serotonin are then centrifuged in plastic tubes (1000 g, 5 min, 4°C) and the supernatant is discarded. The platelet button is suspended in 1.0 ml distilled water, frozen and thawed twice, the proteins are precipitated by the $Zn(OH)_2$ method and serotonin fluorescence may then be determined in the clear supernatant solution.

There are several limitations on the use of isolated platelet preparations for the determination of blood levels of serotonin. The first is related to the administration of reserpine which depletes serotonin from the platelet storage granules. As a result platelet bound serotonin is markedly reduced and "free serotonin" may appear in the plasma (68). In such a circumstance the "whole blood" assay would be necessary. A potential shortcoming of the determination of platelet serotonin expressed in terms of platelet protein is raised by the findings of Stacey (72) and Born and Gillson (9) that serotonin "leaks out" of human platelets when they are washed in isotonic saline. This loss of serotonin on exposure to saline does not appear to occur in platelets from the dog (84).

C. URINE

In normal adults the daily excretion of serotonin in urine is less than 0.1 mg. Although in patients with serotonin producing carcinoids the values are elevated to 1–2 mg/24 hr, it still constitutes less than 1% of the total urinary 5-hydroxyindoles (68), but in the 5-HTP producing cases the urinary values may be as high as 50 mg/24 hr representing 20–25% of the total (50).   The most practical and efficient method for separating serotonin from urine is with the weak cation exchange resin Amberlite IRC-50 (49).

An aliquot of urine is filtered, diluted to 10 ml with water (if necessary) and adjusted to pH 7.5 by the dropwise addition of $0.2N$ NaOH.   (For a normal human 24-hr urine sample this would require an aliquot of 2–5% of the total volume with proportionately smaller aliquots required when serotonin excretion is increased.)   The sample is then passed through a 0.5×5.0 cm column of IRC-50 (NH_4^+ form), the column is washed with 10 ml $0.02N$ ammonium acetate buffer, pH 7.5, and the serotonin is then eluted with 5.5 ml $1.0N$ HCl.   The eluate is adjusted to 6.0 ml with $1.0N$ HCl and a 1.0 ml aliquot is acidified with 0.3 ml conc. HCl to determine the serotonin fluorescence.   Recovery of serotonin through the procedure is $95 \pm 5\%$ and yields are linear in the range of 1.0–20 μg per sample.

2. 5-Hydroxyindoleacetic Acid (5-HIAA)

The major end metabolite of serotonin metabolism in man and other animal species is 5-HIAA.   This compound may normally be found in the urine, but its presence in tissues or plasma is generally measurable only when levels are increased as in patients with malignant carcinoid syndrome or following the administration of the 5-HIAA precursors 5-HTP or serotonin.   Very small amounts of 5-HIAA may also be detected in cerebrospinal fluid (CSF).

A. URINE ASSAYS

a. Qualitative (Screening) Test for Malignant Carcinoid.   This test is useful in the screening of patients suspected of having the carcinoid syndrome (69).   Since the biochemical hallmark of this disease is the increased urinary excertion of 5-hydroxyindole compounds (mainly 5-HIAA) it has been found that a direct nitrosonaphthol reaction on a small urine aliquot is useful in diagnosing the majority of cases (69).

A 0.2 ml urine aliquot is mixed with 0.8 ml water and 0.5 ml of nitrosonaphthol reagent (0.1% nitrosonaphthol in ethanol); 0.5 ml nitrous acid reagent (freshly made by the addition of 5.0 ml $2N$ H_2SO_4 to 0.2 ml 2.5% Na NO_2) is added and the mixture is allowed to stand at room temperature for 10 min.   Five milliliters of ethylene dichloride are added, the tube is shaken vigorously and the two phases are allowed to separate.   A positive test is indicated by a purple color in the aqueous (top) layer. Normal urines may produce a pale yellow or green color.

There are several precautions which must be exercised in the interpretation of this test clinically. First, urines may be expected to yield positive tests only when the concentration of 5-HIAA or total 5-hydroxyindoles exceeds 30 mg/L. The normal range of urinary 5-HIAA in man is 2–9 mg/24 hours. Some patients with malignant carcinoid syndrome only excrete 15–30 mg 5-HIAA per day and thus may yield negative screening tests. On the other hand, *false positive* tests may be found in patients eating large quantities of serotonin containing foods (79) [banana, tomato, avacado, walnut, plum, pineapple and pineapple juice (fresh and canned), papaya, and eggplant]. In addition guaiacol derivatives contained in a variety of medications (especially cough medicines containing glyceryl guaiacolate) also yield colored products which may be misinterpreted as positive tests.

b. Total 5-Hydroxyindoles in Urine. The qualitative assay should not be used as a quantitative procedure for 5-HIAA, but it is sometimes desirable to measure the total 5-hydroxyindole content of urine in following the course of a patient with carcinoid syndrome or as an index of the absorption of a large dose of a 5-hydroxyindole compound (19). This may be achieved by taking a small aliquot (usually 0.1–0.2 ml) of the urine sample containing 5–50 µg of 5-hydroxyindole and adjusting the volume of all samples to 1.0 ml with water. The procedure is then followed exactly as for the urine screening test, except that the absorbance of the aqueous layer is determined at 540 mµ. Because of variation in the yield of the reaction introduced by differing urine specimens, all samples should be determined in duplicate as well as with duplicate internal recovery samples to which a known amount of the standard solution has been added (usually 10–20 µg).

c. Quantitative Assay for 5-HIAA. This procedure (80) involves the initial extraction of 5-HIAA from an acidified aliquot of urine into diethyl ether, the return to an aqueous phase and then reaction with nitrosonaphthol reagent to form a colored derivative whose absorption may be determined on a spectrophotometer. In rabbits and humans with metabolic disorders (phenylketonuria) where large quantities of keto acids are excreted, these compounds must first be removed by precipitation with dinitrophenylhydrazine since they interfere with the final color determination. Preliminary extraction of all urines must be performed with $CHCl_3$ to remove indoleacetic acid which will also yield a colored product with nitrous acid.

Six ml of urine are mixed with 6.0 ml of 2,4-dinitrophenylhydrazine reagent (0.5% dinitrophenylhydrazine in 2.0N HCl) in a glass-stoppered bottle or shaking tube (45–60 ml capacity) and allowed to stand at room temperature for 30 min.   The precipitate is removed by centrifugation and then 25 ml CHCl₃ is added and the mixture is shaken for 15 min, centrifuged, the organic layer is removed, and the extraction with CHCl₃ is repeated.   In normal human urine where large quantities of keto acids are not expected to be present the precipitation step may be omitted and 6.0 ml 2N HCl are substituted for the dinitrophenylhydrazine reagent before a single CHCl₃ extraction.   After removal of the CHCl₃ layer 10 ml of the aqueous phase is transferred to another glass-stoppered vessel containing sufficient NaCl (5–10 g) for salt saturation; 25 ml of freshly prepared diethyl ether are added and the tube is shaken for 5 min and centrifuged.   (The ether must be shaken just before using with a dilute solution of FeSO₄ in distilled water followed by a distilled water wash to remove peroxide impurities which will oxidize and destroy the 5-HIAA in the sample.)   A 20-ml aliquot of the ether layer is then transferred to another shaking tube, 1.5 ml of 0.5M phosphate buffer, pH 7.0, are added and the mixture is shaken for 5 min and centrifuged. The ether layer is removed and 1.0 ml of the aqueous phase is transferred to a tube containing 0.5 ml nitrosonaphthol reagent; 0.5 ml nitrous acid reagent is added and after a 5-min incubation at 37°C the solution is shaken with 5 ml ethylene-dichloride.   The optical density of the aqueous phase is then determined at 540 mμ.

Standards are carried through the assay by adding 10 μg and 20 μg of 5-HIAA to 6.0 ml of water, in the initial step; a reagent blank is made by substituting 6 ml H₂O for a urine sample.   The colored product should yield about 0.125 O.D. units/20 μg 5-HIAA carried through the assay. Recovery of 5-HIAA ranges from 50–90%; due to the fact that it is variable with different urines an internal recovery should be run on a duplicate of each sample for highly accurate results.   The assay is suitable for urine samples containing 5–75 μg 5-HIAA and the aliquot of urine used must be adjusted accordingly, especially in patients with carcinoid syndrome.   The same limitations must be considered as with the qualitative procedure (vide supra).   Certain drugs such as phenothiazines or their metabolities may interfere with the development of the nitrosonaphthol color reaction, but if internal recoveries are run with each sample it is possible to correct for this interference.

B. 5-HIAA IN TISSUES AND PLASMA

Normally the 5-HIAA content of tissues or body fluids other than urine and brain is very low or not measurable by conventional methods.   Levels are increased in patients with malignant carcinoid syndrome or following administration of 5-HTP or serotonin. The 5-HIAA may be extracted from tissue or plasma after protein precipitation and then assayed as in urine samples.

Two milliliters of plasma or a portion of tissue homogenate in $0.1N$ HCl equivalent to at least 1.0 g of tissue is diluted to 6.0 ml with H_2O; 1 ml of 10% $ZnSO_4$ and 0.5 ml of $1.0N$ NaOH are added with vigorous mixing and the resulting precipitate is sedimented by centrifugation. Four milliliters of the supernatant are transferred to a glass-stoppered centrifuge tube containing 0.3 ml $6N$ HCl and sufficient NaCl to effect salt saturation; 20 ml of freshly washed ether (vide supra) are added and the tube is shaken for 5 min. After brief centrifugation 15 ml of the ether layer are transferred to a tube containing 1.5 ml $0.5M$ phosphate buffer, pH 7.0, shaken for 2 min and then the ether layer is removed. A 1.0 ml-aliquot of the aqueous phase is then acidified with 0.3 ml conc. HCl and the 5-hydroxyindole fluorescence is determined spectrophotofluorometrically (excitation 295 mμ, emission 550 mμ). Standards and reagent blanks must be carried through the procedure and meaningful values for 5-HIAA may only be determined if the sample fluorescence is at least 2 or 3 times the blank reading. Most specimens from normal tissues and plasma will yield nonmeasurable quantities.

C. 5-HIAA IN CEREBROSPINAL FLUID

Serotonin synthesis rates in brain have been studied by studying the rate of accumulation of its primary metabolite, 5-HIAA, in the brains of animals where its major means of elimination, the active transport of 5-HIAA from brain to plasma, is blocked with probenecid (48,76). Since 5-HIAA is thought to leave brain only slowly into the CSF by a simple diffusion process (48) it has been proposed that the rate of serotonin synthesis in brain could be determined indirectly by the rate of rise of 5-HIAA in the CSF after probenecid treatment (51). Since 5-HIAA is thought to be the only 5-hydroxyindole present in CSF (2) assays in this body fluid have been based upon the fluorescence of the acidified supernatant after protein precipitation. Three milliliters CSF are added to 0.5 ml 10% $ZnSO_4$ and 0.25 ml $1N$ NaOH. Any precipitated proteins are removed by centrifugation and 1.0 ml of the supernatant is acidified with 0.3 ml conc. HCl before determining its fluorescence. It should be emphasized that this procedure measures total 5-hydroxyindoles since no separatory techniques have been employed, and furthermore the fluorescence of the samples is so low under normal conditions as to provide a reasonable doubt of the quantitative accuracy of the method. Investigators have claimed that the CSF content of 5-HIAA is markedly increased (0.3 ± 0.05 μg/ml) in patients with hydrocephalus and low in patients with depressive psychosis (0.01 μg/ml) (2) and Parkinson's disease (76) and that the rate of serotonin synthesis is reduced in this latter condition. Improvement on current methods presumably using one of the more sensitive fluorescent techniques should result in more meaningful data.

3. 5-Hydroxytryptophan

5-Hydroxytryptophan, the immediate metabolic precursor of serotonin, is not normally present in measurable quantities in animal or human tissues. It is possible, however, to detect 5-HTP in tissues, urine and plasma following administration of large quantities or in the malignant carcinoid syndrome either following the administration of an aromatic amino acid decarboxylase inhibitor (67) or in those rare patients whose tumors produce 5-HTP rather than serotonin (50). Two methods have been developed to assay 5-HTP. One is not specific since it depends on the prior extraction of serotonin and 5-HIAA with the assumptions that extractions are quantitative and that the only 5-hydroxyindole compound which remains is 5-HTP. The second method is quite specific though more laborious and depends on the conversion of 5-HTP to serotonin by a decarboxylase enzyme preparation.

Extraction Procedure (81). One millilitor of a tissue homogenate or plasma is diluted with 2 ml 0.1N HCl and the proteins are precipitated by the addition of 1 ml 40% trichloroacetic acid. (It is not necessary to precipitate proteins in urine samples.) The supernatant solution is extracted twice with 10 ml ether to remove 5-HIAA and the excess trichloroacetic acid. Enough 20% Na_2CO_3 is added to achieve a pH of 10.0 and the samples are adjusted to an equal volume; 15 ml n-butanol are added and the mixture is shaken to remove the serotonin. To 1.0 ml of the aqueous phase is added 0.6 ml conc. HCl and the fluorescence is determined (excitation 295 mμ, emission 550 mμ). Both reagent blanks and standards must be carried through the entire assay. It should be noted that all the fluorescence in the sample is not due to 5-HTP since tissues contain amounts of substances which yield a small fluorescent emission equivalent of up to 5 μg 5-HTP/g of tissue. Furthermore, since neither of the extraction steps yields 100% removal of serotonin or 5-HIAA, in the presence of large amounts of these substances there may be a significant contribution of residual compound to the "apparent" 5-HTP fluorescence.

Enzymatic Procedure. This procedure was developed for the assay of 5-HTP in the urine of patients with the malignant carcinoid syndrome (67). The serotonin in urine is first removed by adsorption on a cation-exchange column and the effluent containing both 5-HTP and 5-HIAA is collected, incubated with an aromatic-L-amino acid decarboxylase preparation and the serotonin formed by this reaction is then isolated on a cation-exchange column.

An aliquot of urine, plasma or the supernatant of a tissue homogenate calculated to contain 5–50 μg of 5-HTP should be adjusted to pH 8.4 in a final volume of up to 10 ml and applied to a column of Amberlite IRC-50 (NH_4^+ form, 8 $\times$ 40 mm, buffered with 0.2M Na acetate buffer, pH 8.4). The column should be washed with 2 ml 0.02M Na acetate buffer, pH 8.4, and the combined effluent collected and adjusted to 12 ml for all samples. To a 5-ml aliquot of the sample is then added 0.8 ml of 0.5M tris buffer pH 8.5, 0.25 ml pyridoxal phosphate ($2 \times 10^{-3}M$), 0.2 ml guinea pig kidney decarboxylase preparation (see below Section IV for details of preparation) and 0.5 ml pargyline ($10^{-3}M$) to make a final volume of 6.75 ml. The mixture is incubated at 37°C for 30 min and then a 5.0-ml aliquot is diluted to 10 ml with water and passed through an Amberlite IRC-50 column (NH_4^+ form, 8 $\times$ 50 mm buffered at pH 7.5 with 0.5M NH_4 acetate buffer). The column is washed with 10 ml 0.02M NH_4 acetate buffer, pH 7.5, and then the serotonin is eluted with 6.0 ml 1N HCl. One milliliter of the eluate is added to a cuvet containing 0.3 ml conc. HCl and the serotonin fluorescence is measured. Reagent blanks and standards (D,L-5-HTP or L-5-HTP) should be carried through the procedure. Only L-5-HTP is measured by the assay and recoveries from urine average 95%.

4. Bufotenin (N,N'-Dimethylserotonin)

As its name implies bufotenin (N,N'-dimethylserotonin) was first isolated from the venom of a toad (31,85). Its major importance today stems from its presence in "cohoba" (74), a hallucinogenic snuff of plant origin (21), and from the controversy whether bufotenin is (11,23) or is not (62,66) found in the urine of schizophrenic patients. Nonetheless, studies to date have generally been based on the presence or absence of a spot corresponding to authentic bufotenin on the chromatogram of large urine extracts. No specific biochemical assay for bufotenin has been developed since the extraction and fluorometric properties of bufotenin and serotonin are similar. Assays based on the differential extraction into butanol (81) or ether (63) yield relative separations of these indole amines, but bidirectional chromatographic systems are necessary for absolute separation. Because of variable recoveries in different steps it would be desirable to incorporate a sample recovery based on the addition of a known amount of radioactive labelled bufotenin to the original sample. Quantitation at the final step may be achieved by any of the methods used for serotonin.

5. Melatonin (N-Acetyl-5-methoxytryptamine)

Though very small quantities of melatonin (N-acetyl-5-methoxy-tryptamine) have been found in mammalian peripheral nerves (6,36) the content of this indole amine as well as the presence of the biosynthetic enzyme hydroxyindole-O-methyl transferase is essentially confined to the pineal gland in amphibians, birds and mammals (86). The importance of this trace indole amine derives from its biologic effects. Though initially thought to be primarily the frog skin lightening substance of pineal origin (37) melatonin has been shown more recently to be intimately related to pineal function and the effects of light and dark on gonadal function in a variety of mammalian species (86).

It is possible to assay melatonin biochemically when it is synthesized in large quantities (3). However, its concentration, even in pineal gland, is so small that unless thousands of glands were pooled it would be impossible to measure except by bioassay. These bioassays depend on the ability of melatonin to blanch the color of the darkened isolated skin of Rana pipiens (38) or the intact Xenopus laevis larvae (55). The content of melatonin in pineal gland ranges from 0.2–0.7 μg/g and the average pineal gland in a rat weighs 1 mg.

IV. ENZYMES RELATED TO SEROTONIN METABOLISM

1. Tryptophan Hydroxylase

Tryptophan hydroxylase is the first and rate-limiting enzyme in the conversion of tryptophan to serotonin. This enzyme is present only in very low concentrations in mammalian tissues. An exception is the malignant mouse mast cell tumor which is a rich source and can be assayed using the nitrosonaphthol reaction to measure the formation of hydroxy-indoles (64). In general however it is necessary to use an isotopic assay. Theoretically the release of tritium from 5-tritiotryptophan would be the ideal assay and Renson et al. have shown this to be practical in certain situations (59). Because of the instability of highly labelled tryptophan and variable blanks (42) this assay has not proved feasible for measuring low levels of activity in brainstem and pineal tissue.

Carbon-labelled tryptophan has been used as substrate by several laboratories (24,26,27,40,47) and has the advantage of giving a relatively low blank but presents a distinct problem in separation of the product from the substrate. We have used this latter approach and solved the separation problem by converting the 5-hydroxytryptophan formed to serotonin which is easily separated on a small ion exchange column. The assay is made quantitative by including a relatively large amount of unlabelled 5-hydroxytryptophan in the incubation, and determining both radioactivity and serotonin content in the final eluate.

A. ENZYME PREPARATION

Tryptophan hydroxylase can be assayed in extracts of either rat brainstem or the pineal gland of several species. The tissues are homogenized in $0.05M$ tris·HCl pH 7.4 (3 volumes) and the homogenate is centrifuged at $30,000 \times g$ for 20 min. The supernatant fraction is then dialyzed overnight against 100 volumes or more of a solution containing $0.1M$ 2-mercaptoethanol and $0.05M$ tris·HCl, pH 7.4. These extracts can be used directly for tryptophan hydroxylase studies or can be further fractionated with ammonium sulfate (34).

B. REAGENTS

Methylene labelled tryptophan can be obtained from either Amersham-Searle Inc. or New England Nuclear Corp. The radioactive amino acid (100 μc) is purified by dissolving in 1 ml of water and passage through a very small 1×0.5 cm column of Permutit which has been washed extensively with water. The Permutit used for this and in the assay is obtained from Fisher Scientific Co (P-64). The reduced cofactor, 2-amino-4-hydroxy-6,7-dimethyl-5,6,7,8-tetrahydropteridine ($DMPH_4$) is obtained from either Aldrich Chemical Co. or Calbiochem Inc. N-Methyl, N-benzyl-propynylamine (Pargyline) is a gift of Abbott Laboratories Inc. Unlabelled L-tryptophan, L-5-hydroxytryptophan and serotonin are obtained from Calbiochem Inc., 2-mercaptoethanol from Eastman Organic Chemicals Inc. and omnifluor from the New England Nuclear Corp. Other materials were obtained from the usual commercial sources. Aromatic L-amino acid decarboxylase was prepared through the alumina C γ eluate stage by the previously published procedure (14). The following standard solutions are prepared and can be stored for a period of weeks: $3M$ tris·acetate pH 7.5, $1M$ tris·HCl pH 9.0, $2 \times 10^{-3}M$ pyridoxal phosphate, $0.02M$ pargyline, $2.5M$ 2-mercaptoethanol, $0.02M$ L-5-hydroxytryptophan, $0.015M$ L-tryptophan, and $9 \times 10^{-4}M$ ^{14}C-L-tryptophan 52 μc/μM. Just prior to each experiment solutions of ferrous ammonium sulfate, 1.06 mg/ml and $DMPH_4$, 7 mg/ml are prepared in $0.01M$ 2-mercaptoethanol.

C. ENZYME ASSAY PROCEDURE

The incubation mixture for a typical experiment consist of 200 μl of the enzyme solution, 20 μl buffer, 5 μl 2-mercaptoethanol, 5 μl pargyline, 10 μl L-5-hydroxytryptophan, 10 μl iron solution. It is generally convenient to make a cocktail of the above ingredients to be added to the enzyme solution. These ingredients are placed in a 20 ml screw cap Kimax tube and preincubated for 10 min. The reaction is started by the addition of 30 μl ^{14}C-L-tryptophan, 10 μl cold L-tryptophan and 10 μl DMPH$_4$ solution usually prepared as a cocktail. The incubation is normally allowed to proceed for 60 minutes at 37° during which time the reaction has been found to be linear (41). The reaction is terminated by the addition of 0.5 ml of 1M tris·HCl pH 9.0 and dilution to 2 ml with water. Next 50 μl of the L-5-HTP solution, 100 μl of the pyridoxal phosphate solution and sufficient decarboxylase to decarboxylate about 50% of the L-5-HTP during a 20-min incubation at 37° are added. At the end of the second incubation the mixture is diluted to 20 ml and applied to a column of Permutit (0.5 × 3 cm) which is held in a glass column obtained from Scientific Glass Inc. (Cat. No. JT-7390). The exchange resin is washed thoroughly with water before use and is washed with 20 ml of boiling distilled water after application of the sample. The latter wash removes all the tryptophan and 5-HTP from the column. The column is next eluted with 4.2 ml of 2N NH$_4$OH. This serotonin containing eluate is divided into two parts, 2.0 ml are placed in a counting vial and 2.0 ml are placed in a chilled tube containing 0.3 ml glacial acetic acid. To the former fraction is added 10 ml of omnifluor solution and the radioactivity determined in a Packard Tricarb scintillation counter. A 0.4 ml aliquot of the latter fraction is diluted to 2 ml with H$_2$O, 0.6 ml of concentrated HCl is added and the amount of serotonin in the eluate determined by the standard fluorescent technique. From this data the specific radioactivity of the serotonin becomes known and is the same as the specific activity of the 5-HTP. Since the total amount of 5-HTP in the system is known, the total amount of radioactivity entering this pool can be calculated, and from this the amount of tryptophan hydroxylated. The following equation can be used to calculate the results:

μM tryptophan hydroxylated

$$= \frac{\text{dpm}/\mu\text{M serotonin} \times \mu\text{M L-5-HTP carrier}}{\text{dpm}/\mu\text{M L-tryptophan}}$$

The level of enzyme activity in several different tissues are shown in Table III.

TABLE III
Tryptophan Hydroxylase Activity of Various Mammalian Tissues[a]

Tissue	Specific activity	Tissue	Specific activity
Mouse mast cell	16.5	Dog hypothalmus	0.050
Beef pineal	0.49	Rat liver	0.008
Rat pineal	0.53	Rat kidney	0.017
Carcinoid tumor	0.23	Human platelets	0.050
Rat brainstem	0.060	Rat intestinal mucosa	<0.002
Rabbit brainstem	0.058	Guinea pig intestinal mucosa	0.015
Dog brainstem	0.025	Rat spleen	<0.002
Dog caudate nucleus	0.015	Rat heart	<0.002

[a] This table is taken from Ref. 39. Specific activity is expressed as $m\mu M/mg$ protein per hour.

2b. Aromatic L-Amino Acid Decarboxylase

This enzyme catalyzes the second step in serotonin synthesis and is widely distributed in the animal body particularly in tissues containing serotonin. It is also unusually high in kidney. The enzyme shows broad specificity decarboxylating many aromatic amino acids and has often been termed either 5-HTP decarboxylase or DOPA decarboxylase, although most evidence indicates these activities are associated with a single enzyme species. Numerous approaches to the assay of this enzyme have been used. In tissues having appreciable activity it is feasible to measure CO_2 evolution manometrically using either 5-HTP or DOPA as a substrate (14). Other systems have been devised to measure CO_2 production by using 1-^{14}C-labeled substrate and trapping the $^{14}CO_2$ for counting (7). We have found that it is convenient to measure the activity of the decarboxylating enzyme using 5-HTP as substrate with subsequent isolation of the serotonin formed on a small Permutit column. The serotonin is measured by its native fluorescence affording considerable sensitivity. The small ion exchange columns are easily regenerated for rapid reuse.

A. ENZYME PREPARATION

To study the enzyme in crude tissue extracts a 35,000 × g supernatant fraction of a 25% homogenate in water is prepared. It is extremely important, when using crude extracts to include a preincubation in the presence of a monoamine oxidase inhibitor. Pargyline is routinely used in our laboratory, since it is essential to completely inhibit monoamine oxidase in order to get quantitative recoveries of serotonin. A procedure for obtaining the partially purified enzyme of guinea pig kidney has recently been detailed (39). It is possible to get a 70-fold purified enzyme by ammonium sulfate fractionation adsorption and elution from alumina C γ gel, and chromatography on a DEAE cellulose column. This type of preparation is particularly useful when studying the nature of potential enzyme inhibitors.

The previously described procedure (39) is as follows:

"a. Ammonium Sulfate Fractionation. Forty guinea pig kidneys weighing about 1 g., each are homogenized in 120 ml of water with a ground glass conical homogenizer. The homogenate is then centrifuged in a Sorval refrigerated centrifuge at 35,000 × g for 30 min. Four ml of 1N tris·HCl, pH 7.3, are added to the supernatant fraction which is then adjusted to 37% saturation with solid ammonium sulfate. The precipitate formed is removed by centrifugation and discarded. The supernatant fraction is then adjusted to 55% saturation with ammonium sulfate. The resulting precipitate which contains the enzyme is separated by centrifugation and redissolved in 40 ml of 0.05M tris·HCl, pH 7.3. The enzyme solution is desalted by chromatography on Sephadex G-25.

"b. Adsorption on Alumina C γ Gel and Elution. The protein content of the above enzyme solution is estimated by absorbancy at 280 mμ and the solution is adjusted to pH 5.8 with 0.02M acetic acid. One mg of alumina C γ gel as a suspension is added per milligram protein. After stirring for 15 min the gel is removed by centrifugation, washed once with water and eluted by 3 successive washes with 10 ml of 0.1M KPO$_4$ buffer, pH 6.3. The three elution fractions which contain most of the enzyme activity are combined.

"c. DEAE Cellulose Chromatography. The enzyme is precipitated from the alumina C γ gel eluate by addition of ammonium sulfate (60% saturation). The precipitate is redissolved in 5 ml 0.001M phosphate buffer, pH 7.0, and dialyzed overnight against the same buffer. This enzyme is applied to a 1 × 18 cm column of DEAE cellulose which has been prepared by the method of Peterson and Sober (53) and equilibrated with 0.001M phosphate buffer, pH 7.0. A linear elution gradient is established between 0.001 and 0.05M phosphate buffer, pH 7.0, using 300 ml of buffer in each of two chambers of a Varigrad mixer. Five milliliter fractions are collected. The majority of the activity is eluted in fractions 37 to 53, at a phosphate concentration of about 0.008M. By combining the tubes with the peak activity the enzyme is obtained in an approximately 22% yield with a purification of 66-fold when compared to the supernatant of the homogenate."

B. ENZYME ASSAY

a. **Reagents.** L-5-Hydroxytryptophan and serotonin are obtained from Calbiochem Inc. Pargyline is obtained from Abbott Laboratories Inc. Permutit (P-64) is purchased from Fisher Scientific Company and pyridoxal phosphate from Sigma Chemical Corporation. All other reagents are the best commercial grades. The following solutions are prepared prior to performing the assay: tris·HCl pH 9.0, $0.5M$; pargyline $10^{-3}M$; DL-5-hydroxytryptophan $9 \times 10^{-3}M$; pyridoxal phosphate, $2 \times 10^{-3}M$; and serotonin 0.1 mg/ml.

b. **Procedure.** Incubations are carried out in air in 30-ml beakers at 37°. The incubation mixture contains enzyme, 0.3 ml pargyline, 0.1 ml pyridoxal phosphate, 0.5 ml tris buffer and water to make a total volume of 2.5 ml. After a 5-min preincubation 0.5 ml of the 5-HTP solution is added to start the reaction. Aliquots (0.5 ml) are removed at the start of the reaction and at subsequent times (usually 10 and 20 min) and placed in test tubes which are contained in a boiling water bath. After heating for 1 min the tubes are removed and 5 ml of water are added to each tube. At the conclusion of the incubation the samples are centrifuged if necessary and applied to small Permutit columns 0.5 × 3.0 cm. The columns are next washed with 20 ml of boiling water and the serotonin is eluted with 3 ml of 20% NaCl. To quantify the reaction, incubation beakers are usually prepared containing no enzyme, but 0, 20, and 40 μg of serotonin. A single aliquot (0.5 ml) is removed from each of the standard beakers and is diluted with 5 ml of water, and are subsequently treated in a manner similar to enzyme samples. The native fluorescence of the serotonin containing eluates is determined in a spectrophotofluorometer. The uncorrected excitation and fluorescent wavelengths are 290 and 350 mμ, respectively. The Permutit columns can be used repeatedly by simply washing with 20 ml of boiling water after each experiment.

The unit of enzyme activity has been defined (39) as the amount of enzyme required to form 1 μM of serotonin per hour under optimal conditions. The enzyme is specific for the L-isomer of 5-HTP, and although it decarboxylates a variety of L-aromatic amino acids, 5-HTP exhibits the lowest Km of those examined (43).

3. Monoamine Oxidase

This enzyme initiates the major catabolic route for serotonin. Monoamine oxidase appears to be primarily in the mitochondria and catalyzes the oxidative deamination of many physiologically active amines. Details are presented elsewhere in this volume; however a brief description of the means to measure serotonin deamination used in our laboratory follows (61): 25% tissue homogenate in $0.05M$ KPO_4 buffer, pH 7.5, is used. One-tenth of a milliliter or less of the enzyme solution is incubated in 0.5 ml of $0.05M$ KPO_4 buffer, pH 7.5, containing $5 \times 10^{-4}M$ 2-^{14}C-serotonin (100,000 to 200,000 dpm/μM, obtained from New England Nuclear Corp.) in test tubes in air at 37° for 10 to 30 min. At the conclusion of the incubation the mixture is quickly poured through a small column of IRC-50* (0.5 × 3.0 cm) contained in a Pasteur pipet. The column is washed with 1 ml of water and the complete effluent collected in a counting vial to which is added 10 ml of omnifluor scintillation solution. In this assay the unmetabolized serotonin adheres to the ion exchange column whereas the 5-hydroxyindolacetaldehyde and 5-Hydroxyindolacetic acid do not and from the amount of radioactivity in the counting vial the amount of serotonin metabolized can be calculated. Appropriate 0 time tubes and blanks must be subtracted. The assay is rapid and the Pasteur pipet containing the unmetabolized serotonin and the IRC-50 can be discarded.

4. Hydroxyindole O-Methyl Transferase

This enzyme has been studied by Axelrod and his coworkers (4,87). The methylation of N-acetyl-serotonin by the transfer of a methyl group from S-adenosyl methionine appears to be the rate-limiting step in the synthesis of melatonin in the pineal gland. The activities of this enzyme undergo a diurnal variation which is reflected in a diurnal variation on the melatonin content of the pineal gland. A procedure for the assay of this enzyme in rat pineal has been developed by Axelrod et al. (4).

A. ENZYME PREPARATION

The upper portion of the skull is removed from the heads of rats immediately after decapitation. The pineal gland, a small round white body, usually adheres to the top of the skull and can be removed with a pair of forceps. The pineal gland, usually about 1 mg in weight, is placed in a small conical glass homogenizer with 0.5 ml of ice cold $0.05M$ phosphate buffer, pH 7.9, and homogenized by hand. The homogenate is used as the enzyme source.

* The IRC-50 (Amberlite CG-50, 100–200 mesh) is obtained from Mallinckrodt Chemical Works and prepared for use by the procedure of Pisano (54).

B. REAGENTS

N-Acetylserotonin is obtained from Regis Chemical Co. and methyl ^{14}C-S-adenosylmethionine (40 μc/mole) from New England Nuclear Corp.

C. ENZYME ASSAY PROCEDURE

Two hundred microliters of enzyme solution are incubated with 50 μg of N-acetyl-serotonin and 50 mμM of labeled S-adenosylmethionine in a total volume of 300 μl in a 15-ml stoppered tube. The incubation is conducted for 1 hr at 37° in air at which time 1.0 ml of 0.2M borate buffer, pH 10, and 8 ml of chloroform are added. The tube is shaken for 5 min and the aqueous phase is removed and replaced with 1 ml of fresh borate buffer. After centrifugation, 5 ml of the chloroform phase is transferred to a counting vial and evaporated to dryness under a stream of nitrogen. Ten milliliters of a scintillator solution are added and the radioactivity determined. The result is corrected for the small amount of radioactivity found in the blank (no enzyme).

References

1. N. E. Andén and T. Magnusson, *Acta Physiol. Scand.*, *69*, 87 (1967).
2. G. W. Ashcroft, and D. F. Sharman, *Nature*, *186*, 1050 (1960).
3. J. Axelrod and H. Weissbach, *J. Biol. Chem.*, *236*, 211 (1961).
4. J. Axelrod, R. J. Wurtman, and S. H. Snyder, *J. Biol. Chem.*, *240*, 949 (1965).
5. J. D. Barchas, (Stanford University School of Medicine), Personal communication.
6. J. D. Barchas and A. B. Lerner, *J. Neurochem.*, *11*, 489 (1964).
7. J. D. Barchas, and S. Udenfriend, Unpublished observations.
8. D. F. Bogdanski, A. Pletscher, B. B. Brodie, and S. Udenfriend, *J. Pharm. Exp. Ther.*, *117*, 82 (1956).
9. G. V. R. Born and R. E. Gillson, *J. Physiol.*, *146*, 472 (1959).
10. G. Brecher and E. P. Cronkite, *J. Appl. Physiol.*, *3*, 365 (1950).
11. G. G. Brune, H. H. Hohl, and H. E. Himwich, *J. Neuropsychiat.*, *5*, 14 (1963).
12. B. S. Bull, M. A. Schneiderman, and G. Brecher, *Amer. J. Clin. Pathol.*, *44*, 678 (1965).
13. R. F. Chen, *Proc. Nat. Acad. Sci., U. S.*, *60*, 598 (1968).
14. C. T. Clark, H. Weissbach, and S. Udenfriend, *J. Biol. Chem.*, *210*, 139 (1954).
15. S. F. Contractor, *Biochem. Pharm.*, *13*, 1351 (1964).
16. G. Curzon and J. Giltrow, *Nature*, *173*, 314 (1954).
17. C. E. Dalgliesh, *Biochem. J.*, *64*, 481 (1956).
18. C. E. Dalgliesh, C. C. Toh, and T. S. Work, *J. Physiol.*, *120*, 298 (1953).
19. K. Engelman, Unpublished data.
20. V. Erspamer, *Arch. Int. Pharmacodyn. Ther.*, *93*, 293 (1953).
21. D. Fabing and J. R. Hawkins, *Science*, *123*, 886 (1956).
22. H. M. Fales and J. J. Pisano, *Anal. Biochem.*, *3*, 337 (1962).
23. E. Fischer, T. A. Fernandez Lagravere, A. J. Vazqnez, and A. O. DiStefano, *J. Nerv. Ment. Dis.*, *133*, 441 (1961).

24. E. M. Gal, J. C. Armstrong, and B. Ginsberg, *J. Neurochem.*, *13*, 643 (1966).
25. S. Garattini and L. Valzelli, *Serotonin*, Elsevier, Amsterdam, 1965, p. 241.
26. D. G. Grahame-Smith, *Biochim. Biophys. Acta*, *86*, 176 (1964).
27. H. Green and J. L. Sawyer, *Anal. Biochem.*, *15*, 53 (1966).
28. J. P. Greenstein and M. Winitz, *The Chemistry of the Amino Acids*, Wiley, New York, 1961, p. 2324.
29. R. Håkanson and G. Hoffman, *Biochem. Pharm.*, *16*, 1677 (1967).
30. M. G. Horning, *Theory and Application of Gas Chromatography*, H. S. Kroman and S. R. Bender, Eds., Grune and Stratton, New York, 1968, p. 135.
31. H. Jensen and K. K. Chen, *J. Biol. Chem.*, *116*, 87 (1936).
32. J. B. Jepson, *Chromatographic and Electrophoretic Techniques*, Vol. I, I. Smith, Ed., Wiley-Interscience, New York, 1960, p. 183.
33. J. B. Jepson and B. J. Stevens, *Nature*, *172*, 772 (1953).
34. E. Jequier, D. S. Robinson, W. Lovenberg, and A. Sjoerdsma, *Biochem. Pharm.*, *18*, 1071 (1969).
35. R. Kuntzman, P. A. Shore, D. Bogdanski, and B. B. Brodie, *J. Neurochem.*, *6*, 226 (1961).
36. A. B. Lerner, J. D. Case, W. Mori, and M. R. Wright, *Nature*, *183*, 1821 (1959).
37. A. B. Lerner, J. D. Case, Y. Takahashi, T. H. Lee, and W. Mori, *J. Amer. Chem. Soc.*, *80*, 2587 (1958).
38. A. B. Lerner and M. R. Wright, *Methods of Biochemical Analysis*, Vol. 8, D. Glick, Ed., Wiley-Interscience, New York, 1960, p. 295.
39. W. Lovenberg, *Methods of Enzymology*, Vol. 17, S. P. Colowich and N. O. Kaplan, Eds.; H. Tabor and C. W. Tabor, Volume Eds., Academic Press, New York. In press.
40. W. Lovenberg, E. Jequier, and A. Sjoerdsma, *Science*, *155*, 217 (1967).
41. W. Lovenberg, E. Jequier, and A. Sjoerdsma, *Advan. Pharm.*, *6A*, 21 (1968).
42. W. Lovenberg and D. S. Robinson, Unpublished data.
43. W. Lovenberg, H. Weissbach, and S. Udenfriend, *J. Biol. Chem.*, *237*, 89 (1962).
44. R. P. Maickel, R. H. Cox, Jr., J. Saillant, and F. P. Miller, *Int. J. Neuropharm.*, *7*, 275 (1968).
45. R. P. Maickel and F. P. Miller, *Anal. Chem.*, *38*, 1937 (1966).
46. R. P. Maickel and F. P. Miller, *Advan. Pharm.*, *6A*, 71 (1968).
47. S. Nakamura, A. Ichiyama, and O. Hayaishi, *Fed. Proc.*, *24*, 604 (1965).
48. N. H. Neff, T. N. Tozer, and B. B. Brodie, *J. Pharm. Exp. Ther.*, *158*, 214 (1967).
49. J. A. Oates, *Methods in Medical Research*, Vol. 9, J. H. Quastel, Ed., Year Book Medical Publishers, Chicago, 1961, p. 169.
50. J. A. Oates and A. Sjoerdsma, *Amer. J. Med.*, *32*, 333 (1962).
51. R. Olsson and B. E. Roos, *Nature*, *219*, 502 (1968).
52. I. H. Page and A. A. Green, *Methods in Medical Research*, Vol. 1, V. R. Potter, Ed., Year Book Publishers, Chicago, 1948, p. 123.
53. E. H. Peterson and H. A. Sober, *Methods of Enzymology*, Vol. 5, S. P. Colowich and N. O. Kaplan, Eds., Academic Press, New York, 1962, p. 3.
54. J. J. Pisano, *Clin. Chim. Acta*, *5*, 406 (1960).
55. W. B. Quay and J. T. Bagnara, *Arch. Int. Pharmacodyn. Ther.*, *150*, 137 (1964).
56. M. M. Rapport, *J. Biol. Chem.*, *180*, 961 (1949).
57. M. M. Rapport, A. A. Green, and I. H. Page, *J. Biol. Chem.*, *176*, 1243 (1948).
58. D. v. Redlich and D. Glick, *Anal. Biochem.*, *29*, 167 (1969).
59. J. Renson, J. Daly, H. Weissbach, B. Witkop, and S. Udenfriend, *Biochem. Biophys. Res. Commun.*, *25*, 504 (1966).

60. J. Renson and S. Udenfriend, Unpublished observations.
61. D. S. Robinson, W. Lovenberg, H. Keiser, and A. Sjoerdsma, *Biochem. Pharm.*, *17*, 109 (1968).
62. T. M. Runge, F. Y. Lara, N. Thurman, J. W. Keyes, and S. H. Hoerster, Jr., *J. Nerv. Ment. Dis.*, *142*, 470 (1966).
63. E. Sanders and M. T. Bush, *J. Pharm. Exp. Ther.*, *158*, 340 (1967).
64. T. L. Sato, E. Jequier, W. Lovenberg, and A. Sjoerdsma, *Eur. J. Pharm.*, *1*, 18 (1967).
65. P. J. Schechter and K. Engelman, Unpublished observations.
66. M. Siegel, *J. Psychiatr. Res.*, *3*, 205 (1965).
67. A. Sjoerdsma, J. A. Oates, P. Zaltzman, and S. Udenfriend, *New Eng. J. Med.*, *263*, 585 (1960).
68. A. Sjoerdsma, H. Weissbach, L. L. Terry, and S. Udenfriend, *Amer. J. Med.*, *23*, 5 (1957).
69. A. Sjoerdsma, H. Weissbach, and S. Udenfriend, *J. Amer. Med. Ass.*, *159*, 397 (1955).
70. I. Smith, *Chromatographic and Electrophoretic Techniques*, Vol. I, I. Smith, Ed., Wiley-Interscience, New York, 1960, p. 40.
71. S. H. Snyder, J. Axelrod, and M. Zweig, *Biochem. Pharm.*, *14*, 831 (1965).
72. R. S. Stacey, *Symposium on 5-Hydroxytryptamine*, G. P. Lewis, Ed., Pergamon, London, 1958, p. 125.
73. E. Stahl, *Thin Layer Chromatography*, E. Stahl, Ed., Springer, Berlin, 1962, p. 292.
74. V. L. Stromberg, *J. Amer. Chem. Soc.*, *76*, 1707 (1954).
75. E. W. Sutherland, C. F. Cori, R. Haynes, and N. S. Olsen, *J. Biol. Chem.*, *180*, 825 (1949).
76. T. N. Tozer, N. H. Neff, and B. B. Brodie, *J. Pharm. Exp. Ther.*, *153*, 177 (1966).
77. B. M. Twarog and I. H. Page, *Amer. J. Physiol.*, *175*, 157 (1953).
78. S. Udenfriend, D. F. Bogdanski, and H. Weissbach, *Science*, *122*, 972 (1955).
79. S. Udenfriend, W. Lovenberg, and A. Sjoerdsma, *Arch. Biochem. Biophys.*, *85*, 487 (1959).
80. S. Udenfriend, E. Titus, and H. Weissbach, *J. Biol. Chem.*, *216*, 499 (1955).
81. S. Udenfriend, H. Weissbach, and B. B. Brodie, *Methods of Biochemical Analysis*, Vol. 6, D. Glick, Ed., Wiley-Interscience, New York, 1958, p. 117.
82. S. Udenfriend, H. Weissbach, and C. T. Clark, *J. Biol. Chem.*, *215*, 337 (1955).
83. J. W. Vanable, Jr., *Anal. Biochem.*, *6*, 393 (1963).
84. H. Weissbach, D. F. Bogdanski, and S. Udenfriend, *Arch. Biochem. Biophys.*, *73*, 492 (1958).
85. H. Wieland, W. Konz, and H. Mittasch, *Ann. Chem.*, *513*, 1 (1934).
86. R. J. Wurtman, J. Axelrod, and D. E. Kelly, *The Pineal*, Academic Press, New York, 1968.
87. R. J. Wurtman, J. Axelrod, and L. S. Phillips, *Science*, *142*, 1071 (1963).

Determination of Amine Oxidases

R. KAPELLER-ADLER, *Department of Pharmacology, University of Edinburgh Medical School, Edinburgh, Scotland*

I. INTRODUCTION

The group of amine oxidases comprises enzymes which oxidatively deaminate amines, mono-, di-, and polyamines with the stoichiometric formation of one molecule each of aldehyde, ammonia, and hydrogen peroxide (1–3).

$$RCH_2NH_2 + O_2 + H_2O \rightarrow RCHO + NH_3 + H_2O_2$$

All the components of this fundamental amine oxidase-substrate reaction, with the exception of water, have been utilized for the estimation of the activity of amine oxidases.

Most of the natural substrates of amine oxidases belong to the group of biogenic amines which are formed *in vivo* in various metabolic processes taking place in microorganisms as well as in higher forms of life. Many of those biogenic amines appear to be involved in important *in vivo* regulatory enzymic mechanisms.

With regard to the classification of amine oxidases, two main groups of these enzymes, based on differences in inhibitor specificities, have been independently proposed by Blaschko and his colleagues (4) and Zeller and his associates (5,6).

Group I comprises amine oxidases resistant to carbonyl reagents (4) and to semicarbazide (5,6), and Group II contains amine oxidases inhibited by carbonyl reagents (4) and sensitive to semicarbazide (5,6). Group I includes the classical monoamine oxidase, an intracellular, mainly mitochondrial, insoluble enzyme present in many vertebrate and invertebrate tissues (1,7), and mouse liver histaminase (8). Amine oxidases of this group act on primary, secondary, and tertiary amines, and on long-chain aliphatic amines.

Group II contains histaminase, diamine oxidase (2,3,9,10), preponderantly an intracellular, probably mitochondrial, soluble enzyme (11) and plasma enzymes, which are extracellular enzymes and include the ruminant plasma amine oxidase, spermine oxidase (12–15), and the non-ruminant plasma amine oxidase, benzylamine oxidase (15–18). Furthermore, Group II comprises rabbit liver amine oxidase, mescaline oxidase (16,19–22), the pea seedlings amine oxidase (23), and the bacterial amine oxidase, the polyamine oxidase (24).

Amine oxidases belonging to Group II oxidize primary amines but do not attack secondary amines.

Only methods for the determination of amine oxidases in animal tissues will be described in this presentation. In a recent monograph (25) the historical background of amine oxidases, their relationship to the biogenic amines, especially in brain tissue, and their biological significance is discussed, and a description of estimation methods of amine oxidases in pea seedlings and bacteria, besides that in animal tissues, is given.

II. METHODS OF ISOLATION, SOLUBILIZATION, PURIFICATION, AND ESTIMATION OF MONOAMINE OXIDASE (MAO)

Cotzias and Dole (26) and Hawkins (27) independently demonstrated the intimate adherence of MAO in most organ tissues to the insoluble structures of mitochondrial membranes. This behavior of MAO may explain why attempts at the purification of this enzyme have not been very successful until recently. Partially purified MAO has been, however, recently prepared by various workers by treatment of the mitochondrial membranes with detergents, or with ultrasonic waves, or by disruption of the mitochondria by homogenization and subsequent treatment with Triton X-100.

1. Isolation of Tissue Mitochondria

A. METHOD OF OSWALD AND STRITTMATTER (28)

Tissues from adult male Wistar rats, freshly excised, are homogenized in an ice-cold mixture of $0.25M$ sucrose and $0.001M$ EDTA, at pH 7.0, using a Potter Elvehjem homogenizer. On centrifugation of a 20% homogenate at $100,000g$ for 1 hr and subsequent washing the sediment once by its resuspension in the original volume of the sucrose-EDTA mixture, and recentrifugation, a "total particulate preparation" is obtained which essentially contains all the MAO activity of the homogenate, tyramine being used as substrate. The sediment is resuspended in the sucrose-EDTA mixture and stored at $-10°C$. In the frozen state MAO activity is stable for several weeks.

B. METHOD OF SEIDEN AND WESTLEY (29)

Male albino rats (Holtzman, Sprague-Dawley line), 60 to 75 days old, are decapitated. The brain, removed from each animal, is transferred to a tissue grinder (Tenbroek) containing 9 vol of $0.25M$ sucrose, and homogenized using a motor at 1750 rev/min. The homogenate is centrifuged at $1500g/6$ min and the sediment of cell debris and nuclei is discarded. The supernatant which contains suspended mitochondria, microsomes, and the "white fluffy layer" (Burkard et al., 30) is placed on top of 15 ml of $0.88M$ sucrose and centrifuged for 20 min at $18,000g$. The "white fluffy layer" is thereby retained at the interface of the $0.25M$ and $0.88M$ sucrose solutions, while the mitochondria sediment to the bottom of the tube. The supernatant and the "white fluffy layer" are both removed by aspiration. The mitochondria are lysed by their resuspension and gentle agitation in a glass homogenizer with a volume of $0.05M$ phosphate buffer (pH 7.4), equal to the threefold weight of the intact brain. The suspended, lysed mitochondria may be frozen and stored at $-30°C$ for at least 2 months without significant loss of MAO activity.

C. METHOD OF GIORDANO, BLOOM, AND MERRILL (31)

On surgical removal, rabbit or dog kidney is sliced, weighed, and then homogenized with 4 vol of $0.25M$ sucrose. The pH is then adjusted to 7.0 with potassium hydroxide, and the brei is placed in a 50-ml vitrex tube, and is further homogenized for 1 min with a motor-driven pestle. After centrifugation of the brei at $700g$ for 10 min, the sediment is discarded. The supernatant is recentrifuged at $700g$ for 10 min and the sediment is again discarded. The resulting supernatant is then centrifuged at $11,000g$ for 20 min and, after removal of the supernatant fluid, the sediment is washed with 10–20 ml of $0.25M$ sucrose and centrifuged at $11,000g$ for an additional 20 min. All the above operations are carried out in a cold room. The final precipitate of mitochondria is weighed and suspended in $0.067M$ phosphate buffer (pH 7.4) whereby 1.0 ml of buffer is used for 50 mg of wet sediment. The suspension obtained is quick-frozen and lyophilized. The resulting powder is stored at $-10°C$. The lyophilized mitochondria thus obtained may serve as a stable source of monoamine oxidase, its MAO activity according to Giordano and his colleagues (31) remaining essentially unchanged from the initial enzymic values for 13 months. In the hands of the authors this method has proved to be a rapid, reliable, and a simple one to perform.

D. METHOD ACCORDING TO SCHNEIDER AND HOGEBOOM (32); HAWKINS (27); AND KOBAYASHI AND SCHAYER (33)

To remove liver glycogen, adult male (Sprague-Dawley) rats are fasted overnight. The animals are then sacrificed by decapitation, the livers excised, chilled over cracked ice, and freed in a mesher from connective tissue. The liver pulp is weighed, placed in a Potter Elvehjem homogenizer, and homogenized with a Teflon pestle in 9 vol of ice-cold isotonic $(0.25M)$ or hypertonic $(0.88M)$ sucrose in distilled water at the low speed of $600g$ for about 5 min. The first supernatant is centrifuged at $22,000g$ for 45 min in the refrigerated centrifuge, whereby the soluble, enzymically inactive protein is separated from the sediment of formed elements, mitochondria and microsomes, containing the total MAO activity of the first supernatant. This sediment is resuspended in 3–4 vol/g of original liver weight of isotonic or hypertonic sucrose solution and is recentrifuged at $22,000g$ for 45 min. All operations are performed at 4°C. The mitochondrial MAO preparation can be stored frozen for 14 days without a significant loss of activity.

2. Solubilization and Purification of Mitochondrial MAO

A. METHOD OF GUHA AND KRISHNA MURTI (34)

Rat liver mitochondria are prepared in $0.25M$ (isotonic) sucrose solution according to the method of Schneider and Hogeboom (32; see Section II-1-D) and washed once with $0.01M$ phosphate buffer (pH 7.6). Resuspended in $0.01M$ phosphate buffer, pH 7.6, the mitochondria are exposed for 25 min in a Mullard Magnetostrictor Generator to ultrasonic waves at 25 kHz (output amperage, 2.5). Transfer the suspension to a stainless steel flask, fitted to the transducer element. During the exposure to sound waves keep the content of the steel flask chilled and, thereafter, centrifuge the suspension in a Servall refrigerated centrifuge at maximum speed at 2°C. The resulting supernatant contains over 90% of the original mitochondrial MAO activity.

For further purification, apply aliquots of the supernatant, containing 40 mg of protein, directly to a DEAE-cellulose column (30 × 2.5 cm), which had been previously equilibrated with $0.01M$ phosphate buffer (pH 7.6) including sodium chloride in an $0.01M$ concentration. The MAO is eluted by stepwise addition of four fractions of 100 ml each of

$0.01M$ phosphate buffer (pH 7.6) containing 0.01, 0.05, 0.1, and $0.5M$ sodium chloride, respectively. Carry out the stepwise elution at 8°C, collecting 10 ml fractions. The active eluates are water clear and pale yellow and the entire MAO activity is recovered in the early fractions at a low sodium chloride concentration. Pool and lyophilize the active fractions, and rechromatograph them on DEAE-cellulose. Fractions emerging on rechromatography at a $0.01M$ NaCl concentration usually contain monoamine oxidase of a specific activity indicating a 350-fold over-all purification.

B. METHOD OF NARA, GOMES, AND YASUNOBU (35)

Beef liver mitochondria were prepared according to the method of Schneider and Hogeboom (32; see Section II-1-D) and were subjected to a procedure which involved a disruption of the mitochondria by homogenization and Triton X-100 treatment, followed by ammonium sulfate fractionation, adsorption on alumina gel Cγ, and chromatography of the eluate on DEAE-cellulose. A 58-fold purification of MAO was achieved by the authors.

About 120 g of purified beef liver mitochondria were suspended in 500 ml of $0.1M$ phosphate buffer (pH 7.4) and were homogenized in batches, for 2 min each, in a Potter Elvehjem homogenizer (25×200 mm) using a stirring motor at 3000 rpm. The homogenate was adjusted by the addition of $0.1M$ phosphate buffer (pH 7.4) to a protein concentration of 30 mg/ml (*Fraction I*).

To achieve a better extraction of MAO, 73 ml of 20% Triton X-100, adjusted to pH 7.4, were added to a homogenate-aliquot of 900 ml, whereby a final Triton X-100 concentration of 1.5% was attained. The mixture was stirred gently for 3 hr. The homogenate was then centrifuged by means of the No. 21 rotor in the Spinco Model L ultracentrifuge at 20,000 rpm for 45 min. A second aliquot of 900 ml of the homogenate was treated in the same fashion and the supernatant of both treated homogenates were combined to yield about 1530 ml of *Fraction II*.

To the clear reddish-brown solution 162 g of solid ammonium sulfate were added to achieve a 20% ammonium sulfate concentration and the mixture was centrifuged. A further addition of 219 g of solid ammonium sulfate brought the solution to a 45% saturation with ammonium sulfate. The solution was then centrifuged in the ultracentrifuge at 20,000 rpm for 25 min. The insoluble enzyme protein rose to the top and was collected by removing the solution with a hypodermic syringe. The enzyme (7.43 g) was dissolved in 450 ml of $0.1M$ phosphate buffer (pH 7.4) yielding *Fraction III*. At this stage the enzyme proved to be rather stable in the frozen state and could be stored frozen for several days.

The concentration of protein of the Enzyme Fraction III was adjusted by the addition of $0.1M$ phosphate buffer (pH 7.4) to that of 10 mg/ml. To this solution were added 20.8 ml of 20% sodium cholate, corresponding to 0.56 mg of cholate/mg of protein, about 137 g of solid ammonium sulfate to achieve a 25% saturation. The mixture was then centrifuged at 8000 rpm for 15 min in the Servall centrifuge and an additional 115 g of ammonium sulfate were added per liter of solution to 45% saturation. The enzyme floated to the top, and the fluid was again removed with a hypodermic syringe. The enzyme precipitate was dissolved in 240 ml of $0.1M$ phosphate buffer (pH 7.4) (*Fraction IV*).

The enzyme solution of the preceding step was dialyzed with one buffer change against 2 liters of $0.01M$ phosphate buffer (pH 7.4), each time for 2 hr. The dialyzed solution was then adjusted by the addition of 52.8 ml of $0.01M$ phosphate buffer, pH 7.4, to a protein concentration of 20 mg/ml. To this solution 162 ml of alumina gel Cγ (18 mg/ml, dry weight), prepared according to the method of Willstätter and Kraut (36) were added so that the ratio of protein to alumina became 2:1. The mixture was then stirred gently for 5 min, after which it was centrifuged at 8000 rpm for 15 min and the precipitate discarded. A further addition of 486 ml of alumina gel Cγ made the protein to alumina ratio 1:2. The mixture was allowed to stand overnight. After centrifugation, the gel obtained was washed with approximately 200 ml of $0.01M$ phosphate buffer (pH 7.4) and eluted with 200 ml of $0.1M$ phosphate buffer (pH 7.4) and this elution process was repeated two or three times depending on the yield of the enzyme. The brown eluates were combined (470 ml) and concentrated by precipitation with 120 g of solid ammonium sulfate to 45% saturation. The floating reddish-brown precipitate obtained after centrifugation was dissolved in 53 ml of $0.1M$ phosphate buffer (pH 7.4) (*Fraction V*).

To remove the ammonium sulfate, this enzyme solution was passed through a Sephadex G-25 column (30 × 3.5 cm) and was, for further purification, thereafter applied to a DEAE-cellulose column (45 × 2.0 cm) which had been equilibrated with $0.01M$ phosphate buffer (pH 7.4). Elution was started with $0.01M$ phosphate buffer (pH 7.4), but at Tube 20 the buffer concentration was increased to $0.1M$ (pH 7.4), and from Tube 40 onwards gradient elution was used by mixing 1000 ml of a 1.0% Triton X-100/$0.1M$ potassium phosphate buffer (pH 7.4) into 1000 ml of $0.1M$ potassium phosphate buffer (pH 7.4). At Tube 175, the buffer concentration was increased to $0.3M$ potassium phosphate buffer (pH 7.4) containing 1.0% Triton X-100. Fractions of 14 ml were collected. The active enzyme was eluted at a final Triton X-100 concentration of 0.2–0.3%. The slightly brownish solution was concentrated by precipitation with solid ammonium sulfate to 45% saturation, and after centrifugation the solid phase was dissolved in 15 ml of $0.1M$ potassium phosphate buffer (pH 7.4) to yield *Fraction VI*.

The purified enzyme could be stored in the frozen state. There was an initial loss of 20% in enzyme activity, but after a period of 2 weeks the activity remained constant. Different results were obtained by the authors with different batches of mitochondria. In about 50% of their experiments the authors obtained enzyme preparations in which the MAO appeared to be solubilized, shown by the fact that the enzyme sedimented to the bottom after the addition of ammonium sulfate and, furthermore, that the enzyme could be dialyzed against buffer which did not contain any detergent. Such enzyme preparations proved to be stable for indefinite periods when stored in the freezer. Tested for purity in the ultracentrifuge, enzyme preparations with a specific activity of 4760 showed a single peak at pH 7.4. Determinations of the sedimentation coefficient of an 0.5% solution of the beef liver enzyme in $0.1M$ phosphate buffer (pH 7.4) gave a value of $10.5S$ (uncorrected). It may be taken that until further detailed criteria of the purity of mitochondrial MAO are presented, this method yields a highly purified enzyme preparation.

C. METHOD OF YOUDIM AND SOURKES (37)

From livers of male (Sprague-Dawley) rats mitochondria are prepared by the method of Hawkins (27; see Section II-1-D) and are suspended in $0.0125M$ phosphate buffer (pH 7.4) containing $0.003M$ benzylamine. After sonication for 100 min at 20 Hz, cholic acid to a final concentration of 1% is added. Stand the mixture for 30 min and then centrifuge at $160,000g$ for 90 min. The supernatant contains up to 85% of MAO activity; three quarters of this activity can be precipitated with ammonium sulfate at a saturation between 30 and 55%. This precipitate dissolves readily in water. For further concentration, this partially purified enzyme is adsorbed on a Sephadex G-200 column and is eluted with $0.05M$ phosphate buffer (pH 7.4). Pool fractions containing active MAO and fractionate them with ammonium sulfate as before. The enzyme is then chromatographed on a column of DEAE-Sephadex, A-50, equilibrated with $0.05M$ phosphate buffer (pH 7.4) and is desorbed from the column with an increasing gradient of concentration of sodium chloride. Subsequently, the active eluate is chromatographed on a column of hydroxylapatite, and eluted from the column by phosphate buffer (pH 6.8) in concentrations increasing from 0.005 to $0.5M$. Four protein peaks are thereby obtained but only the last one contains MAO activity. Fractions of this peak are pooled and are used for further investigations. Up to a 208-fold purification as compared with the crude homogenate can be achieved by this technique with a recovery of about 25% of the original activity. In the pure form the enzyme is yellow in color and can be stored in $0.05M$ phosphate buffer (pH 7.4) at 4°C for at least 2 weeks with little loss of activity. The enzyme is stable in the pH range between 5.5 and 9.5; outside this range, however, it is readily inactivated. The active protein, when precipitated, dissolves easily in water without use of dispersing agents.

This method has been very recently modified by Youdim, Collins, and Sandler (38) and adapted to the preparation of soluble MAO from human placental mitochondria. Placental mitochondria are first subjected to sonic oscillation and the enzyme is then purified by treatment with Triton X-100, ammonium sulfate fractionation, and column chromatography on Sephadex G-200 and on DEAE-Sephadex. The resulting enzymic preparation shows a specific activity of 3500 and a 400-fold purification over the original homogenate. On polyacrylamide gel electrophoresis the human placental MAO displays two active bands, whereas solubilized and partly purified MAO from rat liver (37), similarly treated, showed three bands of enzyme activity.

D. EXTRACTION AND PURIFICATION OF HOG BRAIN MITOCHONDRIAL
MONOAMINE OXIDASE ACCORDING TO THE METHOD OF TIPTON (39)

Reports in the literature on the purification of brain monoamine oxidase are rather scanty. Nagatsu (40) was the first to demonstrate that MAO can be extracted from beef brain mitochondria by sonication in the presence of a detergent. Tipton (39) has recently designed a simple, if rather tedious, technique of repeated sonication, freezing, and thawing by which, in the absence of a detergent, pig brain mitochondrial MAO can be extracted and concentrated about 1000 times.

From freshly killed hogs the brains are removed, freed from fat, and suspended in 9 vol (w/v) of $0.25M$ sucrose. After adjustment with $1.0M$ K_2HPO_4 to pH 7.6, the suspension is homogenized in a Waring Blender at medium speed for 1 min and the pH readjusted with $1.0M$ K_2HPO_4 to 7.6. Mitochondria are prepared from this homogenate by a method similar to that described by Brody and Bain (41). Thus, the homogenate is centrifuged at $18,000g$ for 20 min, the residue is resuspended in half the original volume of $0.25M$ sucrose, adjusted to pH 7.6 with $1.0M$ K_2HPO_4, recentrifuged at $18,000g$ for 20 min, and resuspended in the sucrose phosphate medium.

To liberate the MAO from the mitochondria, the above suspension is centrifuged at $18,000g$ for 20 min. The residue is suspended in about one-tenth of the original volume of the homogenate in ice-cold water. The suspension is homogenized by hand in a Potter-Elvehjem homogenizer with an all-nylon pestle and centrifuged at 18,000g for 45 min. The residue is taken up in a similar volume of 0.01M phosphate buffer (pH 7.6) and stored frozen for 3 days. The preparation is allowed to thaw out and is diluted with 0.01M phosphate buffer (pH 7.6) to give a final protein concentration of 10–15 mg/ml.

The suspension is then subjected to sonication for 45 min in a Dawe Soniprobe, fitted with a $\frac{1}{2}$-in. probe at an output amperage of between 6 and 7. During this treatment the suspension is cooled in ice. The sonicate is then centrifuged at 108,000g for 2 hr, and the supernatant, which should be a cream-colored opalescent solution, is carefully decanted. Should the color of the supernatant show a distinct brown, the solution must be recentrifuged. The residue is resuspended in 0.01M phosphate buffer (pH 7.6), homogenized, diluted to give a protein concentration of 10–15 mg/ml, and sonicated as before. The sonicated suspension is stored frozen, thawed out at room temperature, and centrifuged as before. This procedure of sonication, freezing, and thawing is carried out 6 times, whereby 60 to 80% of MAO activity is liberated into the supernatants. (Tipton has obtained, however, an excellent purification of MAO using only two sonication steps.) The active supernatants are then pooled.

The combined active supernatants from the sonication steps are then exposed to low pH. After cooling in ice, add 1.0N HCl with continuous stirring until the pH falls to 3.0. Then add immediately 1.0N NaOH to return the solution to pH 7.0, whereby a white precipitate forms. After standing at 4°C for 20 min, centrifuge the suspension at 35,000g for 20 min. Discard the residue, adjust the supernatant to pH 7.6, and subject it to a treatment with DEAE-cellulose (Whatman DE 52). For this purpose, equilibrate DEAE-cellulose with 0.01M phosphate buffer (pH 7.6) and remove the excess liquid by means of a sintered glass funnel and gentle suction with a water pump. Add the DEAE-cellulose cake, thus obtained, to the supernatant using 0.5 g DEAE-cellulose/1 mg of protein. Stir the suspension and let it stand at 4°C for 20 min. Remove the DEAE-cellulose by centrifugation at 35,000g/30 min.

The clear supernatant, which should contain between 0.5 and 1.5 mg protein/ml, is finally subjected to an alcohol fractionation. After cooling in ice, add an equal volume of absolute ethanol to the supernatant slowly, within 60 min, and with continuous stirring, and stand the mixture in ice for another 2 hr. The fine precipitate which forms is discarded after centrifugation at 35,000g for 45 min at −5°C. Then add half the original volume of absolute ethanol slowly to the cooled supernatant and stand the mixture at $-10°C$ overnight. The fine precipitate which forms contains the active enzyme. It is separated by centrifugation at 35,000g at $-5°C$ for 45 min and is then taken up in a minimum volume of ice-cold 0.01M phosphate buffer (pH 7.6). Insoluble material is removed by centrifugation and the solution is stored frozen. A further yield of MAO can be obtained by adding the original volume of absolute ethanol to the alcoholic supernatant, allowing the mixture to stand overnight at $-10°C$ and centrifuging it as before. The residue is again taken up in 0.01M phosphate buffer (pH 7.6) and the insoluble material is removed as before. The two active solutions are pooled and stored in the frozen state for 2 days. The precipitate which forms is discarded after centrifugation at 35,000g for 45 min. The supernatant contains MAO which appears to be purified about 1000-fold over the original mitochondrial activity.

3. Methods for the Quantitation of Monoamine Oxidase Activity

A. INTRODUCTION

In 1964 Werle (42) reviewed the methods then available for the quantitative estimation of MAO activity. They include diverse techniques, i.e., manometric, diffusion, and spectrophotometric, based mainly on the consumption of oxygen, production of ammonia, concomitant analyses of oxygen uptake, and ammonia evolution as well as determination of substrate disappearance in the presence of MAO.

Until recently the most commonly applied methods for the estimation of MAO activity depended upon the measurement of oxygen uptake by the standard Warburg technique (43) or the determination of substrate disappearance (44,45).

The application of Warburg's manometric technique for the determination of MAO activity has been severely criticized by many workers. Zeller et al. (46) have recently stressed the fact that about a 40- to 50-fold amount of tissue homogenate is required for the manometric procedures as compared with the quantities of homogenate necessary for a sensitive spectrophotometric method. This makes it impossible to analyze the distribution of MAO in organs with low enzymic activity by means of the O_2-uptake technique. Furthermore, Zeller (46) points out that the manometric method does not lend itself to investigations of the inhibitory effect of various drugs on MAO, since large amounts of homogenate, as required for the Warburg technique, seem to overcome most of the inhibitory effects of such drugs and thus render screening tests useless. Moreover, an apparent time lag phase is often observed with preparations of low MAO activity so that the readings become erratic and the determination of initial velocity cannot be accurately carried out. Finally, the O_2-consumption may be caused not only by the MAO reaction but also by an O_2-uptake, due to the other reactions with endogenous substrates. Contrary to Zeller, Slater (47) emphasizes that the manometric technique, if properly used under correctly chosen conditions, is still a very accurate method. It is essential, however, to measure the rate of O_2-consumption during steady-state conditions. When, after the addition of mitochondria, a constant rate of O_2-consumption prevails for at least 15 min, the total O_2-consumption can be accurately calculated.

In 1938 Petering and Daniells (48) were the first to measure the respiration of diverse cell suspensions and rat liver homogenates by means of a dipping-mercury electrode, and this polarographic technique has been applied since to measurements of the O_2-uptake by microorganisms, fibroblasts, and mitochondria. Clark (49) has greatly improved this method of determination of O_2-uptake by shielding the electrodes from the incubation medium with thin polyethylene films and by ensuring that the electrodes have a constant ionic environment. The "Clark-type electrode" has been very recently used by Tipton and Dawson (50) for the estimation of MAO activity in hog brain (see Section II-3-B-b).

In the analytical procedures based on the estimation of substrate disappearance, serotonin or tyramine have been used as substrates, for both these substances can be readily measured. However, this method of determination of MAO activity is very cumbersome as it involves multiple extraction procedures (51). Since most of the methods so far mentioned proved to be insufficiently sensitive for estimations of MAO in organs with low MAO activity, such as dog heart, or in minute amounts of tissue (e.g., in a simple sympathetic ganglion), and furthermore, since attention is being focused on the pharmacological actions of MAO inhibitors in such sites, some time ago it became clear to many scientists that more sensitive and accurate methods would have to be designed for the study of a direct correlation of biochemical and pharmacological effects. Modern spectrophotometric, fluorometric, radiometric, and histochemical procedures have been recently published. Some of these methods are described here in detail.

B. ESTIMATION OF MAO ACTIVITY BY MEASUREMENT OF O_2-UPTAKE

a. **Standard Warburg Technique of Creasey** (43). Monoamine oxidase activity is measured manometrically at 38°C by following the O_2-consumption during the deamination of the substrate tyramine in the presence of $10^{-2}M$ semicarbazide and $10^{-3}M$ cyanide. Under these circumstances only MAO and catalase appear to be active, spontaneous oxidations being inhibited, and O_2 being absorbed at the rate of 1 atom/ molecule of tyramine deaminated.   Tyramine is rapidly oxidized, and the O_2-uptake remains linear over the first 30 min of the reaction and is directly proportional to the concentration of MAO.

Reagents required are as follows: $0.24M$ sodium phosphate buffer (pH 7.0); $0.01M$ KCN solution; $0.1M$ semicarbazide; $0.1M$ tyramine; and $2.0M$ KCN.

Place in the main compartment of a Warburg vessel 1.0 ml of the enzyme preparation, and add 0.2 ml of $0.24M$ sodium phosphate buffer (pH 7.0), 0.2 ml of $0.01M$ KCN, 0.2 ml of $0.1M$ semicarbazide, and 0.2 ml of distilled water.   Pipet into the side arm 0.2 ml of $0.1M$ tyramine and place in the center well 0.1 ml of $2M$ KCN and a filter paper strip. Adjust all solutions except that of $2M$ KCN to pH 7.0.   The Warburg flasks are then gassed with O_2 for 2 min and equilibrated for 10 min before tipping in the substrate.   For the determination of MAO activity, readings are taken at 0, 10, 20, and 30 min.   For the measurement of the total O_2-uptake, readings are taken at 30 min intervals and, thereafter, until O_2 ceases to be absorbed.   A reaction mixture containing no substrate is used as a control, and any O_2 absorbed by the latter is subtracted from that absorbed in the test reaction mixtures.

b. Determination of MAO Activity by a Polarographic Method According to Tipton and Dawson (50).   The activity of monoamine oxidase was determined in homogenates as well as in the mitochondrial fraction, prepared from individual regions of pig brain, by measuring O_2-uptake with a Clark oxygen electrode.

Pig brains, obtained from an abattoir within 15 min of the animals being killed, were kept in ice for approximately 60 min.   The brains were then dissected at 5°C, portions weighed and homogenized by hand in 9 vol (v/w) of $0.25M$ sucrose, and adjusted with $1.0M$ K_2HPO_4 to pH 7.6. The brain regions were homogenized in a Potter-Elvehjem homogenizer using an all-nylon pestle.   Samples of the homogenate were removed and stored at $-10°C$ until assayed.   Mitochondria were prepared from the remainder of the homogenate by the method of Brody and Bain (41). The homogenate was centrifuged at $1500g$ for 20 min at 4°C, the supernatant carefully decanted and recentrifuged at $18,000g$ for 20 min at 4°C. The resulting residue was resuspended in half the original volume of a $0.25M$ sucrose-K_2HPO_4 medium (pH 7.6) and again centrifuged at $18,000g$ for 20 min at 4°C.   The residues were taken up in $0.02M$ sodium phosphate buffer (pH 7.6) and stored frozen until assayed.

Enzyme assays were carried out in an apparatus similar to that devised by Dixon and Kleppe (52). The O_2-consumption was measured with a Clark oxygen electrode (Yellow Springs Instrument Co. Inc., Yellow Springs, Ohio) connected via a voltage divider to a Honeywell-Brown 1 mV strip-chart recorder. All assays were performed at 30°C, and the apparatus was standardized with distilled water saturated with air at 30°C. Proportionality of reaction velocity and enzyme concentration was tested and verified in each assay.

The assay mixture contained in a total volume of 2.4 ml: 200 μmoles of sodium phosphate buffer (pH 7.0), 50 units of catalase, 20 μmoles of semicarbazide, 2 μmoles of KCN, and enzyme. The mixture was equilibrated with air at 30°C and the reaction was started by the addition of 100 μl of 0.2M tyramine hydrochloride. Separate controls were determined in the absence of enzyme. For the sake of comparison Tipton and Dawson (50) express the MAO activity as mμg atoms of oxygen consumed per minute and per milligram of protein. These authors used 6 samples in parallel in each of the investigated cases and results for MAO activities are given as means ±S.E.M.

Protein concentrations were determined by the microbiuret method of Goa (55).

C. DETERMINATION OF MAO ACTIVITY BY MICROESTIMATION OF AMMONIA USING THE WARBURG MANOMETRIC APPARATUS [METHOD OF BRAGANCA, QUASTEL, AND SCHUCHER (53)]

The principle of this method is essentially the same as that of the microdiffusion procedure of Conway and Byrne (54) in which the ammonia liberated from the alkaline solution is allowed to diffuse into a closed compartment containing acid. In the method of Braganca and his colleagues (53) both the liberation and the diffusion of the ammonia takes place in Warburg manometric vessels.

Procedure. Place in the center well of a Warburg manometric flask a small roll of filter paper soaked with 0.2 ml N H_2SO_4, and put 0.3 ml of a saturated potassium carbonate solution into the side arm. In the main compartment the MAO activity of the homogenate is assayed in a total volume of 3 ml in the presence of 0.05M phosphate buffer (pH 7.6), and 30 μmoles of tyramine hydrochloride, neutralized before use. Incubation is carried out for 30 min at 37°C. The potassium carbonate solution present in the side arm is then tipped into the main compartment which raises the pH of its content to 10.5. This arrests the course of the enzymic reaction. The ammonia formed diffuses into the center well where it is absorbed by the acid on the filter paper. For complete diffusion and absorption of amounts of ammonia, ranging from 10 to 500 μg, the Warburg vessels are shaken at 37°C for 3 hr. The vessels are then taken off the bath, and the small rolls of filter paper are removed from the center well with forceps and placed in graduated tubes. The content of the center well is quantitatively removed and added to the graduated tube by washing five times with distilled water using a micropipet. The content of the graduated tube is made up to a definite volume, and aliquots corresponding to between 20 and 30 μg NH_4^+ are assayed by nesslerization. For this purpose 1 ml of Nessler's solution and 2 ml of $2N$ NaOH are added to each of these aliquots, and the mixture is made up with distilled water to a final volume of 10 ml. The intensity of the color produced is estimated at 425 nm and all values are corrected for blanks.

This method has recently been used successfully by Guha and Krishna Murti (34) for MAO estimations in rat liver mitochondria.

D. SPECTROPHOTOMETRIC METHODS FOR THE ESTIMATION OF MAO ACTIVITY

a. **Method of Weissbach, Smith, Daly, Witkop, and Udenfriend** (45). This assay of MAO activity is based on the rate of disappearance of kynuramine used as a substrate of MAO. Kynuramine can now be obtained from Regis Chemical Co., Chicago, Illinois.

Procedure. Tissues are homogenized in 5 vol of cold distilled water and passed through a layer of cheese cloth. To remove cellular debris, the homogenate is centrifuged at 500 rpm for 15 min. In a 3-ml silica-experimental cuvet place 0.1 or 0.2 ml of tissue homogenate, add 0.3 μmole of kynuramine and 0.3 ml of $0.5M$ phosphate buffer (pH 7.4), and make up with distilled water to a total volume of 3 ml. The initial extinction of this reaction mixture at 360 nm is approximately 0.5. Since the incubation is run at room temperature, only the enzyme solutions are kept cold. A control cuvet is prepared in which the kynuramine is replaced with water. Mix by inversion and take an initial reading at 360 nm. Further readings are then recorded at suitable intervals depending on the activity of the MAO preparation. Activity is expressed as the change in extinction at 360 nm per unit of time.

With crude preparations, the reading observed in the first minute or two may be erratic because of settling of particles in the cuvet. After this period, however, no difficulty is encountered. A Beckman model DU spectrophotometer or the Model 14 Cary Recording spectrophotometer was used by Weissbach et al. (45).

Kynuramine disappearance remains linear with time until the extinction falls below 0.150. Enzyme activity has been found to be pro-

portional to enzyme concentration. Furthermore, this assay has proved
to be extremely sensitive to iproniazid, the known very good inhibitor of
MAO. Rat liver mitochondria, an excellent source of MAO, rapidly
metabolise kynuramine, as do partially purified MAO preparations from
guinea pig liver (56).

Although Weissbach and his colleagues (45) emphasize that their
spectrophotometric assay of MAO not only allows a rapid determination
of the disappearance of kynuramine but also a quick and exact spectrophotometric estimation (between 310 and 335 nm) of the appearance of
4-hydroxyquinoline, a metabolite of kynuramine, no experimental data
for such a procedure are given in their publication, which is regrettable.

The spectrophotometric estimation of MAO activity by Weissbach
et al. (45) has been recently used by Barbato and Abood (57) in their
kinetic studies on the effect of various inhibitors upon purified beef liver
mitochondrial MAO, and the results obtained were compared by the
authors with those which they found with the spectrophotometric technique devised by Tabor and his associates (13). In the latter method,
benzylamine is used as a substrate and the benzaldehyde formed in this
reaction is read at 250 nm. Barbato and Abood (57) maintain that
kynuramine is a better substrate for a spectrophotometric MAO estimation than is benzylamine, particularly when working with carbonyl
reagents such as KCN, which may react with the benzaldehyde formed,
or when applying the inhibitor phenanthroline, the extinction coefficient
of which is extremely high at 250 nm. With $10^{-4}M$ kynuramine as substrate, however, the extinction is measured at 360 nm (45), a wavelength at which the extinction caused by phenanthroline is negligible.

Zeller and his colleagues (46) have recently adversely criticized the
spectrophotometric kynuramine method of Weissbach et al. (45) as
follows: Upon the action of MAO, kynuramine is deaminated and the
resulting aldehyde condenses intramolecularly to form 4-hydroxyquino-
line (58). This metabolite shows a much lower extinction at 360 nm
than the starting material, kynuramine. Since the initial optical readings with 0.1 μmole kynuramine are relatively high, readings of small
differences in optical density may thus be of doubtful value. This
difficulty could be at least partly overcome by following the increase
in optical density between 310 and 335 nm, due to the formation of
4-hydroxyquinoline.

**b. A Rapid Spectrophotometric Method for the Determination of MAO
Activity and MAO Inhibition According to Zeller, Ramachander, and
Zeller** (46). This method is based on the high extinction at 253 nm by
the *m*-iodobenzaldehyde formed during the action of MAO on *m*-iodo-

benzylamine, when used as its substrate. m-Iodobenzylamine is readily degraded by MAO and its primary product, the m-iodobenzaldehyde, does not seem to undergo further enzymic oxidation by oxido-reductases. In solubilized mitochondrial homogenates this method yields well-defined reaction curves during the first minute of incubation. Furthermore, within a wide range a linear relationship between enzyme concentration and increase in extinction is obtained as well as between the reciprocal values of substrate concentrations and reaction rates (Linneweaver-Burk relationship). This technique has been used by Zeller and his colleagues (46) for the determination of MAO levels in the liver and brain of various species. Moreover, these authors used this rapid technique for the analysis of progressive inhibition of MAO by the drugs pargylamine and o-chloropargylamine.

The compound m-iodobenzylamine was chosen by Zeller et al. (46) as a substrate, for much higher reaction rates were obtained in its presence than in that of benzylamine, the classical substrate of MAO. It should be noted that the rate of degradation of m-iodobenzylamine by MAO is much slower in the presence of air than in that of pure oxygen (46).

Synthesis of m-*Iodobenzylamine.* On photobromination of m-iodotoluene (59), m-iodobenzylbromide (mp 50°C) is formed, which is converted to m-iodobenzylamine by means of hexamethylene tetramine (60). m-Iodobenzylamine hydrochloride is purified by recrystallization from absolute ethanol (mp 193°C). At 253 nm the molar extinction coefficient of m-iodobenzylamine in $0.067M$ phosphate buffer, pH 7.4, is 780, and an $0.33 \times 10^{-3}M$ solution, which is the final substrate concentration in the standard test mixture of this method, gives an extinction of 0.26.

Procedure.   Liver tissues or mitochondria are homogenized with a cold solution containing 1.92 ml of $0.067M$ phosphate buffer (pH 7.2) and 0.08 ml of Cutscum (isooctylphenoxy-polyethoxy-ethanol, Fisher Scientific Co.) per 100 mg of tissue. The material is homogenized with 50 up-and-down strokes in a Teflon pestle tissue grinder (Thomas, Philadelphia, Pa.) and then centrifuged for 30 min at $12,000g$. The thin lipid top layer is gently pierced so that 1–2 ml of the clear aqueous layer can be withdrawn. This enzyme solution is diluted with an equal volume of cold phosphate buffer (pH 7.2) and is kept in an ice bath before the assay. When, however, tissues poor in MAO, e.g., brain, are investigated the enzyme solution is used undiluted.

The *Standard Test solution* contains 0.2 ml of the enzyme solution and 2.8 ml. of $0.36 \times 10^{-3}M$ m-iodobenzylamine in $0.067M$ phosphate buffer (pH 7.2), giving a final substrate concentration of $0.33 \times 10^{-3}M$. The control consists of 0.2 ml of enzyme in phosphate buffer only.

Assay. Vessels containing the substrate m-iodobenzylamine in $0.067M$ phosphate buffer (pH 7.2) are shaken under oxygen at 38°C for at least 10 min and are then sealed with Parafilm while the enzyme solution is kept in an ice bath and is not flushed with oxygen. After breaking the seal of the tube containing 2.8 ml. of the buffered substrate solution, 0.2 ml of the enzyme solution is added and the reaction velocity of the first 1 or 2 min from the start of recording is measured. The recording is usually started 10–20 sec from the time the enzyme is pipetted into the buffered substrate solution. In very active liver preparations the reaction rate decreases after 2 min and becomes nonlinear after this period. The increase in optical density at 253 nm which takes place upon the addition of the enzyme solution to the substrate-buffer solution, is measured in a Model 14 Cary Recording spectrophotometer at a compartment temperature of 38°C. The Beckman ultraviolet spectrophotometer as well as the Guilford four-channel photometer are likewise suitable instruments for the above determinations. Results are expressed as differences in optical density/gram of enzyme/minute and pertain to milligrams or grams of enzyme present in 1 ml.

This microspectrophotometric method seems to lend itself for kinetic studies on MAO activity in various tissues, especially in the presence of inhibitors.

E. FLUOROMETRIC METHODS FOR THE DETERMINATION OF MAO ACTIVITY

It is well known that even an average fluorometer will allow the determination of quantities of fluorescent material as small as 0.1 to 0.001 μg/ml. With the more sensitive modern instruments and with compounds showing a high extinction and a high quantum yield of fluorescence, a millimicrogram (nanogram) of a fluorophor may emit enough fluorescence to permit measurement (Udenfriend, 61). Such amounts are much smaller than required for other methods of assay. Therefore fluorometry offers an additional tool for the assessment of specificity.

a. A Sensitive Fluorometric Assay of MAO activity *in vitro* **Based on the Rate of Formation of Indole Acetic Acid According to Lovenberg, Levine, and Sjoerdsma** (62). Tryptamine is a good substrate for MAO (7). When tissue homogenates are incubated with tryptamine, indole acetaldehyde is first formed which, in the presence of aldehyde dehydrogenase and NAD, is oxidized to indole acetic acid (IAA). Whereas measurements of the disappearance of small amounts of tryptamine are often not possible, corresponding quantities of IAA formed can be easily determined since the fluorescent properties of IAA permit the quantitation of microgram amounts. Furthermore, it should be emphasized that the endogenous quantities of IAA present in tissues are below the level of detection.

Procedure. Tissues are frozen on removal and thawed prior to homogenization in $0.25M$ sucrose by means of a motor-driven glass homogenizer. In studies on small samples, the homogenate was often diluted 50 to 100-fold.

Preparation of Aldehyde Dehydrogenase. Guinea pigs are pretreated 18 to 24 hr before the experiment by an intraperitoneal injection of the irreversible MAO inhibitor, β-phenylisopropylhydrazine (JB-516, Catron, Lakeside Laboratories), using 5 to 10 mg/kg. The animals are then sacrificed, and a 20% kidney homogenate is prepared and centrifuged in a Spinco Model L centrifuge for 1 hr at 78,000g. The resulting supernatant shows a high aldehyde dehydrogenase activity but a negligible MAO effect. It can be stored at $-5°C$ for periods up to 2 weeks without any enzymic loss. The activity of aldehyde dehydrogenase is tested by determining the rate of $NADH_2$ formation during its action on acetaldehyde (Racker, 63).

Assay of MAO. Reagents required: Stock solutions of tryptamine hydrochloride, corresponding to a content of 5.6 mg free base/ml of nicotinamide (24.4 mg/ml), and of indole acetic acid (100 μg/ml) are prepared, and they are stable for several weeks when stored at 2°C. NAD (23.2 mg/ml) must be prepared fresh daily.

Aliquots of the homogenate, equivalent to 10 to 200 mg of tissue, are pipetted into 20 ml-beakers containing 0.3 ml (60 μmoles) nicotinamide, 0.4 ml (14 μmoles) of NAD, 250 μmoles of phosphate buffer (pH 7.4), and 0.2 ml of the aldehyde dehydrogenase preparation, and the mixture is made up with distilled water to a total volume of 2.8 ml. Incubate in air at 37°C in a Dubnoff metabolic shaker. After a few minutes of equilibration start the reaction by adding 0.2 ml (7 μmoles) of tryptamine. Remove aliquots of 0.5 ml at various intervals and place them in tapered, glass-stoppered 50 ml-centrifuge tubes, each containing 3 ml of 0.5N HCl to stop the enzymic reaction. Then add 15 ml of toluene to extract the IAA formed, and agitate the tube in a mechanical shaker for 5 min. Centrifuge and transfer 10 ml of the organic layer to another glass-stoppered tube containing 1.5 ml of 0.5M phosphate buffer (pH 7.0). Shake the tube again for 5 min and thus extract the IAA into the aqueous phase. Measure the fluorescence of the IAA in the phosphate buffer solution directly in an Aminco Bowman spectrofluorometer at 280 nm excitation and 370 nm fluorescent wavelength (uncorrected wavelengths). The direct calculation of results will be facilitated by putting appropriate blanks and internal standards through the entire procedure. Lovenberg and his co-workers (62) found on comparison with internal aqueous standards that in this procedure 95 to 100% of IAA can be recovered in the presence of tissue. In all the experiments these authors carried out, a control assay on the aldehyde dehydrogenase was done to correct for traces of MAO activity which might be present in this enzyme preparation.

This fluorometric assay of MAO determination seems to be very well suited for problems such as mapping of MAO activity in specific parts of the central and peripheral nervous system, studying the action of cardiovascular drugs on heart MAO, and for the classification of relationships between pharmacological and biochemical effects of MAO inhibitors.

Lovenberg and his colleagues (62) emphasize, however, that certain precautions must be taken when studies of the effects of MAO inhibitors with this fluorometric method are proposed. Thus it must first be ascertained that the effects obtained are due exclusively to inhibition of MAO and not to inhibition of aldehyde dehydrogenase or, possibly, to a reaction between the inhibitor and the first metabolite of tryptamine, indole acetaldehyde. Moreover, since in this method small samples of tissue require considerable dilution for handling, it is possible that in the case of reversible inhibitors such as harmine, the results obtained do not represent an accurate reflection of the *in vivo* inhibition.

b. A Rapid Microfluorometric Technique for the Determination of MAO of Krajl (64). This method is a microfluorometric adaptation of the spectrophotometric MAO assay designed by Weissbach and his associates (45; see Section II-3-D-a) in which, however, not the disappearance of the substrate, kynuramine, but the appearance of its metabolite 4-hydroxyquinoline (4HOQ) is recorded. The latter compound is formed by a spontaneous cyclization reaction of the intermediate aldehydic metabolite of the MAO action on the substrate kynuramine (45).

Procedure. Add to 1.0 ml of an enzyme solution, i.e., to an aqueous homogenate of tissue containing the desired amount of wet weight of tissue, 0.5 ml of kynuramine (100 μg of kynuramine dihydrobromide), 0.5 ml of $0.5M$ phosphate buffer (pH 7.4) and make up with distilled water to a total volume of 3.0 ml. Incubate at 37°C with air as the gas phase for 30 min and then add 2.0 ml of 10% trichloracetic acid. Separate the precipitated proteins by centrifugation and add 1.0 ml of the supernatant to 2.0 ml of $1N$ NaOH in a silica cuvet. In an Aminco Bowman spectrofluorometer, coupled to an Electro Instruments Model 101 X-Y recorder, activate the solution at 315 nm and measure the fluorescence at 380 nm, or record by scanning from 320 to 450 nm. Appropriate blanks and 4HOQ standards must be carried through the entire procedure. All wavelengths reported are uncorrected values. On activation at 315 nm, 4HOQ in $1N$ NaOH exhibits an intense fluorescence maximum at 380 nm. Under these conditions a tenfold excess of kynuramine exhibits no appreciable fluorescence. At the sensitivity settings employed, slit system 3, meter multiplier 0.03, sensitivity 50 (American Co. Service Manual No. 768 A, p. 15, 1960), full-scale galvanometer deflection is obtained with solutions containing between 0.1 and 0.2 μg of 4HOQ/ml. At the highest sensitivity settings, as little as 1 mμg (1 ng) of 4HOQ can still be detected.

A 4HOQ standard of 10 mμmoles (1.99 μg of 4HOQ) carried through the entire procedure gives an arbitrary fluorescence value of 77 ± 2 units, with blank values of 5 ± 1 units recorded under these conditions. Duplicates agree very well, the average error being less than 2%. The manipulations of this method are very simple and not time consuming.

This very sensitive microfluorometric method of MAO estimation has been successfully applied to different tissues, e.g., rat brain, guinea pig stria, and cat ganglion. A great advantage of this method is that it does not depend on the tissue sample containing aldehyde dehydrogenase. For, as mentioned above, the aldehyde produced by the oxidative deamination of kynuramine undergoes a spontaneous and coupled cyclization to 4-hydroxy quinoline. Hence, 4HOQ production depends only on the activity of MAO.

F. RADIOMETRIC TECHNIQUES FOR THE QUANTITATION OF MAO IN MICROGRAM AMOUNTS OF TISSUE

Fluorometric techniques of MAO estimation require milligram quantities of tissue despite their great sensitivity. Moreover, due to potential variation in endogenous levels of fluorescent material, these methods may be subject to error. Many radiometric assays which permit accurate quantitative determinations of MAO in microgram amounts of tissue have recently been developed by various workers.

a. A Sensitive and Specific Radiometric Assay for the Estimation of MAO of Wurtman and Axelrod (65). This simple method measures the radioactive metabolite formed on oxidative deamination by MAO of ^{14}C-tryptamine and identified as ^{14}C-indole acetic acid.

Procedure. Tissues are homogenized in chilled isotonic KCl, and 1 to 100 μl (10 μg to 1 mg tissue) of the homogenate are assayed. Tryptamine-2-^{14}C-hydrochloride (New England Nuclear Co., 1.3 mCi/mmole) is dissolved in water and stored at $-4°C$. A typical assay contained in a 15-ml glass-stoppered centrifuge tube a mixture of 25 μl of tissue homogenate, 25 μl (6.25 mμmoles, 10,000 cpm) of ^{11}C-tryptamine, and 250 μl of $0.5M$ phosphate buffer (pH 7.4). The mixture was incubated at 37°C for 20 min; to stop the reaction, 0.2 ml of $2N$ HCl was added and the deaminated radioactive material was extracted by shaking with 6 ml of toluene. After centrifugation, a 4-ml aliquot of the toluene layer was transferred to a vial which contained 10 ml of phosphor [0.4%, 2,5-diphenyloxazole and 0.005% of 1,4-di(2,5-phenyloxazole)benzene in toluene]. Counting was performed for 1 to 5 min in a liquid scintillation spectrophotometer. A minute amount of ^{14}C-tryptamine (i.e., less than 0.3%) is extracted by this procedure. By incubating ^{14}C-tryptamine with boiled enzyme a correction is made for the blank value, amounting to approximately 30 to 50 cpm.

The reaction is linear with time for at least 20 min and with enzyme concentrations ranging from 5 to 1000 μg of liver. Duplicate assays of the MAO activity of 250 μg of many liver specimens differed by less than 2%.

The enzyme specificity of this method was examined by Wurtman and Axelrod (65) in a study of the effect on hepatic MAO of a pretreatment *in vivo* or preincubation *in vitro* with tranylcypromine (SKF 385), a potent inhibitor of MAO, at doses which are without effect on diamine oxidase (66). Tranylcypromine produced a decrease in MAO activity of 97 to 99% in both *in vivo* and *in vitro* experiments.

b. A Radioisotopic Assay for MAO Determinations According to Otsuka and Kobayashi (67). This method is based upon the formation from ^{14}C-tyramine, used as substrate, of a radioactive metabolite which is soluble in anisole. After completed incubation with MAO, the reaction product is extracted into anisole, the aqueous phase is frozen, and the anisole containing the radioactive metabolite and phosphor is placed in a counting vial and is assayed in a liquid scintillation spectrometer. By this method radioactive metabolites from as little as 1.4 ng of tyramine can be detected.

[14]C-Tyramine, 4.6 mCi/mmole, was obtained from California Corporation for Biochemical Research and was diluted with nonisotopic tyramine to yield a solution with an activity of 50,000 disintegrations per minute (dpm)/4 μg tyramine/0.1 ml solution. MAO was prepared by the method of Satake (68). Hog kidney acetone powder (65 g) was twice

extracted with 30 ml of $0.2M$ phosphate buffer, pH 7.2. The residue was suspended in 20 ml of water and the suspension was dialyzed in a Visking cellulose bag first against tap water for 4 hr and then against distilled water for 4 hr. The dialyzed suspension contained 1.12 mg of nitrogen ($\equiv 7.00$ mg protein)/ml. and showed a specific activity of 44 μl O_2 consumed/hr/mg protein, using tyramine as subatrate.

Liquid Scintillation Counting Method. A Packard Tri-Carb liquid scintillation spectrometer, Model 314 EX, was used for this method. The *extraction solvent* consisted of reagent-grade anisole, containing 0.6% 2,5-diphenyloxazole (PPO). This solution, anisole-PPO, yielded 53% counting efficiency after its use in the extraction of end products.

^{14}C-Tyramine was assayed at a counting efficiency of 50% using a toluene-ethanol mixture (70:25, v/v) containing 0.37% PPO and 0.01% 1,4-bis-2(4-methyl-5-phenyloxazolyl)benzene (dimethyl-POPOP).

For the estimation of radioactivity in protein solutions, 1 ml of methanol containing $1M$ hyamine hydroxide (Packard Instrument Co.) was added to 0.2 ml of the protein solution, and was heated at 60°C for 4 hr to dissolve the protein, and was assayed thereafter with the addition of 10 ml of toluene-PPO (toluene containing 0.4% of 2,5-diphenyloxazole). The counting efficiency was 32%.

Counting efficiencies of various solvents were examined by the internal standard method using a standard solution of ^{14}C-toluene with 10,000 dpm/0.1 ml. All the countings were performed in low potassium 5-dram glass vials (Wheaton Glass Co.).

Incubation and Extraction Procedure. The enzyme assays were carried out in a screw-cap culture tube (10 × 1.5 cm) in a final volume of 2.0 ml in air at 37°C, and in a Dubnoff metabolic shaker. On completed incubation of 60 min, 0.4 ml of $2M$ citric acid was first added to stop the enzymic reaction, followed by the addition of 10 ml of anisol-PPO solution. After vigorous shaking for 1 min the reaction mixture was centrifuged in an International Centrifuge, Model PR-1, at 1000 rpm, and was left standing at -20°C until the lower aqueous phase was frozen. The upper layer was then transferred into a counting vial and assayed in a liquid scintillation counter.

Standard Assay for Extraction Studies. Incubate 1 ml of hog kidney MAO preparation with 40 μg of ^{14}C-tyramine and $10^{-3}M$ EDTA in 20 ml of $0.1M$ sodium phosphate buffer (pH 7.5) at 38°C for 60 min. Then heat the reaction mixture in a boiling water bath for 5 min and thereafter centrifuge. Aliquots of the supernatant are used for extraction studies. EDTA accelerates the enzymic reaction and prevents further oxidation of the reaction product.

Since in extraction studies Otsuka and Kobayashi (67) have found that the anisole-PPO mixture was a solvent superior to the toluene-PPO system for the metabolites of the MAO-tyramine reaction, they abandoned the use of the latter extraction mixture as a solvent for the isotopic MAO assay.

With regard to the stoichiometry of the MAO reaction using tyramine as substrate, these workers determined the excess of tyramine as well as the amount of the end-product formed and the O_2-uptake during incubation. A typical assay consisted of a mixture of 4 μg of ^{14}C-tyramine, 10 μmoles of nonisotopic tyramine, 0.25 or 0.50 ml of enzyme, and $10^{-3}M$ EDTA made up with $0.1M$ sodium phosphate buffer (pH 7.5) to a total volume of 2.2 ml. After 60 min incubation at 38°C in air, the metabolite formed was extracted 8 times with 10-ml portions of anisole-PPO, and the quantity of this metabolite was calculated from the total radioactivity of the combined extracts. The result obtained indicated an average formation of 1.88 moles of the metabolite per mole of O_2 consumed. This finding seems to approximate the expected 2:1 ratio for the MAO reaction in the absence of aldehyde oxidase but in the presence of catalase in the enzyme preparation (Blaschko, 7). Furthermore, Otsuka and Kobayashi (67) established that the formation of the end-product of the MAO-tyramine reaction was roughly proportional to the incubation time during the first hour, but deviated from linearity with longer incubation periods.

The sensitivity of this method seems to depend on the specific activity of the tyramine available. In the diluted preparations the authors could easily determine the metabolism of 17 ng of tyramine. Without dilution the metabolism of 1.4 ng of tyramine (100 cpm at 50% efficiency) could still be detected. This method seems to lend itself to the *in vivo* study of the effectiveness of MAO inhibitors in the circulation as well as for determining the biological half-life of these inhibitors (67).

c. Isotopic Method for the Microdetermination of MAO Activity in Microgram Quantities of Tissue of McCaman, McCaman, Hunt, and Smith (69). This method allows the quantitative estimation of MAO activity in 2 to 5 μg of nervous tissue (dry weight) using ^{14}C-labeled substrates of different specific activities such as 5-HT, 3-hydroxytyramine, or tyramine.

Procedure. In a pointed microtube (2.5 mm $\times$ 4 cm) place the sample equivalent to 1 to 20 μg of dry brain tissue or 1 μl of homogenate (1 g tissue/25 ml H_2O) and add 10 μl of the ice-cold buffer substrate solution containing $0.1M$ potassium phosphate buffer (pH 7.2) and 0.8 mM [3-^{14}C] serotonin, which had been stored at $-20°C$. Place all the tubes in a tray with ice water, mix the contents of each tube without warming, and incubate at 38°C for 30 min.   Then stop the reaction by adding 1 μl of 3N HCl, add 50 μl of ethyl acetate to each tube, mix the samples thoroughly, and centrifuge to separate the phases.   Remove from each tube 40 μl of the ethyl acetate layer, and transfer it to another tube containing 30 μl of 0.3N HCl.   By this "washing step" the last trace of the radioactive substrate is removed.   Mix thoroughly, centrifuge, and transfer 35 μl of the ethyl acetate layer to a counting vial (20 ml) and mix with 1 ml of absolute methanol, followed by 15 ml of scintillator-toluene solution [4 g of 2,5 diphenyloxazole and 0.1 g of 1,4-bis-2(5-phenyloxazolyl)-benzene per liter of toluene].

Radioactivity is determined in a Packard scintillation counter and the enzyme activity is calculated from the known specific activity of the substrate.   The same protocol is used with ^{14}C-labeled 3-hydroxytyramine (2.5 mM in 0.1M phosphate buffer, pH 6.75), and ^{14}C-labeled tyramine (3.0 mM in 0.1M phosphate buffer, pH 7.95).   The radioactivity extracted into ethyl acetate from the blanks is usually equivalent to less than 0.5% of the quantity of 5-HT or tyramine present in the buffer-substrate.   "One backwash" reduces this amount to less than 0.005% (i.e., 40 cpm), or to no detectable activity, which depends upon the specific activity.   Similar blank values are obtained in tubes containing either the buffer-substrate solution alone, or the latter in conjunction with tissue homogenate, added after incubation, or finally in tubes with boiled homogenate.   For every mμmole of 5-HT showing a specific activity of 1 mCi/mmole, metabolized in the above procedure by brain, a delta of approximately 2000 cpm with a blank of 5 cpm is obtained.

In experiments of McCaman et al. (69) the isotopic measurement of the compounds extracted into ethyl acetate following incubation with ^{14}C-tyramine indicated that the ratio of acid to aldehyde was 4:1.   On the other hand, determinations of the end-products of the brain enzyme reaction with ^{14}C-5-HT showed that both the indole acetaldehyde and indole acetic acid were present in approximately equal quantities.

The optimum substrate concentrations during incubation of MAO containing tissue are 0.8 mM for 5-HT and 2.5 mM for 3-hydroxytyramine.   At higher concentrations of both substrates a marked inhibition of the MAO reaction is observed.   Thus a 50% MAO inhibition is found with an 8 mM concentration of 5-HT, or with a 14 mM amount of 3-hydroxytyramine, used as substrates.   With tyramine as substrate, virtually the same activity is obtained with concentrations ranging from 2.5 to 10 mM.

The amount of the radioactive end-product of the MAO reaction was found to be proportional to the quantity of brain tissue over the range of

5 to 170 μg of wet brain at an incubation for 30 min at 38°C. The MAO activity in brain homogenates decreases with time after an incubation of 15 min, so that the values at 30 min seem to fall off by about 15% and those after 60 min by 30%. Using this isotopic method, McCaman and his associates (69) tested the effect of some well-known MAO inhibitors on the activity of this enzyme in rabbit brain homogenates. These workers preincubated three noncompetitive MAO inhibitors (trans-2-phenylcyclopropylamine, SKF 385; 1-phenyl-2-hydrazino-propane, JB-516; and isonicotinic acid 2-isopropylhydrazine, Marsilid) with a buffered rabbit brain homogenate for 15 min at 38°C before the addition of the substrate. No inhibitory effect on the extraction of the labeled products of the MAO reaction was observed with any of these drugs. Although simple unsubstituted amines such as tryptamine and phenylethylamine turned out to be the most potent inhibitors of the oxidative deamination of tyramine and 3-hydroxy-tyramine, α-methyl or N-methyl substituted derivatives were more potent inhibitors of 5-HT degradation than non-substituted analogs.

d. A Rapid and Sensitive Method for the Microradiometric Assay of MAO in Submicrogram Quantities of Tissue of Aures, Fleming, and Hakanson (70). Thin-layer chromatography has recently been used with much success for the rapid isolation and identification in tissues of minute amounts of biogenic amines. Thus the presence of nanogram amounts of these amines can be discovered by exposing the thin-layer chromatogram either to an o-phthalaldehyde spray or to paraform-aldehyde gas, whereby the amines are converted into highly fluorescent derivatives (71). This technique has been recently adapted as a basis for simple microradiometric estimations of MAO activity (70).

Assay of MAO Activity in the Pineal Gland of the Rat (70). One rat pineal gland (approximately 1 mg of pineal tissue) is homogenized in 50 μl of 0.1M phosphate buffer (pH 6.8). The homogenate is incubated with 2 μg ^{14}C-5-HT (free base, 39.7 mCi/mmole, Radiochemical Centre, Amersham) in a total volume of 60 μl with oxygen as gas phase at 37°C for 60 min. Incubation is stopped by the addition of 1 ml of a mixture of ethyl acetate-acetic acid (95:5). The precipitated proteins are separated by centrifugation and 0.5 ml of the clear supernatant is evaporated to dryness *in vacuo*. The dry residue is extracted with a drop of the ethyl acetate-acetic acid mixture, which extracts the formed 5-hydroxy-indole acetic acid quantitatively but only a trace of 5-HT. The extract is submitted to thin-layer chromatography on silica gel.

A microscope slide (2.5 × 7.5 cm) is coated with a thin layer (250 μ) of silica gel (Kieselgel G and Kieselgel H, Merck, Darmstadt) whereby the layer is applied in the form of a slurry consisting of 30 g of silica gel suspended in 70 ml of redistilled water. The chromatoslide is then activated by drying in an oven at 100°C for 1–2 hr, and the ethyl acetate-acetic acid extract of the pineal gland is spotted onto the chromatoslide by means of a glass capillary. The reaction product, 5-hydroxyindole acetic acid, is separated from residual 5-HT in less than 15 min with the solvent system n-butanol–acetic acid (1:1). The thin-layer material containing 5-hydroxyindole acetic acid is scraped off and transferred to a counting vial containing 0.1 ml distilled water as eluant. The radioactivity is recorded using a liquid scintillation counter.

Estimation of MAO Activity in Rat Brain Caudate Nucleus Slices. Aures and her colleagues (70) investigated the uptake and metabolism of dopamine in the rat corpus striatum which is known to contain a high amount of this amine. Slices of rat caudate mucleus, weighing approximately 50 mg, are incubated with 0.1 μCi of ^{14}C-2-dopamine (22 mCi/mmole, Nuclear, Chicago) in 2 ml of Krebs-Ringer bicarbonate solution, containing 1% D-glucose, pH 7.4, at 37°C for 30 min in an atmosphere of 95% O_2 and 5% CO_2. On removal from the incubation medium, the slices are rinsed and homogenized in acidified methanol. The protein precipitate formed is spun down, and an aliquot of the supernatant is chromatographed on a cellulose thin layer along with reference compounds. The cellulose thin layer is again applied as a slurry (15 g of cellulose powder, MN 300 HR, Macherey, Nagel & Co., Düren, suspended in 90 ml of redistilled water). The chromatogram is developed in a two-dimensional solvent system consisting of mixture I: n-butanol saturated with 0.1N HCl; and mixture II: isopropanol-5N NH_4OH-H_2O (8:1:1); the spots are visualized by exposure to paraformaldehyde gas and are subsequently scanned in uv light. The spots are then scraped off and the radioactivity determined by liquid scintillation counting.

Using this microtechnique, Aures and her colleagues (70) obtained results which compare well with earlier findings of Goldstein et al. (72).

G. HISTOCHEMICAL DEMONSTRATION OF MAO ACTIVITY IN VARIOUS TISSUES BY THE STANDARD TRYPTAMINE-TETRAZOLIUM METHOD OF GLENNER, BURTNER, AND BROWN, JR. (74)

Francis (73a,b) was able to demonstrate that MAO can be localized histochemically when applying tyramine as substrate and a tetrazolium compound as hydrogen acceptor. As is well known, hydrogen peroxide is formed during the oxidative deamination of tyramine by MAO, which indicates that in this reaction dehydrogenation occurs. Instead of molecular oxygen, neotetrazolium (p,p'-diphenylene-bis-2-[3,5-diphenyl]-

tetrazolium chloride) is applied as hydrogen acceptor. The tetrazolium compound is precipitated as blue formazan on reduction, and its formation is the histochemical demonstration upon which MAO activity depends.

In 1957 Glenner, Burtner, and Brown, Jr. (74) modified the histochemical method of Francis (73a,b) by using tryptamine as a substrate instead of tyramine in the presence of tetrazolium salts. In experimental MAO studies comprising substrate specificity, effects of inhibitors, pH optimum, and relative reactivity of various tetrazoles, selected tissues of young adult guinea pigs, rabbits, and rats were subjected by Glenner and his co-workers (74) to this histochemical technique. The tissues were freshly removed and 20 μ thick frozen sections were cut using the sliding microtome according to the technique of Adamstone and Taylor (75).

Assay procedure. Place in a 20-ml polyethylene jar 25 mg of tryptamine hydrochloride, 4 mg of sodium sulfate, 5 mg of nitro-blue tetrazolium, 5 ml of 0.1M phosphate buffer (pH 7.6), and 15 ml of distilled water. Warm to 37°C and incubate in this medium tissue sections for 30 to 45 min at 37°C. After incubation the tissue sections are washed in running water, fixed in buffered 4% formaldehyde for 24 hr, dehydrated, cleared in graded acetone-xylene solution, and mounted in Permount.

Control. Glenner and his colleagues (74) slightly modified the technique of Koelle and Valk (76) and used it for the controls to compare sites of staining with the tryptamine-tetrazolium method. In this modified method, tissue sections are incubated in the presence of tryptamine with the previously prepared hydrochloride of 2-hydroxy-3-naphthoic acid hydrazide. This derivative is more soluble than the hydrazide itself which was used in the original method. Moreover, the addition of sodium sulfate in a high concentration is omitted in the modified technique. The aldehyde formed from tryptamine on the action of MAO condenses with the hydrazide and the product is converted to a blueish-purple pigment by coupling with tetrazolized o-dianisidine. This modification of the rather cumbersome method of Koelle and Valk (76) reduces the incubation time to 60 min and obviates the necessity for continuous aeration of the incubation medium.

Inhibitor studies. Preincubate mounted tissue sections with 0.01M concentrations of potential inhibitors in phosphate buffer (pH 7.6) for 15 min at 37°C. Wash the slides thoroughly in running water and place them in a tryptamine-tetrazolium (INT) incubating solution for 30 min at 37°C. For concomitant incubation-inhibition experiments, various potential inhibitors in 0.1M concentration are incorporated in the trypta-mine-tetrazolium incubating solution with immersion of slides for 30 min at 37°C. Controls without potential inhibitors are used throughout all assays.

Substrate specificity. Replace in the standard medium tryptamine with various compounds in 0.005M concentration. Incubate in the presence of tetrazole components such as INT, neotetrazolium, or nitro-blue tetrazolium (nitro-BT) for periods from 30 min to 2 hr at 37°C. Visible reactions with similar localization during incubation for 30 to 120 min are observed only with tryptamine and 5-HT. No visible reaction is obtained with 5-methoxy-1-methyltryptamine, 5-methoxy-2-methyl-tryptamine, gramine, bufotenine, indole-3-acetic acid, DL tryptophan, indole, skatole, tyramine, histamine, spermine, and acetaldehyde. Dopa, dopamine, and epinephrine cause a spontaneous dye reduction in the absence of tissue sections. N-N-α-Dimethyltryptamine yields a faint reaction.

Optimal activity, with the tryptamine-tetrazolium method is obtained at pH values between 7.2 and 7.6 at 37°C. No reaction takes place below pH 6.8 and above pH 8.0. The reaction is inhibited by heating the slides for 20 min at 60°C. An 0.005M concentration of tryptamine is optimal for the tetrazolium reaction, while lower amounts of this sub-strate require a longer incubation time. An incorporation of a 20% sodium sulfate solution in the incubating mixture causes a moderate reduction in staining intensity, but a slightly sharper localization of the formazan dye.

In 1962 Cunningham and his colleagues (77) developed a controlled temperature freeze-sectioning technique which allows unfixed tissue to be sectioned at 8 μ. This facilitates the presentation of the cellular structure to a degree adequate for accurate histochemistry.

A histochemical coupled peroxidatic oxidation technique of MAO estimation has been recently reported by Graham and Karnovsky (78). In the presence of peroxidase its substrate, 3-amino-9-ethylcarbazole, is oxidized by the H_2O_2 generated during the oxidation of tryptamine catalyzed by MAO. The resulting insoluble red oxidation product is apparently deposited at sites of MAO activity. Graham and Karnovsky have been able to demonstrate histochemically the presence of MAO in liver and kidney of rats and guinea pigs (78) with this technique. This coupled oxidation method of MAO determination may offer a useful histochemical tool alternative to the tetrazolium technique.

H. DIRECT MEASUREMENT OF MAO INHIBITION IN HUMANS BY THE
METHOD OF LEVINE AND SJOERDSMA (79)

Most methods that are currently applied to the determination in man
of MAO inhibition by drugs are indirect.   Certain errors may be inherent
in indirect investigations, e.g., in measurements of the urinary excretion
of a substrate amine.   Urinary levels of the excreted amines depend, of
course, not only upon variations in the intake of their precursors but also
on the rate of their formation or the rate of their release from tissues.
Furthermore, the relationship between the degree of inhibition of a MAO
substrate to the amount of the urinary excretion of the latter is not yet
known.
    A direct approach to the measurement of MAO inhibition in man was
attempted by Levine and Sjoerdsma (79) in specimens of mucosa of the
human jejunum obtained by peroral biopsy.   These authors determined
the mucosal levels of MAO activity before, during, and after the administration of two different MAO inhibitors.   Mucosal tissue MAO levels
were then correlated with urinary levels of excreted tryptamine, which
had been used as substrate, and the relative competence of oral and
parenteral ways of drug application was compared.   The relationship of
MAO inhibition in the intestine to that in other organs in laboratory
animals was also studied.

Procedure.   Nine adult hospitalized patients suffering from essential
hypertension were chosen for this study.   In 5 subjects, specimens of
jejunal mucosa were obtained before treatment by peroral biopsy (Smith
and co-workers, 80) and were tested for MAO activity by the fluorometric
method of Lovenberg, Levine, and Sjoerdsma (62; see Section II-3-E-a).
Two patients were then treated with isocarboxazid, a hydrazine-type of
MAO inhibitor, and three other patients were subjected to a treatment
with MO-911 (pargyline hydrochloride), a nonhydrazine-type of MAO
inhibitor.   The drugs were administered once daily in doses adequate to
produce postural hypotension.   Between the 10th and the 60th day of
treatment, specimens of jejunal mucosa were again obtained and tested
for MAO activity.   To avoid errors, biopsies during treatment were
performed only 20 to 24 hr after the application of the inhibitory drug.
Findings in this study have indicated that determination of MAO activity
in specimens of jejunal mucosa, obtained by biopsy before and after the
application of the inhibitory drug, represents a fairly satisfactory means
for the direct demonstration in man of MAO-inhibiting characteristics of
a drug.   The results obtained in these experiments have further shown
that both MAO-911 and isocarboxazid in doses commonly used in clinical

practice bring about marked inhibition of MAO activity in the mucosa of human jejunum.

However, there are certain limitations of the direct method for the estimation of MAO inhibition. This technique can be used only for the study of drugs producing an irreversible MAO inhibition because tissue specimens as small as those obtained by jejunal biopsy are homogenized at such high dilutions that the concentrations of reversible inhibitors may fall below their effective levels. Finally, because of different degrees of inhibition produced in various tissues by MAO inhibitors, the authors think it advisable to test new MAO inhibitors first in experimental animals in order to establish the relationship of MAO inhibition in the intestinal mucosa to that in other organs (79).

III. METHODS OF PURIFICATION, CRYSTALLIZATION, AND ESTIMATION OF HISTAMINASE, DIAMINE OXIDASE (DAO)

In 1932 McHenry and Gavin (10) were the first to attempt a purification of hog kidney histaminase. These workers prepared an acetone powder which was stable at room temperature for about 12 months from hog kidney cortices. Subsequently, partial purification of hog kidney extracts was achieved by salt fractionation (81,82), by thermal denaturation of the concentrated inert proteins at 60 to 62°C (83), or by isoelectric precipitation (84). A considerable (300-fold) purification of the hog kidney enzyme was obtained by Tabor (85a) with a method involving fractionation of hog kidney cortex by acetone, sodium sulfate, heating at 60°C, adsorption on Cγalumina, and pH precipitation. Swedin (85b), by substituting Tabor's Cγalumina batch procedure with alumina column chromatography, achieved a further high degree of purification of the hog kidney enzyme. More extensively purified homogeneous preparations of this enzyme have been obtained lately by Goryachenkova (86a,b), Kapeller-Adler and MacFarlane (87), and Mondovi and his colleagues (88a,b). Yamada and his co-workers (89) have recently described a method by which a crystalline preparation of diamine oxidase was obtained from hog kidney cortices. This method will be discussed here in detail.

1. Procedure for the Purification and Crystallization of Diamine Oxidase from Hog Kidney According to Yamada, Kumagai, Kawasaki, Matsui, and Ogata (89)

Procedure. All operations are carried out at 5°C.

Step 1. In a Waring Blender fresh hog kidney cortices (about 1100 g) suspended in 1100 ml of $0.03M$ phosphate buffer, pH 7.0, are homogenized for 2 min and the suspension is then centrifuged.

Step 2. The supernatant is subjected to a fractionation with ammonium sulfate at 30 to 60% saturation, followed by dialysis against $0.03M$ phosphate buffer (pH 7.0).

Step 3. The dialyzed enzyme solution is applied to a DEAE-Sephadex A-50 column (6 × 60 cm) which had been equilibrated with $0.03M$ phosphate buffer (pH 7.0). The enzyme is desorbed from the column by stepwise elution with $0.07M$ and $0.1M$ phosphate buffers (pH 7.0). The eluted active fractions are pooled and concentrated by precipitation with ammonium sulfate at 60% saturation. The precipitate is collected and dialyzed against $0.1M$ phosphate buffer (pH 7.0).

Step 4. The dialyzed enzyme solution is again submitted to a fractionation with ammonium sulfate at 35 to 55% saturation, followed by dialysis against $0.1M$ phosphate buffer (pH 7.0).

Step 5. The dialyzed enzyme fraction is adsorbed on a hydroxylapatite column (5 × 15 cm), equilibrated before use with $0.1M$ phosphate buffer (pH 7.0). The column is then washed with $0.1M$ phosphate buffer containing $0.2M$ ammonium sulfate, and the enzyme is eluted from the column with $0.1M$ phosphate buffer (pH 7.0) containing $0.5M$ ammonium sulfate. The active eluates with a specific enzymic activity greater than 0.200 are combined and concentrated by the addition of ammonium sulfate at 60% saturation.

Step 6. The precipitate is dissolved in $0.1M$ phosphate buffer (pH 7.0) and passed through a Sephadex G-150 column (2 × 100 cm), equilibrated with $0.1M$ phosphate buffer (pH 7.0). Active enzyme fractions showing a specific activity greater than 0.500 are combined and concentrated by the addition of ammonium sulfate at 60% saturation. The precipitate is collected and dissolved in $0.1M$ phosphate buffer (pH 7.0).

Step 7. Finely powdered ammonium sulfate is then very carefully added to the enzyme solution until the latter has become slightly turbid, and the reaction mixture is placed in an ice bath. After 3 hr, crystallization starts and is virtually completed within a week. Crystals of the enzyme appear as fine, highly refractive needles, showing a pink color. For recrystallization of the enzyme, Step 7 is repeated.

Properties of Crystalline Hog Kidney DAO. On Tiselius electrophoresis, carried out at 4°C in sodium phosphate-sodium chloride buffer (Miller and Golder, 90) of 0.1 ionic strength and pH 7.4, the crystalline DAO preparation migrated as a single symmetric band. In the ultracentrifuge in $0.1M$ potassium phosphate buffer (pH 7.4) at 16°C, the enzyme sedimented as a single symmetric peak. When extrapolating the data obtained from four ultracentrifuge runs to zero protein concentration, the authors arrived at a value for the sedimentation constant of $S_{20,w}^0$ of 9.90×10^{-13} (cm/sec). With increasing protein concentration this sedimentation constant decreased by 0.03×10^{-13} (cm/sec)/mg of protein. Furthermore, a solution of the crystalline enzyme, containing 6.18 mg protein/ml, gave a diffusion constant $D_{20,w}$ of 5.14×10^{-7} (cm^2/sec). Assuming a partial specific volume of 0.75, Yamada and his colleagues (89) calculated a molecular weight of the enzyme of 185,000.

The enzyme is pink and spectrophotometric investigations revealed that the pink color is associated with an absorption maximum at 470 nm. The color is discharged by putrescine as substrate as well as by sodium dithionite and is restored by oxygenation. Investigations for a possible metal content by atomic absorption spectrophotometry and chemical analysis showed that copper is the only metal component of the crystalline enzyme, its content amounting to a value of 11.7 mμ-atoms/mg of the enzyme; this corresponds to 2.17 g-atoms of copper/mole of the enzyme. The crystalline enzyme preparation oxidatively degrades cadaverine, putrescine, 1,6-diaminohexane, histamine, agmatine, and 1,3-diamino-propane at relative rates of 100, 97, 61, 59, 40, and 17, respectively.

The authors estimated the enzymic activity manometrically by measuring the O$_2$-uptake at 38°C in a reaction mixture containing 100 μg of the enzyme, 50 μg of catalase, 10 μmoles of substrate, and 100 μmoles of potassium phosphate buffer (pH 7.4) in a total volume of 3 ml. Under these conditions, the crystalline enzyme with the specific activity of 1.079 oxidized 2.04 μmoles of cadaverine/min/mg. of the enzyme.

Protein was estimated spectrophotometrically measuring the extinction at 280 nm. An E value of 1.63/1 mg/ml, and for 1 cm lightpath, obtained by extinction and dry weight determinations, was used throughout the authors' investigations.

2. Methods for the Histochemical Detection of Histaminase, Diamine Oxidase, Activity in Tissue Sections

A. PROCEDURE OF VALETTE AND COHEN (91).

Feulgen and Rossenbeck (92) suggested that aldehydes might be identified histochemically by the Feulgen reaction, a staining method widely used by histologists and cytologists for cell nuclei and chromosomes. The Feulgen reaction is apparently based upon the fact that aldehydes restore the intense magenta color of basic fuchsin which is the main component of the Feulgen reagent.

Oster and Schlossman (93) were the first workers to use Feulgen's reaction for the detection of aldehyde formation during the action of MAO using tyramine as substrate. Similarly, Valette and Cohen (91) availed themselves of this reaction for histochemical investigations of histaminase activity in various animal tissues.

Procedure. Fresh, frozen sections of various animal tissues are pre-incubated with a 2% (w/v) solution of sodium bisulfite for 24 hr at 37°C. The tissue sections are thoroughly rinsed with distilled water, then immersed in an 0.5% (w/v) solution of histamine dihydrochloride, previously adjusted with a buffer to pH 7.2, and incubated at 37°C for 24 hr. Other tissue sections are similarly incubated as controls, either in histamine-free buffer solutions (pH 7.2), or with histamine in the presence of the histaminase inhibitor, semicarbazide hydrochloride (0.01%, w/v). On completed incubation, all the tissue sections are washed with distilled water, and then plunged for 1 hr into the solution of Feulgen reagent. The tissue sections are then washed twice with a 0.2% solution of sodium bisulfite and finally with freshly boiled distilled water. They are mounted in glycerine. Whereas the control sections show a uniformly distributed rose color, the tissue sections containing histaminase display plaques of cells of deep violet color surrounded by rose-colored tissue free of histaminase.

Preparation of Feulgen Reagent (94). Dissolve 1 g of basic fuchsin in 200 ml of boiling distilled water, filter, cool, and add 2 g of potassium metabisulfite ($K_2S_2O_5$) and 10 ml of 1N HCl. Bleach for 24 hr, then add 0.5 g of activated carbon (Norit), shake for about 1 min, and filter through coarse paper. The filtrate should be colorless.

By means of this histochemical technique Valette and Cohen (91) were able to demonstrate the presence of histaminase in various organs of different animals. Thus in guinea pigs these authors detected hista-minase activity in the proximal convoluted parts of the renal tubules. The rat kidney, which does not contain histaminase, did not give a positive Feulgen test. Histaminase activity was, however, detected in the intestinal epithelium and in the bronchioles of the guinea pig. In human placenta a positive Feulgen reaction was encountered only in the decidual cells, whereas a completely negative result was obtained by Valette and Cohen in fetal membranes. These observations are in agreement with previous findings of Swanberg (95).

B. PROCEDURE FOR THE QUANTITATIVE ESTIMATION OF HISTAMINASE
LOCALIZED IN THE HUMAN UTERUS AND ATTACHED INTACT
PLACENTA, AT TERM, ACCORDING TO GUNTHER AND GLICK (96).

In an effort to localize the source of the markedly increased blood serum levels of histaminase in human pregnancy, the authors subjected fresh-frozen microtome sections, cut serially through the uterus and attached intact placenta, to a quantitative chemical analysis for histaminase activity. For the determination of histaminase activity these authors modified the spectrophotometric technique of Aarsen and Kemp (97), which is based on the measurement of hydrogen peroxide formed during the enzymic oxidation of histamine (see Section I), and adapted it to their studies on histaminase in microtome sections of the human uterus and attached intact placenta at term.

Preparation of Tissue Sections. A human uterus with unseparated placenta from a pregnancy at term was obtained immediately on Caesarean hysterectomy. Blocks of tissue cut from the uterine wall and the attached placenta were rapidly frozen with solid carbon dioxide and stored at $-20°C$. Blocks of human placenta obtained from vaginal deliveries were similarly cut and stored at $-20°C$. Sampling of fresh-frozen cylinders of the tissue and their microtomy in a cryostat at $-15°C$, to obtain circular sections 16 μ thick of 4 mm diameter and 0.2 μl volume, was performed following a conventional procedure used in the author's laboratory.

Analytical Procedure. To obtain serial samples throughout the tissue, place one section on a slide for histological examination by staining with haematoxylin-eosin, the next five sections in a 27 mm-reaction tube for the no-substrate blank, the next one on a slide for histologic study, and the next five in a tube for enzyme reaction, and repeat this pattern.

Constriction pipets are used for all volumes of 200 μl or less.

1. Pipet 50 μl of $0.1M$ sodium phosphate buffer (pH 6.2), freshly prepared before use, into each of the reaction tubes containing the tissue sections for the no-substrate blank.

2. Pipet 40 μl of the same buffer into each of the tubes containing the five tissue sections for the enzyme reaction.

3. Mix well using a vibration mixer to break up the tissue.

4. In a 5-ml Pyrex tube mix 1.6 ml of $0.1M$ sodium phosphate buffer (pH 6.2), 200 μl of horseradish peroxidase stock solution (Boehringer No. POD 11,15302, containing 0.4 mg peroxidase/ml of distilled water, stored at $-20°C$ up to 1 week), and 200 μl of o-dianisidine (Sigma) stock solution (containing 5 mg of o-dianisidine/ml of 95% ethanol, stored at $-20°C$ up to 3 days). Pipet 50 μl of this mixture into each of the tubes containing tissue, and mix.

5. Pipet 10 μl of $0.3M$ histamine solution (110 mg of histamine dihydrochloride/ml of distilled water, containing 70 μl of $5N$ sodium hydroxide for neutralization and diluted twice with an equal volume of water) into each of the tubes for enzyme reaction, but not into the blank tubes.

6. Cap tubes, mix, and incubate at 37.5°C for 8 hr.

7. Stop reaction by addition of 50 μl of 50% sulfuric acid (v/v) to each tube, mix, and centrifuge at 1000*g* for 2 min in an International Clinical Centrifuge, Model CL, or in a Misco Microcentrifuge (Microchemical Specialities, Berkeley, Calif.).

8. Measure extinction of the clear supernatant at 530 nm in a Beckman spectrophotometer, Model DU.

9. Subtract extinctions of no-substrate blanks from those of corresponding enzyme reactions to obtain activity, and express activity as the corrected change in extinction per unit time.

A linear relationship was established by the authors between the number of placental tissue sections (up to nine sections were tested) and enzyme activity with a number of tissue samples obtained from different individuals. The highest histaminase activity was found in the decidua and in the Nitabuch's membrane. These findings compare well with observations of Swanberg (95) in regard to the localization of histaminase in the human placenta, and support the latter authors suggestion that the decidua is a significant source of the increased plasma histaminase levels in human pregnancy.

3. Methods for the Quantitation of Histaminase, Diamine Oxidase (DAO)

A. INTRODUCTION

Most of the commonly used methods for the estimation of histaminase, diamine oxidase, have been recently reviewed by Zeller (98) and Werle (42).

As mentioned above (see Section I), all the components of the fundamental amine oxidase-substrate reaction with the exception of water have been utilized as a basis for the determination of the activity of an amine oxidase. Many different enzymic units have been proposed by various workers for the activity of histaminase, diamine oxidase, and they are still being used to express the results obtained. These units should be brought up to date, however, and replaced by those which comply with the recommendations of the International Union of Biochemistry (99).

To the manometric methods of histaminase, diamine oxidase estimations, measuring the O_2-uptake during the enzymic reaction, the same criticism applies which was mentioned in the discussion of the techniques designed for the estimation of monoamine oxidase activity (see Section II-3-A).

The manometric technique is not sensitive enough to determine adequately the activity of histaminase occurring in very small amounts of many biological materials. Thus, in some instances part of the oxygen consumption may be due not only to an effect of histaminase on its substrate, but to a further oxidation of the primary enzymic product by other enzymes, e.g., in the case of imidazole acetaldehyde produced in the histaminase-histamine reaction, to its transformation in a secondary oxidative process by xanthine oxidase to imidazole acetic acid. Moreover, although in assays of monoamine oxidase activity this extraneous O_2-uptake can be prevented by the addition of cyanide or semicarbazide, this technique cannot be applied, of course, to histaminase, diamine oxidase, since this enzyme is itself inhibited by the substances just mentioned.

Other methods have been designed for the estimation of histaminase, diamine oxidase, which are based on the aldehyde production, or the formation of hydrogen peroxide, or ammonia.

Two types of highly sensitive enzymic methods based upon measurements of the disappearance of substrate have been published thus far. Both methods, however, are applicable only when histamine is used as a substrate. One of these methods involves a biological determination of histamine and the other a spectrofluorometric estimation of the histamine which disappears during the histaminase-histamine reaction. Conversely, methods have been designed which can be used only with aliphatic diamines and not with histamine as substrates since they are based upon a spontaneous cyclization of the aldehyde metabolite of the diamine oxidase-diamine reaction to form a cyclic Schiff base (azomethine), Δ^1-pyrroline being formed from putrescine and Δ^1-piperideine from cadaverine. The production of these compounds can be demonstrated by means of o-aminobenzaldehyde which condenses with the cyclic azomethine to form yellow or orange 1,2-dihydroquinazolinium derivatives.

B. BIOLOGICAL METHOD FOR THE ESTIMATION OF HISTAMINASE ACTIVITY BY MEASURING THE DISAPPEARANCE OF HISTAMINE ACCORDING TO SPENCER (100).

Spencer investigated the effects of enzyme and substrate concentrations upon the rate of inactivation of histamine by histaminase of rat ileum, and found that the time necessary for one-half of the histaminase to be destroyed is inversely proportional to the histaminase content of the incubation mixture.

Procedure. Male Wistar albino rats weighing 120 to 160 g are used. The small intestine is removed from groups of 4 to 5 rats immediately after killing the animals by a blow on the head. After discarding the first 6 and the last 2 in., the tissue is washed with normal saline solution. Grind the bulked tissue in a glass mortar with silver sand, add Tyrode solution (5 ml/g), and extract for 5 min. Centrifuge the supernatant at 2800 rpm for 10 min. The enzyme is present in the supernatant. In a 250-ml conical flask place 95 ml of Tyrode solution (Table I) containing a known amount of histamine and equilibrate to 37°C, add to this solution 5 ml of the supernatant, shake the flask well to mix the contents, remove a first sample immediately, and incubate the remainder with frequent shaking at 37°C. Further samples are removed at 10-min intervals. To stop enzyme activity, bring each sample to the boil, cool, and store at 4°C until it is tested for its histamine content. In assays in which the effect of changes in enzyme or substrate concentration are to be investigated, use different volumes or quantities of extract and histamine so that all incubation mixtures initially amount to 100 ml in volume. Throughout the first 90 min of incubation, the pH of the mixture remains within the limits of 7.0 to 7.4.

The samples are assayed directly on the atropinized ileum of the guinea pig for the estimation of their histamine content. The tests are performed in duplicate, and means are used for the calculation of results.

The results obtained indicate an exponential relationship between histamine concentration and length of incubation time, which becomes linear, however, when plotted semilogarithmically.

For the estimation of histaminase activity Spencer (100) proposes a method as follows: Histamine concentrations are determined at 0 min and various time intervals afterwards, and a plot of $\log_{10} (x_0/x_t)$ against t is made. A linear relationship should prevail over most of the reaction, at least as far as 70 to 80% of histamine degradation. Read the time taken for 50% histamine to be destroyed (or another fixed percentage within the linear part of the graph); this reading represents the DT_{50} value. The ratio of enzyme activity (E) in different tissues is then

$$\frac{E_1}{E_2} = \frac{DT_{50}\ (2)}{DT_{50}\ (1)}$$

The percentage of control tissue activity is expressed as follows:

$$E\ (\text{experimental}) = \frac{DT_{50}\ (\text{control})}{DT_{50}\ (\text{experimental})} \times 100$$

In different incubation mixtures Spencer obtained $\pm 7\%$ maximum error with this method. If an initial histamine concentration within the range of 1.50 to 2.50 μg/ml is applied, then a dilution before the assay is possible throughout the entire range of samples obtained, and this obviates difficulties in assay due to interference by tissue proteins.

The following advantages of this new method are pointed out by Spencer (100):

1. The DT$_{50}$ time depends upon a constant and does not change for a given concentration of enzyme.

2. The value is arrived at from data obtained from all the samples assayed for histamine in a given mixture. Hence one erroneous assay only slightly affects the results, although the initial sample is the most important one in any mixture.

3. Incubation mixtures containing different initial amounts of histamine can be compared, providing they fall within the range between 75 to 150% of the mean initial histamine concentration. Varying amounts of endogenous histamine will not alter the result.

4. A maximum experimental error of less than 10% can be expected.

TABLE I

Preparation of Tyrode solution (*Tyrode*, 118)

Stock solutions	g/liter
Sodium bicarbonate, $NaHCO_3$	40.00
Magnesium chloride, $MgCl_2 + 6H_2O$	8.54
Sodium acid phosphate, $NaH_2PO_4 + H_2O$	2.00
Calcium chloride, $CaCl_2$	8.00
Potassium chloride, KCl	8.00
Sodium chloride, $NaCl$	80.00

For the histamine assay on the guinea pig ileum, place 25 ml each of the first five stock solutions mentioned in a 1000-ml measuring flask and add 100 ml of the sodium chloride stock solution. Add 1 g of glucose and 1 ml of an 0.01% (w/v) aqueous solution of atropine and make up to volume with distilled water.

In regard to the usefulness of highly sensitive methods of histaminase estimation based upon measurement of histamine disappearance, Zeller (98) emphasizes that complementary experiments must be carried out in order to find out whether histaminase alone is responsible for the disappearance of histamine.

C. SPECTROPHOTOMETRIC METHODS FOR THE ESTIMATION OF
HISTAMINASE, DIAMINE OXIDASE, ACTIVITY

a. Spectrophotometric Method for the Determination of Diamine Oxidase Based on the Aldehyde Formation during the Action of DAO on Putrescine According to Holmstedt and Tham (101a), and Holmstedt, Larsson, and Tham (101b). This simple method requires little apparatus and many estimations can be performed simultaneously (101b). It is based upon the formation of γ-aminobutyraldehyde when putrescine is acted upon by diamine oxidase. This aldehyde undergoes a rapid cyclization to Δ^1-pyrroline, which is made to react with o-aminobenzaldehyde to form a yellow-colored product, the 2,3-trimethylene-1,2-dihydroquinazolinium hydroxide (85a,101b; see this Section III-3-A). The colored reaction product shows a strong extinction maximum at 430 nm; its molar extinction coefficient was found to be 1.86×10^3 mole^{-1} cm^{-1} (101a,b).

Method. Reagents required:

1. 0.005M o-aminobenzaldehyde is synthesized by the method of Bamberger and Demuth (102a,b), and stored at $-8°C$ in ampoules containing nitrogen.

2. 0.1M putrescine dihydrochloride dissolved in phosphate buffer, pH 6.8 (Light & Co. Ltd., London).

3. Phosphate buffer (Soerensen) 0.067M, pH 6.8.

4. Trichloracetic acid (10%, w/v).

5. Enzyme preparation; a hog kidney preparation of diamine oxidase of high specific activity, free of catalase and peroxidase, obtained according to the method of Arvidsson and his colleagues (103).

Procedure. In a metabolic shaker at 37°C shake a test tube containing a mixture consisting of 2.5 ml of a solution of 0.005M o-aminobenzaldehyde in 0.67M phosphate buffer (pH 6.8) (stable at 4°C for a week) and the enzyme preparation (2 to 20 mg) dissolved in phosphate buffer (pH 6.8) and made up to 4.5 ml with the same buffer. After completing the temperature equilibration, add 0.5 ml of a 0.1M putrescine dihydrochloride solution freshly prepared in phosphate buffer (pH 6.8) and incubate at 37°C for 3 hr. Controls, which are assays without the addition of substrate, are incubated along with the tests. At the end of incubation add 1 ml of 10% trichloracetic acid to both tubes and thus stop the enzymic reaction. At this stage the tubes can be kept for 24 hr. Centrifuge both, test and control, and read the extinction of both supernatants in a Beckman spectrophotometer, Model DU, at 430 nm and 1 cm lightpath. Should the extinction exceed the reading of 0.8, dilute with distilled water, which is apparently in agreement with ordinary photometric techniques.

Prepare a standard calibration curve which makes it possible to convert the extinction readings of the dye formed in the DAO-putrescine reaction into micromoles of putrescine destroyed per milligram of enzyme in 60 min, in which units Holmstedt and his associates (101a,b) express the DAO activity. This method can be applied in studies of the effect of various inhibitors on DAO. Moreover, in tissue homogenates and in partly purified enzyme preparations this spectrophotometric technique is not impeded by the presence of other oxidative enzymes such as catalase or peroxidase.

b. Spectrophotometric Micromethod for the Estimation of Histaminase, Diamine Oxidase, in Biological Fluids Involving a Peroxidatic Oxidation of Indigo Disulfonate According to Kapeller-Adler (104a,b). This method is a spectrophotometric modification of the microvolumetric technique which was designed in 1951 for the determination of histaminase activity in biological media (Kapeller-Adler, 87, 105,a,b).

The theory underlying this spectrophotometric estimation method of histaminase, diamine oxidase, may be described as follows: According to the fundamental amine oxidase-substrate reaction (see Section I), 1 molecule of hydrogen peroxide is formed from 1 molecule of substrate. In the subsequent coupled oxidation reaction the H_2O_2 thus generated oxidizes the added indigo disulfonate. It is well known that indigo is oxidized by any oxidizing agent to isatin (106) according to the equation:

$$C_{16}H_{10}N_2O_2 + 2O = 2C_8H_5NO_2$$

On the basis of this evidence it is assumed that 2 molecules of H_2O_2 formed from 2 molecules of substrate in the histaminase, diamine oxidase-substrate reaction, will oxidize 1 molecule of indigo disulfonate. Since the molecular weight of indigo disulfonate is 466, it follows that half a micromole of this compound will amount to 233 μg. Furthermore, since 1 molecule of indigo disulfonate requires for its oxidation 2 atoms of oxygen, it is deduced that 1 ml of an $0.002N$ indigo disulfonate solution which contains 233 μg of indigo disulfonate thus corresponds to half a micromole of this compound. Hence it is concluded that 1 ml of $0.002N$ indigo disulfonate, oxidized by H_2O_2 in the coupled oxidation reaction mentioned, corresponds to the destruction of 1 micromole of substrate and, if determined per time unit of 1 min, forms the basis for the International Unit (IU) of histaminase, diamine oxidase. Thus, the consumption of 1 ml of $0.002N$ indigo disulfonate indicates the destruction of 1 micromole each of histamine (111 μg), cadaverine (102 μg), or putrescine (88 μg)/min. Likewise, the consumption of 1 μl of $0.002N$ indigo disulfonate indicates the destruction of 1 millimicromole of histamine (0.111 μg), cadaverine (0.102 μg), or putrescine (0.088 μg)/min, and expresses 1 International milliunit (ImU).

Protein is determined by a modification (Miller, 107) of the method by Lowry et al. (108). The specific activity is defined as enzyme activity per milligram of protein.

Materials and Methods. Partly purified extracts of human placentae (25), amniotic fluids, and pregnancy sera were examined by means of this spectrophotometric method for their enzymic effects on histamine, cadaverine, and putrescine. The amniotic fluids were obtained either by amniocentesis or by puncture of the amniotic sac, and pregnancy sera were taken from women at various stages of pregnancy.

Reagents. *1. Buffers.* (a) 0.175M sodium phosphate buffer (pH 6.8). (b) EDTA-phosphate buffer, 0.175M sodium phosphate buffer (pH 6.8) containing 0.001M EDTA.

2. 0.002N Indigo disulfonate solution. Dissolve 23.3 mg indigo carmine (Analar, BDH) in 100 ml of distilled water. Solution keeps well at 4°C for 2 to 3 weeks.

3. Standard solutions of substrates. Histamine dihydrochloride, BDH. Dissolve 27.6 mg of histamine dihydrochloride in 10 ml of phosphate buffer (pH 6.8). Add a few drops of chloroform for preservation. Solution keeps at 4°C for a few days. *Cadaverine dihydrochloride,* 1,5-diaminopentane dihydrochloride, purissimum (Fluka A.G., Buchs, Switzerland). Dissolve 525 mg of cadaverine dihydrochloride in 20 ml EDTA-phosphate buffer (pH 6.8), and add a few drops of chloroform. Keeps at 4°C for 2 to 3 weeks. *Putrescine dihydrochloride,* 1,4-diaminobutane dihydrochloride, purum (Fluka A.G., Buchs, Switzerland). Dissolve 483 mg of putrescine dihydrochloride in 20 ml EDTA-phosphate buffer (pH 6.8) and add a few drops of chloroform. Keeps at 4°C for 2 to 3 weeks.

Most of the biological fluids seem to contain substances which strongly inhibit the enzymic effect on cadaverine and putrescine but not that on histamine (109). Since this inhibitory effect is completely abolished in the presence of EDTA, phosphate buffer containing 0.001M EDTA is used for the estimation of diamine oxidase activity.

The *optimal substrate* concentrations were found to be 20mM quantities of cadaverine dihydrochloride and of putrescine dihydrochloride, and 1mM of histamine dihydrochloride (104a).

Dialysis. All biological fluids are subjected before being tested for their enzyme activity to a dialysis. About 10 ml of the enzyme solution are dialyzed against 5000 ml of $0.175M$ EDTA-phosphate buffer (pH 6.8) at 4°C for 3 hr using Visking dialyzing tubings which had been pretreated with an $0.001M$ EDTA solution. The dialyzed fluids are centrifuged and the water-clear supernatants are subjected to the enzymic assay.

In all assays air was used as the gas phase.

Incubations were carried out in a thermostatically controlled water bath at 37°C. The extinctions were measured throughout all experiments in an Optica CF4 spectrophotometer.

Procedure. In a Pyrex test tube (2 × 15 cm), marked a, place 0.2 ml of the dialyzed enzyme solution and 0.6 ml of the $0.002N$ indigo disulfonate solution. In another test tube, marked b, place multiples of a mixture consisting of 0.5 ml EDTA-phosphate buffer and 0.2 ml of the standard solution of cadaverine dihydrochloride or of putrescine dihydrochloride. When histamine is used as substrate, place in test tube b multiples of a mixture consisting of 0.6 ml phosphate buffer (pH 6.8) and 0.1 ml of the histamine standard solution. Preincubate all the tubes marked a and b in a thermostatically controlled water bath at 37°C for 15 min. Then pipet carefully into each a-tube 0.7 ml of the substrate buffer mixture present in the b-tube, mix very gently by inversion, and stopper the tubes. Incubate for 60 min at 37°C. Blanks are set up in the same fashion as tests with the exception that, for the blanks, test tube b contains only buffer without the addition of the substrate. The end volume in each blank and test amounts to 1.5 ml. To stop the enzymic reaction on completed incubation, place all the tubes containing tests and blanks in a large beaker with ice-cold water and rapidly dilute the fluid in each tube to 10 ml by adding 8.5 ml of distilled water. Stopper the tubes, mix gently by inversion, and read blanks and tests in a spectrophotometer (Optica CF4), at 613 nm against a dilute phosphate buffer solution, pH 6.8, (1.5 ml phosphate buffer made up to 10 ml with distilled water), using cuvets of 1 cm lightpath.

Calculation. Subtract the extinction reading obtained for the test from that of the blank, calculate the enzyme activity per milliliter of fluid and minute, as discussed above, and express it in ImU. It is convenient to remember that 1 ml of $0.002N$ indigo disulfonate solution, diluted to 10 ml with distilled water, shows an extinction of 1.0 in a 1-cm cuvet at 613 nm.

Assay for Dialyzed Pregnancy Sera. In the Pyrex test tube (2 × 15 cm) marked a, place 0.5 ml of clear dialyzed pregnancy serum and add 0.6 ml of $0.002N$ indigo disulfonate solution. In the test tube marked b, place multiples of a mixture consisting of 0.2 ml of the stock solutions of cadaverine dihydrochloride or putrescine dihydrochloride, and 0.2 ml EDTA-phosphate buffer (pH 6.8). When histamine is to be used as substrate place in test tube *b* multiples of a mixture containing 0.1 ml of the stock solution of histamine dihydrochloride and 0.3 ml phosphate buffer (pH 6.8). Continue as described above for the assays of amniotic fluids or placental extracts. In some pregnancy sera with a very low enzymic effect the incubation time must be extended from 60 to 120 min.

The advantage of this simple spectrophotometric micromethod is that blanks and tests can be examined easily in replicate and that the enzymic effect of different biological media can be studied simultaneously on various substrates.

Previous findings (Kapeller-Adler, 109) suggesting that placental histaminase, due to its significant behavior towards histamine, cadaverine, and putrescine, resembled the diamine oxidase of pea seedlings more than hog kidney histaminase have been confirmed with the spectrophotometric method just described.

c. Spectrophotometric Micromethod for the Determination of Diamine Oxidase Based on an Enzymic Determination of Ammonia of Kirsten, Gerez, and Kirsten (110). In this micromethod the ammonia formed in the primary diamine oxidase-substrate reaction is utilized in a secondary enzymic reaction as substrate of glutamate dehydrogenase, according to the equation:

$$\alpha\text{-oxoglutarate} + NH_4^+ + NADH + H^+ \underset{\text{dehydrogenase}}{\overset{\text{glutamate}}{\rightleftharpoons}} \text{glutamate} + NAD^+ + H_2O$$

This reaction occurs at a neutral pH and thus eliminates the necessity of the strong alkalinization of the method of Conway (54). With the large amounts of α-oxoglutarate applied, ammonia is quantitatively consumed in the enzymic conversion of α-oxoglutarate to L-glutamate.

Method. Reagents required:

1. Glutamate dehydrogenase (Boehringer, Mannheim), free from ammonium ions, 3-times recrystallized, and dialyzed against $0.2M$ sodium phosphate buffer (pH 7.6), dissolved in 50% glycerol (pH 7.3).

2. α-oxoglutaric acid (Boehringer, Mannheim). Five milliliters of $0.08M$ α-oxoglutaric acid in water are neutralized with $2N$ NaOH to pH 7.0 (use indicator paper). The neutral solution keeps for a few days at 4°C.

3. $NADH_2$ solution (Boehringer, Mannheim) in $0.2M$ phosphate buffer (pH 7.6) contains 10 mg of $NADH_2$/ml of buffer.

4. $0.2M$ sodium phosphate buffer (Soerensen) (pH 7.6).

The reagent mixture required for the determination of NH_4^+, formed in the DAO-substrate primary reaction, contains 29 ml of $0.2M$ sodium phosphate buffer (pH 7.0) and 0.6 ml of a $NADH_2$ solution (10 mg $NADH_2$ in 1 ml of $0.2M$ phosphate buffer, pH 7.6).

Assay. Carried out at 25°C. Place 1.50 ml of the reagent mixture in a cuvet with lid and 1 cm light path and mix gently. Read extinction at 366 nm in a spectrophotometer. Then start the reaction by stirring into the solution 20 μl of glutamate dehydrogenase, cover the cuvet with the lid and follow the incorporation of the ammonia, formed in the primary enzymic reaction, into the L-glutamate by measuring the decrease in extinction at 366 nm within 25 min, when the reaction is usually completed. Express the results in micromoles of NH_4^+ formed and calculate from the latter the diamine oxidase activity. Blanks (1.5 ml of reagent mixture and 0.5 ml of water) are set up in parallel with every test reaction.

According to Kristen et al. (110), 1 to 100×10^{-9} moles of ammonia can be determined with an accuracy of $\pm 1\%$.

A better presentation by the authors of the above method is highly desirable.

D. SPECTROFLUOROMETRIC ESTIMATION OF HISTAMINASE ACTIVITY BY MEASURING THE DISAPPEARANCE OF HISTAMINE ACCORDING TO SHORE, BURKHALTER, AND COHN, JR. (111a)

The activity of histaminase is estimated by determining the rate of disappearance of histamine added to tissue homogenates. A simple fluorometric method is used for histamine estimations. In this very sensitive method histamine is first extracted from an alkalinized medium into n-butanol from which it is returned to an aqueous solution. Histamine is then condensed with o-phthalaldehyde (OPT) to yield a product showing a strong and stable fluorescence, which is measured in a spectrofluorometer.

Procedure. Tissues are homogenized in 9 vol of $0.2M$ phosphate buffer (pH 7.2), and after standing for 10 min the homogenate is centrifuged (111b). After preincubation of the supernatant in a Dubnoff metabolic shaker for 15 min at 37°C in an atmosphere of air, histamine is added to give a final concentration of 7.5 μg/ml, and the mixture is incubated at 37°C for 60 min. Samples are then removed for the estimation of histamine. For the extraction of residual histamine to be assayed, add 0.5 ml of $5N$ NaOH, 1.5 g solid NaCl, and 10 ml of n-butanol to a 4-ml sample placed in a 25-ml glass-stoppered shaking-tube. The tube is shaken for

5 min to extract histamine quantitatively into the n-butanol, and after centrifugation, the aqueous phase is removed by aspiration. The n-butanol layer is then shaken for about 1 min with 5 ml of salt-saturated $0.1N$ NaOH. This procedure removes any residual amount of histidine which may be present. Then centrifuge the mixture and transfer an 8-ml aliquot of the n-butanol extract to a 40-ml glass-stoppered shaking-tube containing 2.5 to 4.5 ml of $0.1N$ HCl and 15 ml of n-heptane. Shake for about 1 min and then centrifuge. The histamine is in the aqueous phase and is assayed fluorometrically.

Fluorometric Assay of Histamine. For the histamine estimation in the acid extract, transfer a 2-ml aliquot containing 0.005 to 0.5 μg histamine/ml to a test tube and add 0.4 ml of $1N$ NaOH, followed by 0.1 ml of the OPT reagent (1% w/v of o-phthalaldehyde, Mann Research Laboratories Inc., New York, in reagent-grade absolute methanol). After 4 min, 0.2 ml of $3N$ HCl is added. The contents of the tube are thoroughly mixed after each addition. The solution is then transferred to a cuvet and, on activation at 360 nm, the resulting fluorescence is measured at 450 nm (wavelengths uncalibrated). The fluorescence of the acidified solution is stable for at least 30 min. Over the range of 0.005 to 0.5 μg histamine/ml, the fluorescence intensity was found to be proportional to the concentration of histamine. The small fluorescence found in blanks of tissues or reagents may be corrected by omission of the condensation step. This can be achieved by adding all the reagents to a separate aliquot of a tissue extract but reversing the order of addition of OPT and $3N$ HCl. The resulting solution contains tissue blanks and reagent blanks, but not the fluorophor resulting from the condensation of OPT and histamine, for the condensation reaction does not take place in an acid medium.

Although histidine and histidyl histidine, when treated with OPT, produce fluorophores showing a spectral behavior similar to that of histamine, these compounds do not interfere with the histamine estimation, since they are not extracted into n-butanol from an alkaline solution. Likewise ammonia, which at concentrations greater than 4 μg/ml produces some fluorescence, does not interfere with the fluorometric histamine estimation in tissues or blood, since it is only poorly extracted into n-butanol. Thus, fluorescence readings not exceeding those of reagent blanks can be obtained when amounts of ammonia as high as 150 μg are run through the entire procedure.

In urines, however, this method cannot be used, for the urinary amounts of ammonia are sufficiently high to interfere with the procedure just described.

With this sensitive fluorometric method, amounts of histamine as low as 0.005 µg/ml can be determined.

E. RADIOACTIVE ASSAY METHODS FOR THE ESTIMATION OF DIAMINE OXIDASE ACTIVITY

a. Determination of DAO Activity in Hog Kidneys by Liquid Scintillation Counting According to Okuyama and Kobayashi (112). The technique of diamine oxidase determination by means of liquid scintillation counting is based on the formation of radioactive metabolites of the diamine oxidase-cadaverine-C^{14} reaction. The end products are extracted into toluene and are assayed in a liquid scintillation spectrometer. The method is also applicable to radioactive putrescine used as substrate. This radio assay has provided evidence for the quantitative production of Δ^1-pyrroline and Δ^1-piperideine as metabolites of diamine oxidase action on putrescine and cadaverine, respectively, and thus confirmed an early suggestion by Tabor (85a). This method does not work with radioactive histamine (see Section III-3-A).

Materials and Method. Radioactive compounds:

Putrescine-C^{14} (New England Nuclear Corp.) is diluted with nonisotopic putrescine to yield approximately 130,000 disintegrations/min (dpm)/5 µg base.

Cadaverine-C^{14} picrate synthesized by Schayer and his co-workers (113) was converted on a micro Dowex-1 column to the dihydrochloride and gave between 26,000 and 40,000 dpm/5 µg base.

Enzyme Preparations: Hog kidney diamine oxidase was used. Frozen hog kidneys (233 g) were minced and extracted with 1060 ml of $0.1M$ phosphate buffer (pH 7.6) for 20 min at 59 to 60°C using a mechanical stirrer. The extract was filtered through cheesecloth and fractionated with ammonium sulfate. The precipitate obtained at 40 to 55% ammonium sulfate saturation was dissolved in $0.1M$ phosphate buffer (pH 7.6) and dialyzed against deionized water overnight at 4°C. The enzyme preparation contained 13.48 mg nitrogen/ml and showed an oxygen uptake of 2.59 µl/min/mg nitrogen with cadaverine as substrate. This preparation also showed some catalase activity.

Liquid Scintillation Counting Method. A Packard Tri-Carb Liquid Scintillation Counter, Model 314 X, was used by these workers. The extraction solution consisted of reagent-grade toluene containing 0.35% of 2,5-diphenyloxazole (PPO). The solution gave 59% counting efficiency after being used in the extraction of the end products. For the radioactive assay of cadaverine-C^{14} and putrescine-C^{14} an ethanol-toluene mixture (25:70, v/v) containing 0.35% PPO was applied. The counting efficiency of the latter solution was 36%. All the countings were carried out in low-potassium 5-dram glass vials (Wheaton Glass Co.).

General Extraction Procedure. Into a screw-cap culture tube (10 × 1.5 cm) containing approximately 200 mg of sodium bicarbonate pipet 2 ml of the enzyme reaction mixture. Then add with a rapid constant-volume dispenser 10 ml of the PPO solution, shake the mixture vigorously for about 1 min, and then allow it to stand at −10° until the lower aqueous layer is frozen. Then pour the upper layer into a counting vial and assay in the liquid scintillation spectrometer. Although a single extraction procedure gives satisfactory results in semiquantitative experiments, for quantitative purposes the sample must be extracted 2 to 3 times with 10-ml portions each time of the scintillation (PPO) solution.

Standard Test Solution. This is prepared as follows: 1.0 ml of the hog kidney enzyme preparation is incubated with 25 μg of cadaverine-C^{14} in 20 ml of $0.1M$ sodium phosphate buffer (pH 7.6) for 60 min at 38°C. Aliquots of this mixture are used to determine the optimal extracting conditions. The mixture is stored frozen.

This radio assay was found by the authors (112) to be applicable to inhibition studies of diamine oxidase. Thus, aminoguanidine, a specific inhibitor of this enzyme, was preincubated with the hog kidney enzyme in a final concentration of $10^{-5}M$ for 15 min before the addition of putrescine-C^{14}. After an incubation of 60 min at 38°C the enzyme activity was found to be completely inhibited, i.e., the extracted radioactivity was at the same level as that of the zero time value. When aminoguanidine was added to the reaction mixture after the completion of the enzyme reaction, a slight but insignificant effect on the extraction efficiency was noticed.

The experimental data presented by Okuyama and Kobayashi (112) seem to indicate that this simple, rapid, and quantitative technique for the estimation of diamine oxidase activity compares well with other relevant methods.

b. Measurement of Plasma Diamine Oxidase Activity during Normal Human Pregnancy by an Improved Radioassay Procedure According to Southren, Kobayashi, Brenner, and Weingold (114). The radioassay of Okuyama and Kobayashi (112) was adapted by Southren, Kobayashi et al. (114) to investigations of diamine oxidase activity in plasma during normal human pregnancy.

Method. Ten milliliters of blood is withdrawn from a pregnant woman by an oxalated Vacutainer (Becton, Dickinson & Co., Rutherford, N. J.).

The blood is immediately centrifuged and the plasma is heated at 50°C for 15 min to activate DAO (83). The assay is carried out in a screw-cap culture tube (10 × 1.5 cm) containing 2 ml of heated plasma and 50 μg of putrescine-C^{14}. The incubation mixture (final volume 2.1 ml) is shaken in a Dubnoff metabolic shaker for 2 hr in air at 37°C. Approximately 200 mg of sodium bicarbonate is then added to the reaction mixture and the final product is extracted into 10 ml of toluene containing 0.35% diphenyloxazole and 0.01% of 1,4-bis-2-(5-phenyloxazolyl)-benzene. The reaction mixture is kept in a freezer at −15°C until the aqueous phase is frozen. The toluene extract is then poured into a counting vial for assay in a liquid scintillation counter. The extraction is repeated once. The counting efficiency is approximately 50%. A linear relationship is found between plasma diamine oxidase activity and incubation time up to 150 min. The new unit of diamine oxidase established by Southern, Kobayashi, and their co-workers (114) is defined as the amount of DAO contained in 1 ml of plasma which metabolizes 0.01 μg of putrescine-C^{14} in 2 hr at 37°C in air in a final volume of 2.1 ml.

The range of this radioassay, extended to 5000 DAO units, provides a means by which the plasma diamine oxidase activity may be conveniently determined in the second and third trimester of pregnancy.

Tryding (115) has recently slightly modified the radioassay technique for the estimation of plasma diamine oxidase activity, described above, by increasing approximately 3-fold the concentration of putrescine-C^{14} to 0.001M. Moreover, since the modified technique requires only 0.05 ml of plasma, the analysis can be performed on capillary blood plasma.

It will be, however, essential to establish whether the millimolar concentration of the substrate is sufficient to saturate the very active plasma amine oxidase in human pregnancy.

IV. PROCEDURE FOR THE ESTIMATION OF THE ACTIVITY OF EXTRACELLULAR PLASMA (SERUM) AMINE OXIDASES

SPECTROPHOTOMETRIC METHOD BY TABOR, TABOR, AND ROSENTHAL (13)

This method depends upon the measurement of the increase in absorption at 250 nm occurring on the enzymic formation of benzaldehyde from benzylamine.

The incubation mixture contains 200 μmoles of potassium phosphate buffer (pH 7.2), 10 μmoles of benzylamine, and enzyme in a total volume of 3 ml. The mixtures are incubated for 5 to 15 min in air at 30°C.

84   R. KAPELLER-ADLER

One spectrophotometric unit is defined as the amount of enzyme which causes an increase in the extinction of 0.001/min. This method developed by Tabor et al. (13) for the estimation of spermine oxidase activity from beef plasma was subsequently used with minor modifications by Yamada and Yasunobu (116) for the determination of the activity of crystalline beef plasma spermine oxidase, by Buffoni and Blaschko (18) for assaying the activity of crystalline benzylamine oxidase from hog plasma, and by McEwen (117) for the measurement of benzylamine oxidase activity in human plasma.

Acknowledgments

This article was written during the author's tenure of an award from the Distillers' Co. Ltd., Grants' Committee which is gratefully acknowledged. The author is greatly indebted to Mrs. L. M. Ellis for her very careful preparation of the typescript.

References

1. M. L. C. Hare, *Biochem. J.*, *22*, 968 (1928).
2. C. H. Best, *J. Physiol.*, *67*, 256 (1929).
3. E. A. Zeller, *Advan. Enzymol.*, *2*, 93 (1942).
4. H. Blaschko, P. J. Friedman, R. Hawes, and N. Nilsson, *J. Physiol.*, *145*, 384 (1959).
5. E. A. Zeller, L. A. Blanksma, W. P. Burkard, W. L. Pacha, and J. C. Lazanas, *Ann. N. Y. Acad. Sci.*, *80(3)*, 583 (1959).
6. E. A. Zeller, *Pharm. Rev.*, *11*, 387 (1959).
7. H. Blaschko, *Pharm. Rev.*, *4*, 415 (1952).
8. Y. Kobayashi, *Arch. Biochem. Biophys.*, *71*, 352 (1957).
9. E. A. Zeller, in *The Enzymes*, Vol. *II*, Part *1*, 1st ed., J. B. Sumner and K. Myrbäck, Eds., Academic Press, New York, 1951, p. 536.
10. E. W. McHenry and G. Gavin, *Biochem. J.*, *26*, 1365 (1932).
11. G. C. Cotzias and V. P. Dole, *J. Biol. Chem.*, *196*, 235 (1952).
12. J. G. Hirsch, *J. Exp. Med.*, *97*, 345 (1953).
13. C. W. Tabor, H. Tabor, and S. M. Rosenthal, *J. Biol. Chem.*, *208*, 645 (1954).
14. H. Blaschko and R. Hawes, *J. Physiol.*, *145*, 124 (1959).
15. H. Blaschko and R. Bonney, *Proc. Roy. Soc.*, Ser. B, *156*, 268 (1962).
16. E. Kolb, *Zentr. Veterinaermed.*, *3*, 570 (1956).
17. B. Bergeret, H. Blaschko, and R. Hawes, *Nature*, *180*, 1127 (1957).
18. F. Buffoni and H. Blaschko, *Proc. Roy. Soc.*, Ser. B, *161*, 153 (1964).
19. F. Bernheim and M. L. C. Bernheim, *J. Biol. Chem.*, *123*, 317 (1938).
20. G. Steensholt, *Acta Physiol. Scand.*, *14*, 356 (1947).
21. E. A. Zeller, J. Barsky, E. R. Berman, M. Cherkas, and J. R. Fouts, *J. Pharmacol. Exp. Ther.*, *124*, 282 (1958).
22. E. Werle and E. Pechmann, *Z. Vitamin-, Hormon- Fermentforsch.*, *2*, 433 (1948/49).
23. J. M. Hill and P. J. G. Mann, *Biochem. J.*, *91*, 171 (1964).
24. R. H. Weaver and E. J. Herbst, *J. Biol. Chem.*, *231*, 637, 647 (1958).
25. R. Kapeller-Adler, in *Amine Oxidases and Methods for Their Study*, M. Ancharoff, Ed., Wiley-Interscience, New York, 1970.

26. G. C. Cotzias and V. P. Dole, *Proc. Soc. Exp. Biol. Med.*, *78*, 157 (1951).
27. J. Hawkins, *Biochem. J.*, *50*, 577 (1952).
28. E. O. Oswald and C. F. Strittmatter, *Proc. Soc. Exp. Biol.*, *114*, 668 (1963).
29. L. S. Seiden and J. Westley, *Biochim. Biophys. Acta*, *58*, 363 (1962).
30. W. P. Burkard, K. F. Gey, and A. Pletscher, *Biochem. Pharmacol.*, *3*, 249 (1960).
31. C. Giordano, J. Bloom, and J. P. Merrill, *Experientia*, *16*, 346 (1960).
32. W. C. Schneider and G. H. Hogeboom, *J. Biol. Chem.*, *229*, 953 (1957).
33. Y. Kobayashi and R. W. Schayer, *Arch. Biochem. Biophys.*, *58*, 181 (1955).
34. S. R. Guha and C. R. Krishna Murti, *Biochem. Biophys. Res. Commun.*, *18*, 350 (1965).
35. S. Nara, B. Gomes, and K. T. Yasunobu, *J. Biol. Chem.*, *241*, 2774 (1966).
36. R. Willstätter and H. Kraut, *Ber. Deutsch. Chem. Ges.*, *56*, 1117 (1923).
37. M. B. H. Youdim and T. L. Sourkes, *Can. J. Biochem. Physiol.*, *44*, 1397 (1966).
38. M. B. H. Youdim, G. G. S. Collins, and M. Sandler, *Proc. 5th Meet. Fed. Eur. Biochem. Soc., Prague. 1968, Vol. 18.* D. Shugar, Ed., Acad. Press, New York, 1970, p. 281.
39. K. F. Tipton, *Eur. J. Biochem.*, *4*, 103 (1968).
40. T. Nagatsu, *J. Biochem.*, *59*, 606 (1966).
41. T. M. Brody and J. A. Bain, *J. Biol. Chem.*, *195*, 685 (1952).
42. E. Werle, in *Hoppe-Seyler-Thierfelder's Handbuch d. physiologischen u. pathologisch-chemischen Analyse*, 10th ed., Vol. VI/A, Springer, Berlin, 1964, p. 653.
43. N. H. Creasey, *Biochem. J.*, *64*, 178 (1956).
44. A. Sjoerdsma, T. E. Smith, T. D. Stevenson, and S. Udenfriend, *Proc. Soc. Exp. Biol. Med.*, *89*, 36 (1955).
45. H. Weissbach, T. E. Smith, J. W. Daly, B. Witkop, and S. Udenfriend, *J. Biol. Chem.*, *235*, 1160 (1960).
46. V. Zeller, G. Ramachander, and E. A. Zeller, *J. Med. Chem.*, *8*, 440 (1965).
47. E. C. Slater, in *Methods in Enzymology, Vol. 10*, R. W. Estabrook and B. Pullmann, Eds., Academic Press, New York, 1967, p. 19.
48. G. H. Petering and F. Daniells, *J. Amer. Chem. Soc.*, *60*, 2796 (1938).
49. L. C. Clark, *Trans. Amer. Soc. Art. Int. Org.*, *2*, 41 (1956).
50. K. F. Tipton and A. P. Dawson, *Biochem. J.*, *108*, 95 (1968).
51. S. Udenfriend, H. Weissbach, and C. T. Clark, *J. Biol. Chem.*, *215*, 337 (1955).
52. M. Dixon and K. Kleppe, *Biochim. Biophys. Acta*, *96*, 357 (1965).
53. B. M. Braganca, J. H. Quastel, and R. Schucher, *Arch. Biochem. Biophys.*, *52*, 18 (1954).
54. E. J. Conway and A. Byrne, *Biochem. J.*, *27*, 419 (1933).
55. J. Goa, *Scand. J. Clin. Invest.*, *5*, 218 (1953).
56. H. Weissbach, B. G. Redfield, and S. Udenfriend, *J. Biol. Chem.*, *229*, 953 (1957).
57. L. M. Barbato and L. G. Abood, *Biochim. Biophys. Acta*, *67*, 531 (1963).
58. K. Makino, K. Satch, and Y. Joh, *J. Biochem.* (Tokyo), *42*, 555 (1955).
59. J. F. Brewster, *J. Amer. Chem. Soc.*, *40*, 406 (1918).
60. T. Ukai, Y. Yamamoto, M. Yotsuzuka, and F. Ichimura, *J. Pharm. Soc. Japan*, *76*, 657 (1956); *Chem. Abstr.*, *51*, 280 (1957).
61. S. Udenfriend, "Fluorescence Assay in Biology and Medicine," in *Molecular Biology, Vol. 3*, B. L. Horecker, N. O. Kaplan, and H. A. Sheraga Eds., 4th Printing, Academic Press, New York, 1966, p. 122.
62. W. Lovenberg, R. J. Levine, and A. S. Sjoerdsma, *J. Pharmacol. Exp. Therap.*, *135*, 7 (1962).
63. E. Racker, *J. Biol. Chem.*, *177*, 883 (1949).

64. M. Krajl, *Biochem. Pharmacol.*, *14*, 1684 (1965).
65. R. J. Wurtman and J. Axelrod, *Biochem. Pharmacol.*, *12*, 1439 (1963).
66. P. A. Shore and V. H. Cohn, Jr., *Biochem. Pharmacol.*, *5*, 91 (1960).
67. S. Otsuka and Y. Kobayashi, *Biochem. Pharmacol.*, *13*, 995 (1964).
68. K. Satake, in *Koso Kenkyu-ho (Methods in Enzyme Studies)*, Vol. *2*, S. Akabori, Ed., Asakura Shoten, Tokyo, 1958, p. 534.
69. R. E. McCaman, M. W. McCaman, J. M. Hunt, and M. S. Smith, *J. Neurochem.*, *12*, 15 (1965).
70. D. Aures, R. Fleming, and R. Hakanson, *J. Chromatogr.*, *33*, 480 (1968).
71. R. Hakanson and Ch. Owman, *J. Neurochem.*, *12*, 417 (1965).
72. M. Goldstein, A. J. Friedhoff, C. Simons, and N. N. Procheroff, *Experientia*, *15*, 254 (1959).
73a. C. M. Francis, *Nature*, *171*, 701 (1953).
73b. C. M. Francis, *J. Physiol.*, *124*, 188 (1954).
74. G. G. Glenner, H. J. Burtner, and G. W. Brown, Jr., *J. Histochem. Cytochem.*, *5*, 591 (1957).
75. F. B. Adamstone and A. B. Taylor, *Stain Technol.*, *23*, 109 (1948).
76. G. B. Koelle and A. de T. Valk, Jr., *J. Physiol.*, *126*, 434 (1954).
77. G. J. Cunningham, L. Bitensky, J. Chayen, and A. A. Silcox, *Ann. Histochem.*, *1*, 433 (1962).
78. R. C. Graham and M. J. Karnovsky, *J. Histochem. Cytochem.*, *13*, 604 (1965).
79. R. J. Levine and A. Sjoerdsma, Clin. Pharmacol. and Therap., *4*, 22 (1963).
80. R. B. W. Smith, H. Sprinz, W. H. Crosby, and B. H. Sullivan, *Amer. J. Med.*, *25*, 391 (1958).
81. M. Kiese, *Biochem. Z.*, *305*, 22 (1940).
82. N. R. Stephenson, *J. Biol. Chem.*, *149*, 169 (1943).
83. M. Laskowski, J. M. Lemley, and C. K. Keith, *Arch. Biochem.*, *6*, 105 (1945).
84. L. F. Leloir and D. E. Green, *Fed. Proc.*, *5*, 144 (1946).
85a. H. Tabor, *J. Biol. Chem.*, *188*, 125 (1951).
85b. B. Swedin, *Acta Physiol. Scand.*, *42*, 1 (1958).
86a. V. D. Uspenskaia and E. V. Goryachenkova, *Biokhimiya* (Engl. Transl.), *23*, 199 (1958).
86b. E. V. Goryachenkova, L. J. Stcherbatyuk, and E. A. Voronina, *Biokhimiya*, *32*, 330 (1967).
87. R. Kapeller-Adler and H. MacFarlane, *Biochim. Biophys. Acta*, *67*, 542 (1963).
88a. B. Mondovi, G. Rotilio, A. Finazzi, and A. Scioscia-Santoro, *Biochem. J.*, *91*, 408 (1964).
88b. B. Mondovi, G. Rotilio, M. T. Costa, A. Finazzi-Agrò, E. Chiancone, R. E. Hansen, and H. Beinert, *J. Biol. Chem.*, *242*, 1160 (1967).
89. H. Yamada, H. Kumagai, H. Kawasaki, H. Matsui, and K. Ogata, *Biochem. Biophys. Res. Commun.*, *29*, 723 (1967).
90. G. L. Miller and R. H. Golder, *Arch. Biochem.*, *29*, 420 (1950).
91. G. Valette and Y. Cohen, *C. R. Soc. Bull. Paris*, *146*, 714 (1952).
92. R. Feulgen and H. Rossenbeck, *Hoppe-Seyler's Z. Physiol. Chem.*, *135*, 203 (1924).
93. K. A. Oster and M. C. Schlossman, *J. Cell. Comp. Physiol.*, *20*, 373 (1942).
94. J. Paul, in *Cell and Tissue Culture*, 2nd ed., E. & S. Livingstone, Edinburgh and London, 1961, p. 254.
95. H. Swanberg, *Acta Physiol. Scand.*, *23*, Suppl. *79*, 7 (1950).
96. R. E. Gunther and D. Glick, *J. Histochem. Cytochem.*, *15*, 431 (1967).
97. P. N. Aarsen and A. Kemp, *Nature*, *204*, 1195 (1964).

98. E. A. Zeller, in *The Enzymes*, Vol. 8, P. D. Boyer, H. Lardy, and K. Myrbäck, Eds., Academic Press, New York, 1963, p. 313.

99. *Report of the Commission on Enzymes of the International Union of Biochemistry*, Symposium Series 20, Pergamon Press, London, New York, Oxford, Paris, 1961.

100. P. S. J. Spencer, *J. Pharm. Pharmacol.*, *15*, 225 (1963).

101a. B. Holmstedt and R. Tham, *Acta Physiol. Scand.*, *45*, 152 (1959).

101b. B. Holmstedt, L. Larsson, and R. Tham, *Biochim. Biophys. Acta*, *48*, 182 (1961).

102a. E. Bamberger and E. Demuth, *Ber.*, *34*, 1309 (1901).

102b. E. Bamberger, *Ber.*, *60*, 314 (1927).

103. U. B. Arvidsson, B. Pernow, and B. Swedin, *Acta Physiol. Scand.*, *35*, 338 (1955/56).

104a. R. Kapeller-Adler, Unpublished data (1964).

104b. R. Kapeller-Adler, *Proc. 3rd Int. Pharmacol. Congr.*, *Sao Paulo, Brazil, July 1966, Abstracts*, p. 168.

105a. R. Kapeller-Adler, *Biochem. J.*, *48*, 99 (1951).

105b. R. Kapeller-Adler, *Biochim. Biophys. Acta*, *22*, 391 (1956).

106. H. E. Wagner, *J. Prakt. Chem.*, *89*(2), 377 (1914).

107. G. L. Miller, *Anal. Chem.*, *31*, 964 (1959).

108. O. M. Lowry, N. J. Rosenbrough, A. L. Farr, and R. S. Randall, *J. Biol. Chem.*, *193*, 265 (1951).

109. R. Kapeller-Adler, *Clin. Chim. Acta*, *11*, 191 (1965).

110. E. Kirsten, C. Gerez, and R. Kirsten, *Biochem. Z.*, *337*, 312 (1963).

111a. P. A. Shore, A. Burkhalter, and V. H. Cohn, Jr., *J. Pharmacol. Exp. Ther.*, *127*, 182 (1959).

111b. P. A. Shore and V. H. Cohn, Jr., *Biochem. Pharmacol.*, *5*, 91 (1960).

112. T. Okuyama and Y. Kobayashi, *Arch. Biochem. Biophys.*, *95*, 242 (1961).

113. R. W. Schayer, R. L. Smiley, and J. Kennedy, *J. Biol. Chem.*, *206*, 461 (1954).

114. A. L. Southren, Y. Kobayashi, P. Brenner, and A. B. Weingold, *J. Appl. Physiol.*, *20*, 1048 (1965).

115. N. Tryding, *Scand. J. Clin. Lab. Invest.*, *17*, Suppl. *86*, 197 (1965).

116. H. Yamada and K. T. Yasunobu, *J. Biol. Chem.*, *237*, 1511 (1962).

117. C. M. McEwen, Jr., *J. Biol. Chem.*, *240*, 2003 (1965).

118. M. V. Tyrode, *Arch. Int. Pharmacodyn. Ther.*, *20*, 205 (1910).

The Chemical Determination of Histamine

P. A. SHORE, *Department of Pharmacology, University of Texas Southwestern Medical School, Dallas, Texas 75235*

I. INTRODUCTION

A number of chemical methods have been developed for the determination of histamine, and a detailed review of the techniques available in 1956 was presented in *Methods of Biochemical Analysis* by Code and McIntire (1). At that time, chemical techniques were confined to colorimetric procedures involving the coupling of histamine to a diazotized aromatic amine or to dinitrofluorobenzene. Also available was an isotope dilution technique involving the coupling of histamine with

labeled pipsyl chloride followed by the addition of unlabeled dipipsylhista-
mine and recrystallization to specific activity.

These chemical techniques have been superseded by the fluorometric
method which allows much greater sensitivity and specificity in histamine
measurement as compared to the older colorimetric methods. Also
available now is a new enzymic-isotopic method based on the remarkable
specificity of the histamine methylating enzyme. Accordingly, this
chapter will confine its discussion to the fluorometric method with its
various modifications and to the enzymic-isotopic method.

II. THE FLUOROMETRIC METHOD

1. Principle

The fluorometric method is based on the coupling of histamine with
o-phthalaldehyde (OPT) at a highly alkaline pH to form a fluorescent
product which is rearranged upon acidification to form an even more
highly fluorescent and stable fluorophore (2). The acid stability of the
histamine-OPT fluorophore lends considerable specificity to the procedure, as a number of other primary amines couple with OPT in a basic
medium to form Schiff base products which, however, are broken down to
their constituent parts upon acidification; for example, OPT reacts with
agmatine in a basic medium to form a fluorophore which is destroyed
upon acidification (3). Even so, OPT forms acid-stable interfering
reaction products with a few other substances, thus necessitating special
purification procedures for the estimation of histamine in urine and in
brain. These modifications are described in a later section.

2. Reagents and Standards

A. o-PHTHALALDEHYDE (OPT)

This reagent may be obtained in quite pure form from Mann Chemical
Co., 136 Liberty St., New York, N.Y.; Calbiochem, 3625 Medford St.,
Los Angeles, Calif.; or Nutritional Biochemical Corp., 26201 Miles Rd.,
Cleveland, Ohio. For many purposes the commercial OPT can be used
as supplied, but for minimum reagent blank the OPT should be recrystallized from petroleum ether (30–60° boiling range) by dissolving the
commercial OPT under reflux conditions, decanting the solution from the
residue, and allowing the OPT to crystallize in the refrigerator. The
resulting crystals, after drying, should be stored in a dark bottle.

For histamine analysis OPT is dissolved in reagent grade, acetone-free
methanol, 10 mg/ml. This solution may be kept in a dark bottle in the
cold for up to 2 weeks, but it is the author's practice to prepare small
quantities of OPT solution as needed.

B. OTHER REAGENTS AND STANDARDS

Other reagents, such as NaOH and HCl, should be of the highest purity possible, made up in glass distilled water. A histamine stock solution of 100 μg/ml in 0.1N HCl should be kept in the cold and diluted daily to working standards of 0.005 to 0.5 μg/ml in 0.1N HCl. There is some evidence that histamine may be unstable in glass at subfreezing temperatures, and it has been suggested that if histamine solutions are kept frozen they should be in polyethylene tubes (4).

3. Fluorophore Production and Measurement

To a 2-ml aliquot of histamine solution in a small test tube, 0.4 ml 1N NaOH is added, followed by 0.1 ml OPT reagent (10 mg/ml methanol as described above). After 4 min at room temperature, 0.2 ml 3N HCl is added. The tube is shaken after each addition. The solution is then transferred to a fluorometer cuvet and the fluorescence at 450 mμ resulting from activation at 360 mμ (wavelengths uncalibrated) is measured in a spectrofluorometer or a fluorometer with appropriate filters.

To allow estimation in smaller samples, the procedure may be scaled down proportionately. Fluorescence as measured in a spectrofluorometer is proportional to histamine concentration over the range indicated above. On first use of the method, especially when a filter fluorometer is to be used, it is advisable to re-establish the limits of proportionality, as this may vary somewhat depending on the type of fluorometer used.

4. Reagent Blanks

For most purposes, an ordinary reagent blank (addition of reagents to 0.1N HCl) is satisfactory. In those cases where there exists the possibility of substances with "native" fluorescence being present, a "reversed reagent blank" is valuable. In this case, the 3N HCl addition is made before that of the 1N NaOH. Under these conditions, no reaction of OPT with histamine is possible, and any "native" fluorescence thus noted can be subtracted.

5. Specificity of the OPT Reaction

The acid-stable fluorescence produced by reaction of histamine with OPT at a highly basic pH shows considerable specificity for N-unsubstituted imidazol-ethylamines. Thus no fluorescence is produced under the conditions described above when the following related compounds are substituted for histamine: Acetylhistamine, 1-methyl-4-(β-aminoethyl) imidazol (1-methylhistamine), N-(side chain) alkyl histamines, imidazol, imidazolacetic acid, urocanic acid, carnosine, and anserine. Histidine, histidylhistidine, and other histidyl end group peptides react with OPT, but are readily excluded by purification procedures (see below). Other compounds found in biologic material but not reacting under the conditions described above are serotonin and catecholamines. Although spermidine and ammonia develop relatively little fluorescence with OPT, the former is present in brain and the latter in urine in sufficient quantities that special procedures are required for purification of brain and urinary histamine, as described in a later section.

III. EXTRACTION OF HISTAMINE FROM TISSUES

1. Principle

Histamine is extracted into an organic solvent from an alkalinized, salt-saturated, protein-free tissue preparation and is then returned to an aqueous phase.

2. Solvents

n-Butanol, reagent grade, such as supplied by Merck, may be used without purification if the solvent blank as described below is satisfactory. To reduce possible interference by impurities, the solvent may be washed successively with one-fourth volumes of $1N$ NaOH, $1N$ HCl, and two of distilled water. The excess water remaining may then be removed by the addition of an excess of solid NaCl.

n-Heptane, pure grade, such as the 99 mol % supplied by Phillips Petroleum Co. The solvent may be purified, if necessary, by successive washing with $1N$ NaOH, $1N$ HCl, and water, as described above.

3. Extraction Procedure

Tissues are homogenized in 9 vol of $0.4N$ HClO$_4$ in a glass homogenizer. The homogenate is allowed to stand at room temperature for a few minutes and is then centrifuged. A 4-ml aliquot of the supernatant fluid is transferred to a shaking tube containing 10 ml n-butanol, 0.5 ml $5N$ NaOH, and 1.5 g or more of solid NaCl. The tube is shaken for 5 min or more and centrifuged. If the presence of about as much histidine as histamine is suspected, the organic phase is washed by transferring it to another tube and shaking with 5 ml NaCl-saturated $0.1N$ NaOH. After centrifugation, an aliquot of 8 ml of the organic phase is transferred to another shaking tube containing 2.5 to 4.5 ml of $0.1N$ HCl and 15 ml n-heptane. The tube is shaken for about 1 min, centrifuged, the organic phase removed by aspiration, and 2 ml of the acid phase transferred to a test tube for reaction with OPT as described above.

Internal standards are prepared by adding to $0.4N$ $HClO_4$ known quantities of histamine which are then carried through the entire method. The use of internal standards permits the scaling down of the relative volumes of solvents as desired. Complete solvent and reagent blanks should be prepared by adding the above reagents to $0.4N$ $HClO_4$, and continuing through the entire extraction and coupling method.

4. Modification of the Extraction Method

Little histidine is extracted by n-butanol at a high pH, but if large quantities of histidine are present as in studies on histidine decarboxylase, the n-butanol may be replaced by a mixture of chloroform and n-butanol (2:3, v/v) (5). Another modification for the same purpose calls for the use of n-heptanol as a solvent (4), a procedure which also eliminates the need for heptane addition. Histamine recoveries are lower, however, when n-heptanol is utilized.

IV. SPECIAL PURIFICATION PROCEDURES

The extraction system described in Section III is appropriate to the analysis of histamine in most tissues. In the presence of excess histidine, the solvent extraction procedure may be modified as just described. An additional technique for removal of excess histidine consists of passage through a Dowex-1 carbonate anion exchange resin (6).

The presence of high concentration of spermidine in brain and of ammonia and other interfering constituents in urine has led to the development of special column techniques for the purification of brain and urine histamine prior to coupling with OPT.

1. Analysis of Brain Tissue

A. PRINCIPLE

Histamine and spermidine are extracted from brain with butanol and returned to an aqueous phase as described above. The amines are removed from the aqueous phase by passage through a phosphorylated cellulose column. Histamine is then selectively eluted from the column and coupled with OPT. The following procedure is a modification used in the author's laboratory (7) of the column technique first described by Kremzner and Pfeiffer (8).

B. MATERIALS

Extraction solvents and reagents as described in Section II-2-B.
Cellex-P (Bio-Rad Laboratories, Richmond, Calif.) is washed with

94 P. A. SHORE

successive portions of 0.1M HCl, 1.0M NaCl, H$_2$O, 1.0M NaHCO$_3$, 1.0M Na$_2$CO$_3$, H$_2$O, 0.1M NaOH, H$_2$O, ethanol, H$_2$O, and 0.2M NaHPO$_4$ buffer (pH 6.0). The cellulose is then suspended in 0.03M phosphate buffer (pH 6.0) and transferred to a chromatographic column (Scientific Glass No. JT-7390, 200 mm length, 6 mm bore) to a height of 40 mm and washed with 10 ml 0.03M phosphate buffer (pH 6.0).

C. PROCEDURE FOR ANALYSIS OF BRAIN

Brain tissue is homogenized and extracted with butanol as described in Section III-3. After shaking the butanol phase with heptane and 0.1 HCl, the acid phase is adjusted to pH 6.0 with 0.1N NaOH, diluted to 10 ml with 0.03M phosphate buffer (pH 6.0), and applied to the Cellex-P column. The column is then washed with 5 ml H$_2$O and the histamine eluted with 5 ml 0.2M NaCl. A 2-ml aliquot of the latter is then treated with NaOH, OPT, and HCl as described in Section II-3. Internal standards should be carried through the entire procedure.

2. Analysis of Urine

A. PRINCIPLE

Urine is passed through a column of IRC-50 cation exchange resin at pH 7.5. Histamine is eluted with HCl which is then made basic and subjected to the butanol extraction procedure described above. The procedure described is that of Oates et al. (9).

B. MATERIALS

IRC-50 cation exchange resin (Fisher Scientific Co. CG-50, Type I, 100–200 mesh, hydrogen form) is suspended in 1 vol distilled water. The mixture is adjusted to pH 9 with 10N NaOH. The resin is then washed with water, adjusted to pH 3 or less by the addition of 12N HCl, and after washing again with water, the pH is adjusted to 7.5 with NaOH. The resin is then washed twice with 0.5M sodium phosphate buffer (pH 7.5). Volumes approximately equal to that of the wet resin are used for all washes. The resin may be stored in the pH 7.5 in the refrigerator.

Using the same chromatographic tube specified in the brain assay, the IRC-50 resin is added to a column height of 55 mm. It is then washed with 10 ml of 0.5M sodium phosphate buffer (pH 7.5), followed by 15 ml distilled water.

C. PROCEDURE FOR ANALYSIS OF URINE

Urine samples are collected in bottles containing $6N$ HCl or toluene to avoid bacterial histidine decarboxylation. The urine pH is adjusted to 7.5 with NaOH and the urine is then passed through filter paper. A 20-ml aliquot of the filtered urine is mixed with 10 ml of $0.15M$ sodium phosphate buffer and applied to the IRC-50 column which is then washed with a 10-ml and a 5-ml portion of $0.5M$ sodium acetate buffer (pH 6.5). Histamine is eluted from the column with 5 ml $1N$ HCl, which is then made strongly basic with NaOH and extracted with n-butanol as described in Section III-3. Coupling with OPT is carried out as in Section II-3. Internal standards should be carried through the entire procedure.

V. THE ENZYMIC-ISOTOPIC METHOD

A highly sensitive and specific enzymic-isotopic method has been described which is relatively simple and rapid (10). By the use of labeled reagents of appropriate specific activity, the method can detect as little as 2 ng of tissue histamine. Although its application to measurement of plasma histamine has not been attempted, this technique very likely would be the method of choice because of its potential extreme sensitivity.

1. Principle

Tissues are incubated with ^{14}C (methyl labeled) S-adenosylmethionine and tracer quantities of ^{3}H-histamine in the presence of histamine-N-methyl transferase which specifically transfers the ^{14}C methyl group to histamine. The ratio of the ^{14}C to ^{3}H in the resulting methylhistamine is directly related to the quantity of endogenous histamine present in the tissue sample.

2. Materials

^{3}H-Histamine (about 10 Ci/mM) and ^{14}C-(methyl labeled)-S-adenosyl-methionine (about 40 Ci/M) may be obtained from New England Nuclear Corp. The purity of the ^{3}H-histamine should be checked periodically.

Histamine-N-methyltransferase is partially purified by a modification of the method of Brown et al. (11), as follows: Six male guinea pig brains are homogenized in 10 vol of $0.25M$ sucrose and centrifuged at $78,000g$ for 30 min. To 180 ml of the supernatant fluid, 85 g of $(NH_4)_2SO_4$ is added, followed by centrifugation at $10,000g$ for 20 min. To the decanted supernatant fluid, 33 g of $(NH_4)_2SO_4$ is added, followed by centrifugation at $10,000g$ for 10 min. The resulting precipitate is dissolved in 25 ml of

0.01M phosphate buffer (pH 7.4), and dialyzed overnight against 2 liters of 0.001M phosphate buffer (pH 7.4). The above procedure is carried out at 4°. This partially purified enzyme preparation is stable for several months when stored at −15°.

3. Procedure

Tissues to be analyzed for their histamine content are homogenized in 2 to 10 vol of 0.05M sodium phosphate buffer (pH 7.9). The homogenates are transferred to centrifuge tubes and heated for 10 min at 100° to destroy tissue S-adenosylmethionine and to free any bound histamine. The homogenates are then centrifuged at 10,000g for 20 min. Aliquots of the supernatant fraction (25–500 μl) containing 20–500 ng histamine are mixed in glass-stoppered centrifuge tubes with 200 nCi³H-histamine (1 ng), about 40 nCi (1 nM) of ¹⁴C-S-adenosylmethionine, and 50 μl of the histamine-N-methyl transferase preparation.

After incubation of the mixture for 1 hr at 37°, the reaction is stopped by the addition of 2 ml of 1N NaOH, followed by excess solid NaCl and 6 ml chloroform. The tube is shaken briefly and the aqueous phase removed by aspiration. The chloroform phase, which now contains the methylhistamine, is washed with 2 ml of 1N NaOH. The chloroform is then transferred to a scintillation vial and evaporated to dryness in a stream of air. Then 1 ml of 95% ethanol and 10 ml of an appropriate toluene phosphor solution is added and the radioactivity of ¹⁴C and ³H is measured in a liquid scintillation spectrometer. Standards are prepared by adding known quantities of unlabeled histamine to tissue homogenates before heating and to 0.05M sodium phosphate buffer (pH 7.9). As mentioned above, the histamine content is directly proportional to the ¹⁴C/³H ratio. The sensitivity of the method will vary with the specific activity of the labeled compounds.

VI. FLUOROMETRIC HISTOCHEMICAL TECHNIQUE

Although not directly germane to the quantitative estimation of histamine in the usual sense, mention should be made of a recent adaptation of the OPT-histamine coupling reaction to allow histochemical visualization of histamine in its cellular depots by formation of an OPT-histamine fluorochrome which can be detected microscopically. Details of this histochemical technique will not be given here, but may be found in the original paper (12).

References

1. C. F. Code and F. C. McIntire, in *Methods of Biochemical Analysis*, Vol. 3, D. Glick, Ed., Wiley-Interscience, New York, 1956, p. 49.

2. P. A. Shore, A. Burkhalter, and V. H. Cohn, *J. Pharmacol.*, *127*, 182 (1959).
3. V. H. Cohn and P. A. Shore, *Anal. Biochem.*, *2*, 237 (1961).
4. D. v. Redlich and D. Glick, *Anal. Biochem.*, *29*, 167 (1969).
5. A. Burkhalter, *Biochem. Pharmacol.* *11*, 315 (1962).
6. H. Weissbach, W. Lovenberg, and S. Udenfriend, *Biochim. Biophys. Acta*, *50*, 177 (1961).
7. M. Medina and P. A. Shore, *Biochem. Pharmacol.*, *15*, 1627 (1966).
8. L. T. Kremzner and C. C. Pfeiffer, *Biochem. Pharmacol.*, *15*, 197 (1966).
9. J. A. Oates, E. Marsh, and A. Sjoerdsma, *Clin. Chim. Acta*, *1*, 488 (1962).
10. S. H. Snyder, R. J. Baldessarini, and J. Axelrod, *J. Pharmacol.*, *153*, 544 (1966).
11. D. D. Brown, R. Tomchick, and J. Axelrod, *J. Biol. Chem.*, *234*, 2948 (1959).
12. L. Juhlin and W. B. Shelley, *J. Histochem. Cytochem.*, *14*, 525 (1966).

Determination of Histidine Decarboxylase Activity*

Richard W. Schayer, *Research Center, Rockland State Hospital, Orangeburg, New York 10962*

I. INTRODUCTION

The first demonstrations of enzymatic decarboxylation of L-histidine by extracts of mammalian tissues were made independently in the laboratories of Werle (1) and Holtz (2). Later studies using isotopic L-histidine revealed the presence in rat stomach of a histidine-decarboxylating enzyme with different characteristics (3).

The originally discovered enzyme, which has a low affinity for L-histidine and catalyzes decarboxylation of a number of amino acids, has been

* Supported by U. S. Public Health Service Grant AM 10155.

found to be identical with dopa decarboxylase (4,5). The other histidine-decarboxylating enzyme, which has a high affinity for substrate molecules and which appears to be specific for L-histidine, is probably the major *in vivo* catalyst of histamine formation (4,5). Werle now agrees that the nonspecific and specific histidine-decarboxylating enzymes are different enzymes and not two different centers on the same protein (6).

Recent reviews (7,8) cover the historical development in this field, the sources and properties of histidine decarboxylase, its relationship to histamine formation *in vivo*, the mechanism of action, and inhibitor studies.

Only those methods for determination of the specific histidine decarboxylase in mammalian tissues will be included in this Chapter. Methods for measuring histidine decarboxylase activity in bacteria (9) and plants (10,11) have been published elsewhere.

II. GENERAL PRINCIPLES

Histidine decarboxylase in the presence of the cofactor, pyridoxal phosphate, catalyzes the conversion of L-histidine to histamine plus carbon dioxide. Methods of enzyme assay are based on the amount of histamine or carbon dioxide formed. Disappearance of L-histidine cannot ordinarily be used as a measure of histidine decarboxylase activity since histamine formation is a minor metabolic pathway in mammalian tissues.

Since the amount of histamine formed may be extremely small, and since relatively large quantities of endogenous histamine may be present, isotopic methods are particularly useful.

The methods to be described are of three types: (*a*) nonisotopic; newly-formed histamine is measured, (*b*) isotopic; newly-formed labeled histamine is measured, and (*c*) isotopic; $^{14}CO_2$ is measured. There may be major differences in findings depending on the method used; Section VI, Comments, should be read before a method is selected.

III. PURIFICATION OF HISTIDINE DECARBOXYLASE

1. Method of Hakanson (12)

Pregnant rats are decapitated 15–20 days after mating and the litters removed and pooled. In frozen condition the fetal rat tissue maintained its enzyme activity for several weeks. At least 50 g of whole fetus are taken for each preparation.

Tissue is homogenized in 2 vol of 0.1N NaAc buffer (pH 4.5), centrifuged at 20,000*g* for 20 min at 0°, and the precipitate discarded. The supernatant (I) is warmed at 55° for 5 min, centrifuged at 20,000*g* for 10 min at 0°, and the precipitate discarded. The supernatant (II) contains more than 90% of the initial enzyme activity.

The enzymic material is precipitated with $(NH_4)_2SO_4$ at 25, 40, and 60% saturation. The precipitates are spun down, redissolved in 10–20 ml $0.1M$ phosphate buffer (pH 7.0), and dialyzed against redistilled water at 4° overnight. Sediments develop on dialysis and the extracts are centrifuged before the assay of activity. The fraction which precipitates between 25 and 40% saturation contains most of the enzyme activity. This fraction is again treated with $(NH_4)_2SO_4$ and the material which now precipitates between 28 and 42% saturation contains all enzyme activity. This extract is dialyzed against water for at least 6 hr. This treatment results in a 200-fold purification of the enzyme as compared to supernatant II. More than 50% of the enzyme activity of the initial homogenate is preserved in this extract, which is diluted to contain 0.4% protein before use. In this state the enzyme is stable for several days.

2. Method of Aures (13,14 and unpublished)

A special transplantable mouse mast-cell tumor, extremely rich in histidine-decarboxylating enzyme, is used. Its properties are similar to those of fetal rat-liver histidine decarboxylase but the tumor is a more convenient source. Dopa decarboxylase is also present. The purification steps are shown in Table I and the results in Table II.

The heat treatment does not increase the specific activity of either dopa or histidine decarboxylase under the conditions employed, but decreases both. However, the ammonium sulfate fractionation is less successful if the heat treatment is omitted. The sediment from the 25% saturated ammonium sulfate solution is 20 times as active as the starting material, but the yield is low and the bulk of the decarboxylases appears in the sediment of the 50% saturated ammonium sulfate solution. The latter was therefore used for repeated pH fractionations, and the proteins were simply separated by lowering the pH with diluted acetic acid. A fraction which precipitated between pH 5.5 and 4.5 was refractionated. The most active material was obtained in the fraction between pH 5.1 and 4.7. Under the incubation conditions employed, 12,982 μM CO_2/g protein/hr was formed, measured as initial rates.

This fraction had a great increase in specific activity for histidine; its decarboxylating rate was 150 times as high as that for dopa. The acid treatment also has an activating effect, since quantitatively more activity units were obtained in the sediment than in the starting material. This highly active fraction was stable for a few days, then about 10% of its activity remained for several weeks.

TABLE I
D. Aures, Purification of Histidine Decarboxylase

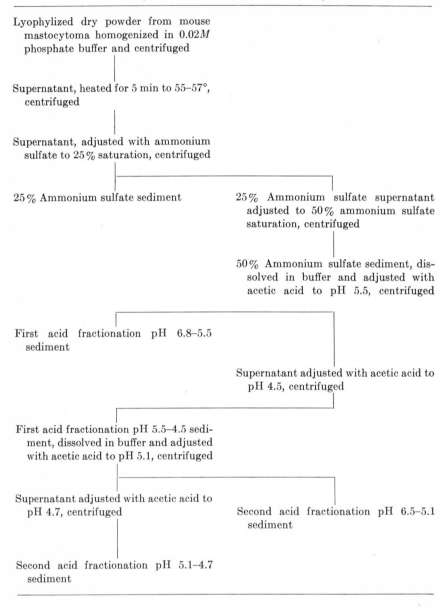

Lyophylized dry powder from mouse mastocytoma homogenized in 0.02M phosphate buffer and centrifuged

Supernatant, heated for 5 min to 55–57°, centrifuged

Supernatant, adjusted with ammonium sulfate to 25% saturation, centrifuged

25% Ammonium sulfate sediment

25% Ammonium sulfate supernatant adjusted to 50% ammonium sulfate saturation, centrifuged

50% Ammonium sulfate sediment, dissolved in buffer and adjusted with acetic acid to pH 5.5, centrifuged

First acid fractionation pH 6.8–5.5 sediment

Supernatant adjusted with acetic acid to pH 4.5, centrifuged

First acid fractionation pH 5.5–4.5 sediment, dissolved in buffer and adjusted with acetic acid to pH 5.1, centrifuged

Supernatant adjusted with acetic acid to pH 4.7, centrifuged

Second acid fractionation pH 6.5–5.1 sediment

Second acid fractionation pH 5.1–4.7 sediment

TABLE II
Purification of Histidine Decarboxylase

Fraction	μmoles CO_2/g protein/hr[a]		Increase in specific activity over the starting material	
	Histidine	Dopa	Histidine	Dopa
Dry powder homogenate	1.64	23.28	—	—
25% Ammonium sulfate sediment	15.06	17.82	9.2	—
50% Ammonium sulfate sediment	6.36	29.26	3.8	1.26
First pH fractionation				
6.8–5.5	3.76	1.74	2.3	—
5.5–4.5	18.90	40.00	12.0	1.72
Second pH fractionation				
6.5–5.1	88.42	2.64	56.5	—
5.1–4.7	12,982.00	1,242.00	7,910.0	53.5

[a]Values obtained from initial rates, 30 min incubation time.

IV. PURIFICATION OF L-HISTIDINE

Commercial L-histidine, isotopic and nonisotopic, is contaminated with a variable amount of histamine. It is important to realize that drastic treatment of histidine at any stage of the assay may cause a serious error through nonenzymatic histamine formation.

1. Method of Schayer (unpublished)

Dissolve the ^{14}C-L-histidine in 1.3% sodium chloride solution equal to about one-half the desired final volume. Then add sufficient $1N$ NaOH solution so that after neutralization and dilution the solution will be in approximately isotonic saline. ^{14}C-Histamine is removed by extracting the alkaline solution six times with butanol-chloroform, 3:1 (volume about that of the aqueous fraction), centrifuging each time if the layers do not separate completely. The organic layer (upper) is discarded. The volume of the aqueous layer (lower), which contains the ^{14}C-L-histidine, is kept constant by addition of water as necessary.

Residual butanol and chloroform are removed by extracting the aqueous layer three times with ether. Residual ether is blown off in a stream of nitrogen, the aqueous solution neutralized with $0.1N$ HCl, and water is added to give the desired final volume. Portions of the ^{14}C-L-histidine solution are transferred to several suitable containers and stored in a freezer.

2. Method of Kahlson et al. (15,16)

The ^{14}C-histidine, not exceeding 4 mg, is dissolved in 1 ml of $0.1M$ sodium phosphate buffer (pH 6.5). A column 30 × 4 mm (0.20 g) of Dowex 50 W-X4 (100–200 mesh), buffered in advance with sodium acetate buffer (pH 6.0), is prepared in a glass tube. The histidine solution is passed through the column, followed by 4 ml of the phosphate buffer. In the collected effluent >96% of the ^{14}C-histidine and <2% of the ^{14}C-histamine are present, compared with the original solution. It should be mentioned that the purchased ^{14}C-histidine never contained >0.1% of the ^{14}C-histamine before purification, and that the size of the column and volume of the buffer must be increased should the histamine content be greater.

In testing this method, by adding radioactive histamine to nonradioactive histidine, a higher degree of purification than that mentioned above was regularly seen, i.e., <0.05% of the ^{14}C-histamine was found in the effluent. It would thus appear that some ^{14}C-histamine was formed on purification of radioactive histidine. However, the procedure, as employed, reduces contamination to such an extent that the activity of blank samples is negligible.

3. Method of Mackay and Shepherd (17)

Nonisotopic L-histidine zwitterion is prepared from commercial L-histidine monohydrochloride monohydrate to eliminate traces of histamine. The hydrochloric acid is removed by adding excess silver carbonate to a warm, concentrated, aqueous solution of the monohydrochloride. After filtration, silver ions are removed from the filtrate by passing in hydrogen sulfide. Filtration and subsequent concentration of the filtrate, under reduced pressure, give a saturated solution of L-histidine zwitterion from which the latter may be precipitated by addition of dehydrated alcohol.

V. METHODS FOR HISTIDINE DECARBOXYLASE ASSAY

1. Isotopic Methods: [14]CO$_2$ Measured

A. METHOD OF KOBAYASHI (18)

Kobayashi has modified his original procedure and currently uses the following method.

Histidine decarboxylase is prepared by excising rat tissues, washing quickly with physiological saline, and mincing with 3 vol (1 g tissue/3 ml buffer) of $0.1M$ phosphate buffer (pH 7.4). The tissue is homogenized in an Omnimixer for 1 min at maximum speed, then frozen and thawed three times. The entire brei is centrifuged at approximately 20,000g for 30 min at 4° and the cell-free solution decanted. The pH of the cell-free preparation is adjusted to 6.9 with 0.01N HCl.

Analyses are made in a single-arm Warburg vessel incubated in a Dubnoff metabolic shaker. The side arm is sealed with a small multiple-dose bottle rubber stopper, and the main chamber sealed with a solid ground-glass stopper. In a typical assay, the flasks are prepared as follows: 2 ml of rat enzyme, 30 μg of pyridoxal phosphate, and 5 μg of ^{14}C-L-histidine (carboxyl-labeled) in the side arm. Hyamine hydroxide (Packard Instrument Co.), used to absorb carbon dioxide, is placed in the center well in one of two ways: 0.2 ml in the center well or 0.01–0.02 ml placed on a strip of Whatman No. 3 filter paper, 7.5 × 25 mm. The filter paper is edged with a thin strip of paraffin wax to prevent the Hyamine from touching the walls of the vessel. The final volume of each flask is usually 2.60 ml. The reaction time is 2 hr at 37°, after which the reaction is stopped by introducing 0.3 ml of 1M citric acid through the rubber stopper with a syringe. The flasks are shaken for an additional hour to ensure quantitative absorption of the carbon dioxide by Hyamine hydroxide. Longer absorption times do not result in increased recovery of carbon dioxide as measured by the amount of radioactivity found.

When Hyamine hydroxide is placed directly into the center well, the Hyamine solution is washed out quantitatively into a counting vial with water such that Hyamine plus the wash water weighs 1.5 g on a triple beam balance; 15 ml of counting solution is then added. The counting solution consists of 6 parts dioxane, 1 part anisole, and 1 part dimethoxyethane: 500 ml of this solvent mixture contained 3.5 g of diphenyloxazole (Pilot Chemical Co.) and 25 mg of p-bis(O-methylstyryl) benzene (Pilot Chemical Co.).

When Hyamine hydroxide on paper is used, it is placed in 10 ml of scintillation solvent consisting of 3 parts ethanol, 7 parts toluene, and 0.4% diphenyl-oxazole. It is essential that the caps of the counting vials be tightly sealed. Samples are counted in a liquid scintillation spectrometer.

More recently Kobayashi has found that the methanol in the Hyamine hydroxide used directly does not inhibit histidine decarboxylase activity. Therefore, the preparation of aqueous Hyamine hydroxide has been abandoned (personal communication).

B. METHOD OF AURES AND CLARK (19)

The incubation mixture consists of 0.1–0.4 ml of enzyme preparation, 0.0102 μM of pyridoxal 5-phosphate, 0.03 μM of aminoguanidine, and 0.2 μCi DL-histidine-1-^{14}C (specific activity 1.05–1.1 μCi/μM), which is adjusted to a total of 0.5 μM of the L-isomer with unlabeled L-histidine. The inhibitors are dissolved in Teorell-Stenhagen buffer (20) or Carbowax-300, a liquid ethylene glycol polymer of 300 cP viscosity, and added in different concentrations. The final mixture is adjusted to pH 6.8 and a volume of 0.5 ml with Teorell-Stenhagen buffer. The blanks are identical except that the enzyme is denatured by heating at 100° for 10 min. The incubations are done in a 5-ml Erlenmeyer flask which was connected with a sleeve-type rubber connector to a short-neck, 20-ml ampule (shown in Fig. 1 of Ref. 19).

After a 2-hr shaking in a Dubnoff shaking incubator, the reaction is stopped by injecting 0.4 ml of $1M$ citric acid with a hypodermic syringe and 23-gauge needle through the rubber connector into the incubation mixture. The whole apparatus is then fastened with rubber bands in a horizontal position on a rotating wheel, as shown in Ref. 19, Fig. 2, and cooled to 0°.

Two milliliters of a phenethylamine, methanol, and toluene-POPOP-PPO mixture (21) (27 ml of phenethylamine, 27 ml of absolute methanol, 0.5 g of PPO, 0.01 g of POPOP, and a sufficient quantity of toluene to make up 100 ml) is injected into the empty 20-ml ampule, and 0.1 ml of $1M$ sodium bicarbonate solution is injected into the Erlenmeyer flask. The diffusion chamber is then rotated on the wheel at a slow speed in a refrigerated ethylene glycol bath at 0°. After 30 min the sodium bicarbonate injection is repeated, and after a total rotation time of 60 min a third sodium bicarbonate addition is made and the rotation continued for a final 30 min. The phenylethylamine is rinsed quantitatively into a counting vial with 10 ml of toluene-methanol scintillator mixture (27 ml of absolute methanol, 0.5 mg of PPO, 0.01 g of POPOP, and a sufficient quantity of toluene to make a total volume of 100 ml). The counting is done in a liquid scintillation spectrometer.

C. METHOD OF AURES, DAVIDSON, AND HAKANSON (22)

Gastric mucosa is scraped off and homogenized in 0.05M phosphate buffer (pH 6.9), to a final concentration of 100 mg (wet weight)/ml. Aliquots of the homogenate (usually 0.4 ml for assay of histidine decarboxylase, 0.1 ml for dopa decarboxylase) are taken for incubation. In one series of experiments the homogenate was centrifuged at 10,000g for 15 min and aliquots of the supernatant (usually 0.5–1.0 ml) were passed through a Sephadex G-25 column (diameter 8 mm; length 200 mm). The protein fraction (absorption at 280 mμ), which had a volume of 3 ml, was collected for assay of histidine decarboxylase. For details see the method of Levine and Watts (Section V-1-D). The activities of DOPA (3,4-

dihydroxyphenylalanine) decarboxylase and histidine decarboxylase are determined radiometrically by estimating the $^{14}CO_2$ released from carboxyl-labeled substrates. All assays are made in duplicate and the enzyme activities expressed as nM CO_2 formed/mg and hour. The results are corrected by using boiled tissue blanks. Apart from the conventional procedure of Aures and Clark (19), a modification designed for the micro-assay of amino acid decarboxylases is employed (for details of the apparatus, see Fig. 1 of Ref. 22). All incubation ingredients, except the radioactive substrate, are mixed in a small glass tube, which is connected with a small glass chamber by a tight-fitting rubber ring. The opening on top of the glass chamber is covered with a rubber cap. The glass chamber contains a small roll of filter paper soaked in NCS-solubilizer (Nuclear Chicago) or Digestin (New England Nuclear). The incubation is started by the injection of the radioactive substrate through a needle inserted in the top rubber cap. The incubation is stopped after 1 hr, the $^{14}CO_2$ produced during the incubation released by acidification (0.1 ml 10% trichloracetic acid) and completely expelled with carrier bicarbonate (20 μl, 1M NaHCO$_3$). After further incubation for 30 min (to permit all $^{14}CO_2$ to be trapped on the filter paper), the top rubber cap is removed and the filter paper immersed in scintillation fluid for determination of radioactivity. Details on the composition of the incubation mixtures and of the incubation conditions are given in Table III.

TABLE III
Incubation conditions[a]

| Incubation ingredients | Histidine decarboxylase | | DOPA decarboxylase | |
	Conventional assay	Micro-assay[b]	Conventional assay	Micro-assay[b]
DL-DOPA-1-^{14}C	—	—	$8 \times 10^{-4}M$ 0.2 mCi/mM	$4.3 \times 10^{-4}M$ 3.75 mCi/mM
L-Histidine-1-^{14}C	$4 \times 10^{-5}M$[c] 10 mCi/mM	$1.5 \times 10^{-4}M$ 13.7 mCi/mM	—	—
Pyridoxal-5'-phosphate	$2 \times 10^{-5}M$	$2 \times 10^{-5}M$	$2 \times 10^{-5}M$	$2 \times 10^{-5}M$
Glutathione	$5 \times 10^{-4}M$	$5 \times 10^{-4}M$	$5 \times 10^{-4}M$	$5 \times 10^{-4}M$
Aminoguanidine	$10^{-4}M$	—	—	—
Phosphate buffer	$0.03M$, pH 6.9	$0.03M$, pH 6.9	$0.03M$, pH 6.9	$0.03M$, pH6.9
Total volume	0.5 ml	0.05 ml	0.5 ml	0.05 ml

[a] Samples were incubated under nitrogen at 37°C, usually for 30 min.
[b] The micro-assays were used exclusively for the determination of enzyme activities in histological sections. Incubation time 1 hr.
[c] The ^{14}C-histidine concentration given was used in most experiments.

D. METHOD OF LEVINE AND WATTS (23)

Specific histidine decarboxylase is prepared from whole fetal rats (19–20 days gestation) using the method of Hakanson (see Section III-1) modified by performing the initial homogenization in $0.1M$ sodium acetate buffer at pH 5.5 rather than 4.5. When the lower pH is used there is occasional complete loss of enzyme activity.

To prepare incubation vessels, a straightened wire paper clip with a narrow loop on one end is forced through the center of the cap of a 25-ml polyethylene vial (Packard Instrument Co.) so that the loop end projects from the lower part of the cap. A rectangular piece of Whatman 3MM filter paper measuring 1 × 3 cm is rolled into a cylinder and clamped firmly into the loop. Next, the paper is dipped into hydroxide of Hyamine and the excess allowed to drip off. It is essential that the loop be so positioned that when the cap is screwed onto the vial the filter paper is entirely in the upper third of the vial and does not touch the edges of the vial. The incubation is carried out at 37° in a Dubnoff metabolic shaker with the vial cap screwed on tightly. Components of the incubation mixture are listed in Table IV.

TABLE IV
Incubation Mixture Components

Component	Final concentration
Enzyme preparation	4–10 mg of protein/2 ml
Pyridoxal-5'-phosphate	$3.7 \times 10^{-5}M$
Streptomycin sulfate	$1.0 \times 10^{-4}M$
Sodium phosphate buffer, pH 6.8	$12.5 \times 10^{-2}M$
L-Histidine	$2.5 \times 10^{-4}M$
^{14}C-DL-Histidine	$0.25 \mu Ci/ml$
Water	To final volume of 2.0 ml

Streptomycin is added to suppress the growth of bacteria that may decarboxylate histidine; it does not affect histidine decarboxylase activity. The protein content varies with the activity of the enzyme preparation.

Blanks are prepared by adding 4-bromo-3-hydroxybenzyloxamine (NSD-1055), final concentration $10^{-4}M$, to incubation mixtures which are duplicates of experimental incubates; under these conditions histidine decarboxylase activity is totally inhibited.

Substrate is prepared in advance as follows. In $10^{-4}N$ HCl, non-radioactive L-histidine and ^{14}C-DL-histidine are dissolved to final concentrations of $5 \times 10^{-3}M$ (concentration is calculated to accommodate contribution of radioactive material to total L-histidine) 5 μCi/ml (2.5 μCi/ml of ^{14}C-L-histidine), respectively. Thus addition of 0.1 ml of this solution to the incubation mixture results in the desired concentration of substrate.

The entire mixture, with the exception of substrate, is preincubated for 10 min. To start the reaction, substrate is added, and the cap of the vial screwed on tightly. $^{14}CO_2$ formation is linear for 2.5 hr. Thirty or 60-min incubations are used for routine assays.

To terminate the reaction, 2 ml of 6N HCl is injected through the side of the vial through a 23-gauge disposable needle; the hole in the vial is sealed promptly with adhesive tape. Incubation is continued for 30 min after acidification to permit quantitative adsorption of CO_2 into the filter paper. Then the vial is opened, the filter paper transferred to another vial containing 10 ml of a scintillation fluorophor solution (24), and the radioactivity determined in a scintillation spectrometer. Recovery of $^{14}CO_2$ is virtually quantitative. Under the conditions used by the authors, with enzyme preparations free from endogenous L-histidine, the recovery of 655 cpm represents the decarboxylation of 1.0 mμM of L-histidine.

Smith and Code have also devised a modified method in which $^{14}CO_2$ is measured (25).

2. Isotopic Methods: Radioactive Histamine Measured

A. METHOD OF SCHAYER

The method is a modification of isotope dilution methods previously described (7,26). It involves incubation of tissue extracts with minute quantities of radioactive L-histidine (diluted by endogenous free L-histidine), and determination of the ^{14}C-histamine formed by addition of carrier which is subsequently extracted, converted to the benzenesulfonyl derivative (BSH), and counted.

Soft tissues are homogenized by hand in cold phosphate buffer, 0.1M (pH 7.2–7.4), containing 0.2% glucose. Enzyme is extracted from tough tissues by repeated freezing and thawing. Either homogenates or cell-free extracts may be used.

If the tissue is significantly active with respect to either dopa decarboxylase, diamine oxidase (histaminase), or histamine methylation, additives are required. Dopa decarboxylase and diamine oxidase can be inhibited by addition of α-methyl dopa and aminoguanidine, respectively. No potent inhibitor of the histamine-methylating enzyme is available. If the tissue contains significant quantities of endogenous histamine there is no significant error; methylation of endogenous histamine occurs in the early phases depleting the supply of the methyl donor, S-adenosylmethionine; ^{14}C-histamine formed during the course of the anaerobic incubation is unaffected. If a tissue is low in endogenous histamine, it is advisable to add histamine to the buffer to give a concentration of about 10 μg/ml.

  Incubations are done in 20-ml beakers to which are added 1.8 ml of crude enzyme preparation, 0.10 ml (50 μg) of pyridoxal phosphate, and 0.10 ml of purified ^{14}C-L-histidine solution. Currently, we use about 0.10 μCi of ^{14}C-L-histidine, weighing approximately 0.4 μg; it is, of course, diluted by the endogenous free L-histidine in the tissue extract unless the latter is removed by dialysis or some other procedure. Blanks are prepared either by substituting 0.10 ml of 0.01M hydroxylamine for pyridoxal phosphate, or by using heat-denatured enzyme preparations. Incubations are carried out in a Dubnoff metabolic shaker for 3 hr at 37° under nitrogen. The reaction is stopped by adding with mixing 1 ml of carrier solution containing 66.4 mg of histamine dihydrochloride (40 mg of histamine base) plus 50 mg of L-histidine monohydrochloride (the latter is used to dilute the substrate) and 3 ml of 0.6M perchloride acid. If the protein precipitate is bulky, it may be removed by centrifugation and washed once with 1 ml of 0.2M perchloric acid. If only a small amount of protein is present, it need not be removed.

  Each incubate-carrier-perchloric acid mixture is transferred through a funnel to an extraction tube of suitable size using 1 ml of 0.2M perchloric acid for rinsing. Then 1 ml of 5N NaOH solution and 20 ml of butanol-chloroform 3:1 (freshly mixed) are added through the same funnel. Through a powder funnel approximately 3 g of solid NaCl is added, the tubes are closed tightly with a Neoprene stopper and shaken mechanically for about 10 min. The solution is centrifuged and most of the organic layer (upper) is transferred to a clean tube (avoid contamination by traces of the lower aqueous layer) and washed by shaking with 5 ml of 0.1N NaOH solution saturated with NaCl for about 5 min. Most of the organic layer is transferred to a third tube and shaken with 10 ml of 0.1N HCl for about 5 min and most of the acid (which contains ^{14}C-histamine plus carrier) is transferred to a 25-ml Erlenmeyer flask, and evaporated to dryness in a stream of warm air.

  The carrier histamine dihydrochloride samples are dissolved in 2 ml of water, and then 3 ml of a solution of benzene sulfonylchloride in dioxane (60 mg of benzenesulfonylchloride/ml of dioxane) and about 0.3–0.4 g of sodium bicarbonate are added. The reaction is allowed to proceed at 37° for 30 min with gentle shaking. Water is added in small quantities over a period of 1–2 hr to a total of about 20 ml. The sides of the flask are scratched to help start crystallization.   The crude dibenzenesulfonyl histamine (BSH) is collected and washed with cold 25% ethanol in water.   Recrystallization is effected by dissolving in a few milliliters of warm acetone, treating with charcoal, filtering through a 2-ml coarse sintered-glass funnel containing filter-aid, and slow addition of small amounts of warm water, avoiding turbidity.   The crystals of BSH are collected, dried, and accurately weighed in vials suitable for liquid scintillation counting.

Since this is an isotope-dilution procedure, losses during the various steps need not be kept uniform for all samples and blanks.

BSH samples are dissolved in 15 ml of a suitable phosphor and counted in a liquid scintillation spectrometer.   After subtracting the background count, samples and blanks are corrected for weight by calculating radioactivity per 100 mg of BSH.   After subtracting blank values, samples are reported as counts per minute per 100 mg of BSH.   Since approximately equal weights of a purified compound are counted, all samples and blanks show almost identical counting efficiency.

In a recent modification, now used routinely in Schayer's laboratory, one-half of each incubate is saved so that samples lost accidentally can be repeated.   At the end of the incubation, volumes of each sample are measured with a 2-ml pipet (approximately 10% of the initial volume is lost by evaporation), one-half aliquots transferred to extraction tubes containing carrier and perchloric acid, and extraction performed as usual. To the remaining one-half is added 1 ml carrier and the mixture frozen and saved.   In this modification the concentration of ^{14}C-L-histidine incubated with enzyme is doubled so that 0.10 ml = 0.20 μCi.

B. METHOD OF KAHLSON, ROSENGREN, AND THUNBERG, CURRENT MODIFICATION (16)

Minced tissue, usually 0.2 g, is incubated for 3 hr at 37° under nitrogen, in beakers containing 40 μg of ^{14}C-L-histidine (base), labeled in the 2 position of the imidazole ring, $10^{-4}M$ aminoguanidine sulphate, $10^{-4}M$ pyridoxal-5-phosphate, and $10^{-1}M$ sodium phosphate buffer (pH 7.4) containing 0.2% glucose, made up to a final volume of 3.2 ml.   After incubation nonradioactive histamine dihydrochloride equivalent to 40 mg of the base is added as carrier, followed by perchloric acid, giving a final concentration of $0.4M$.

After mixing, the sample is allowed to stand for 30 min and is then filtered.   The pH of the filtrate is adjusted to approximately 6.5 with sodium hydroxide.   The sample is transferred to a 50 × 10 mm (2.5 g) column of Dowex 50W-X$_4$ (100–200 mesh), and buffered in advance with

$1N$ sodium acetate buffer (pH 6). The resin is then rinsed with 200 ml of $10^{-1}M$ sodium phosphate buffer (pH 6.5) at a flow rate of 15 ml/hr and finally with 10 ml $1N$ hydrochloric acid. The histamine is then eluted from the column with 10 ml of $10N$ hydrochloric acid and the eluate is evaporated to dryness on a steam bath. The residue, dissolved in 5 ml of water, is treated with activated charcoal and filtered. The histamine in the filtrate is allowed to react with pipsyl chloride (p-iodobenzenesulfonyl chloride) by adding 0.5 g of sodium bicarbonate and 5 ml of acetone, and after mixing, 0.22 g of pipsyl chloride in 4 ml of acetone. Crystals of pipsyl histamine precipitate when distilled water is added slowly. They are collected on a glass filter, rinsed with 25% alcohol, and dissolved in hot acetone. Charcoal is added and, after filtration, the pipsyl histamine recrystallizes on the gradual addition of water. The crystals are again collected on a filter, rinsed with alcohol, transferred to a counting plate, and counted after drying. The samples are repeatedly recrystallized from acetone until they display constant radioactivity.

Measurement of the radioactivity is made at infinite thickness in a flow counter until at least 2000 counts are obtained. Values are corrected for background radiation, usually about 20 cpm and for blanks. Traces of ^{14}C-histamine present as contamination are removed before use (see Section IV-2). Activity of blank samples is a few counts per minute above the background. A fuller description of the various procedures is given by Kahlson, Rosengren, and Thunberg (16).

C. METHOD OF SKIDMORE AND WHITEHOUSE (27)

^{14}C-Histamine formation by gastric mucosa was determined after paper chromatography by a method very similar to that described by Somerville for dopa decarboxylase assay (28). Incubation mixtures are made up as follows: 0.1 ml ^{14}C-histidine (0.5 μCi, 0.14mM) in 0.01N HCl; 0.1 ml nonradioactive substrate (4.8mM) in 0.1M sodium phosphate (pH 6.8); 0.1 ml drug in 0.1M sodium phosphate (pH 6.8); 0.1 ml 0.1M sodium phosphate (pH 6.8); 0.2 ml enzyme solution, containing pyridoxal phosphate; 1.9 ml of the enzyme solution are preincubated at 37° with 0.1 ml 0.8mM pyridoxal phosphate in 0.1M sodium phosphate (pH 6.8) for 10 min. This solution is added last to the incubation mixtures. Ten microliters of this mixture is immediately withdrawn for analysis and chromatographed on Whatman 3 MM paper by the method of Somerville (using isopropanol; 0.88 ammonia; water, 100:5:10 by volume). Ten microliter samples of the reaction mixture are withdrawn at intervals during the incubation and chromatographed in the same way. The portions of the paper corresponding to substrate and product are assayed for radioactivity by scintillation counting in 7 ml of toluene phosphor (containing 4 g PPO and 0.1 g dimethyl POPOP/liter). As controls, parallel incubations are carried out without drugs and also with heat-denatured (boiled) enzyme. All drugs are screened for possible effects on the separation of histamine from histidine by running control chromatograms of a known mixture of the amino acid and the amine in the presence and absence of drugs.

3. Isotopic Methods : Radioactivity Extractible by Butanol-Chloroform Measured

A. METHOD OF REILLY AND SCHAYER (29)

The incubation is similar to that described in Section V-2-A. At the end of the incubation period the incubate is transferred to a tube suitable for extraction (with glass or Neoprene stopper), and the beaker is rinsed with 1 ml of water.   Two milliliters of saturated sodium chloride solution, 1.0 ml of $5N$ sodium hydroxide, 12 ml of butanol-chloroform, 3:1, and approximately 1.5 g of sodium chloride are then added.   This is shaken mechanically and centrifuged.   Ten milliliters of the organic (upper) layer is transferred to another tube and washed by shaking with 5 ml of $0.1N$ sodium hydroxide saturated with sodium chloride and centrifuged. Eight milliliters of the organic layer is transferred to another tube, and shaken with 5 ml of $0.1N$ hydrochloric acid.   Four milliliters of the acid layer is transferred to a counting vial, evaporated to dryness in a steam of warm air, the residue is dissolved in 0.50 ml of water, a water-miscible phosphor is added, and the solution is counted.

4. Nonisotopic Methods : Newly Formed Histamine Determined

Only those methods not included in a previous review (7) will be included in this section.

A. METHOD OF WERLE AND LORENZ (6)

This method was used for thyroid and thymus; modifications may be required for other tissues (30,31). Assays are run in a Warburg apparatus at 37°. To the reaction vessel are added 2.0 ml of crude homogenate (equivalent to 0.35 g of fresh tissue), 0.10 ml of aminoguanidine (final concentration $5 \times 10^{-4}M$), 0.10 ml of tetracycline (20 g/ml of incubate), and 0.3 ml of $0.2M$ phosphate buffer (pH 7.0). For the nonspecific decarboxylase the pH is 8.0, and 20 mg of benzene is added. Additives are dissolved in $0.05M$ phosphate buffer (pH 7.45), and the pH readjusted with $1N$ HCl or $1N$ NaOH.

To the side arm is added 0.5 ml of L-histidine (to produce a final concentration of $10^{-2}M$). The final volume is 3.0 ml and the final buffer concentration is $0.06M$.

Blanks containing acid-inactivated enzyme plus substrate, and blanks containing active enzyme but no substrate, are included.

Assays are gassed for 10 min with pyrogallol-purified nitrogen prior to adding substrate. After 3 hr the reaction is stopped by addition of 0.5 ml of $3N$ perchloric acid. After removal of protein, histamine is extracted, reacted with o-phthalaldehyde, and the product assayed spectrofluorimetrically by Burkhalter's modification (32) of the method of Shore et al. (33).

B. METHOD OF KIM AND GLICK FOR MICROGRAM SAMPLES OF TISSUE (34)

This method was devised to measure histidine decarboxylase activity in microtome sections of rat stomach and in other minute quantities of tissue.

Tissue samples, 50 mg, are homogenized in 5 ml of cold distilled water, centrifuged for 15 min at $2000g$, and the supernate assayed. The preparation of microtome sections is described in the original paper (34).

Reagents are $3.34 \times 10^{-2}M$ histidine (base), $8.27 \times 10^{-5}M$ pyridoxal phosphate, $6.67 \times 10^{-4}M$ aminoguanidine sulfate, and $0.667M$ phosphate buffer (pH 6.4). Equal volumes of the four solutions are mixed just before use. The assay procedure is as follows.

1. Forty microliters of distilled water is added to each reaction tube containing a tissue section and mixed by vibration to break up the tissue, or 40-μl samples of supernatant fluid are pipetted from tissue homogenate into separate reaction tubes. Tubes are 27 mm long and 4 mm bore.

2. To measure the preformed histamine in the sample, 20 μl of $2.4N$ perchloric acid is added, mixed, 60 μl of reagent mixture is added, centrifuged at $1000g$ for 4 min, and 100 μl of the clear supernatant liquid is used for the histamine analysis.

3. To measure the enzyme, 60 μl of reagent mixture is added to each tube (after step *1*), mixed, stoppered, and placed in a water bath at 37°C for 3 hr. The reaction is stopped by adding 20 μl of $2.4N$ perchloric acid, mixing, centrifuging, and using 100 μl for histamine analysis as in step *2*.

4. The enzyme activity is expressed as the difference in the values for preformed histamine and that obtained after the incubation.

Histamine is determined by the fluorometric procedure of von Redlich and Glick (35) with the following two changes.

a. 20 μl of perchloric acid is used in place of 120 μl, and the concentration of the perchloric acid is 2.4N in place of 0.4N.

b. A 3:2 mixture (by volume) of *n*-butanol and chloroform (both redistilled) is used in place of butanol alone.

Von Redlich and Glick have recently published an improved method for fluorometric microdetermination of histamine and serotonin (36).

VI. COMMENTS

Nonisotopic methods can be recommended only for tissues high in histidine decarboxylase activity. These methods are insensitive and there are a number of instances in which nonisotopic procedures have failed to detect activity readily measured by isotopic methods. A number of laboratories have abandoned nonisotopic methods and a trend favoring isotopic procedures exists.

The $^{14}CO_2$ method (Section V-1) has the advantage of simplicity and speed and has been widely used for assay of tissues with high histidine decarboxylase activity. However, its accuracy has now been seriously questioned. Grahn and Rosengren, using a $^{14}CO_2$ assay which was essentially a combination of the method of Kobayashi and its modification by Levine and Watts, compared its results with those of the isotope dilution assay for ^{14}C-histamine (37). They found that reducing agents such as adrenaline and ascorbic acid caused a nonenzymatic formation of $^{14}CO_2$ from carboxyl-labeled histidine without concurrent formation of histamine. Under certain conditions the yield of $^{14}CO_2$ was enormous, in one case being 400 times that of the ^{14}C histamine formed! The reaction mechanism behind the phenomenon is unknown. Grahn and Rosengren conclude that "The radioactive CO_2 method has great limitations and may not be trustworthy for the purpose of measuring the rate of histamine formation, particularly in tissues with low histidine decarboxylase activity."

These findings of Grahn and Rosengren were obtained with minced tissues incubated in air. It is not clear whether cell-free incubations under nitrogen would possess the same potential for error. In any event this method should be used with great care.

The $^{14}CO_2$ method has not as yet been shown to be capable of measuring histidine decarboxylase in tissues with low activity and hence has not been proven generally useful. Formation of $^{14}CO_2$ from ^{14}C-L-histidine may not always be a satisfactory criterion of histidine decarboxylase activity; in metabolically active tissues such as liver, it is possible that histidine is catabolized by the glutamic acid pathway, and $^{14}CO_2$ released only after further degradation.   Blocking of $^{14}CO_2$ liberation by inhibitors of histidine decarboxylase does not ensure that a direct decarboxylation is involved.   If compounds inhibit one pathway of histidine metabolism they may also inhibit others.   Histidine decarboxylase activity of several tissues has been determined simultaneously by the $^{14}CO_2$ method and a nonisotope method; the results agreed reasonably well (38).

Obviously, the $^{14}CO_2$ method can not be used to study *in vivo* decarboxylation of histidine.

Methods requiring isotope-dilution techniques (Section V-2) are tedious.   Nevertheless, they are, in the author's opinion, the only available means for evaluating the physiological significance of histamine through the ability of tissues to produce it.   Radioactive histamine formed during incubation may be destroyed to some extent; however, errors can be minimized by proper precautions.   The isotope dilution methods are used to measure histamine, the only specific product of histidine decarboxylase action, and the substance of biological interest.   When purified ^{14}C-L-histidine is used, blanks are low; blanks in the $^{14}CO_2$ method are said to be high.   The isotope-dilution methods have consistently produced results which are consonant with *in vivo* findings.   Finally, their value is enhanced by the fact that the same procedure can be used to measure histamine formation *in vivo*.

The short method (Section V-3) does not identify the product as histamine.   Nevertheless, it is useful for specialized studies such as testing effects of inhibitors on a single sample of histidine decarboxylase.   It is not reliable for complex *in vitro* experiments or for measuring radioactive histamine formed *in vivo*.

There is no standardization relative to units for expressing histidine decarboxylase activity.   Some laboratories describe enzyme activity as the "amount histamine formed per unit time per unit weight of tissue."   This practice might lead uncritical readers to make the unwarranted assumption that similar rates of histamine formation occur *in vivo*.   For this reason the author prefers to use a unit which shows relative rather than absolute rates of histamine formation.

References

1. E. Werle, *Biochem. Z.*, *288*, 292 (1936).
2. P. Holtz and R. Heise, *Naturwiss.*, *25*, 201 (1937).
3. R. W. Schayer, *Amer. J. Physiol.*, *189*, 533 (1957).
4. P. O. Ganrot, A. M. Rosengren, and E. Rosengren, *Experientia, 17*, 263 (1961).
5. W. H. Lovenberg, H. Weissbach, and S. Udenfriend, *J. Biol. Chem.*, *237*, 89 (1962).
6. E. Werle and W. Lorenz, *Biochem. Pharmacol.*, *15*, 1059 (1966).
7. R. W. Schayer, in *Handbook of Experimental Pharmacology*, Vol. 18, Part 1, M. Rocha e Silva, Ed., Springer, Berlin, 1966, p. 688.

8. D. M. Shepherd and D. Mackay, in *Progress in Medicinal Chemistry*, Vol. 5, G. P. Ellis and G. B. West, Ed., Plenum Press, New York, 1967, p. 199.
9. E. F. Gale, in *Methods of Biochemical Analysis*, Vol. 4, D. Glick, Ed., Wiley-Interscience, New York, 1957, p. 285.
10. G. R. Lloyd and P. J. Nicholls, *Nature*, *206*, 298 (1965).
11. U. von Haartmann, G. Kahlson, and C. Steinhardt, *Life Sci.*, *5*, 1 (1966).
12. R. Hakanson, *Biochem. Pharmacol.*, *12*, 1289 (1963).
13. D. Aures, W. J. Hartman, and W. G. Clark, *Fed. Proc.*, *21*, 269 (1962).
14. D. Aures, W. J. Hartman, and W. G. Clark, *Proc. Intern. Congr. Physiol. Sci.*, *22nd, Leiden*, *2*, 694 (1962).
15. G. Kahlson, E. Rosengren, and R. Thunberg, *J. Physiol.*, *169*, 467 (1963).
16. G. Kahlson and E. Rosengren, *Physiol. Rev.*, *48*, 155 (1968).
17. D. Mackay and D. M. Shepherd, *Brit. J. Pharmacol.*, *15*, 552 (1960).
18. Y. Kobayashi, *Anal. Biochem.*, *5*, 284 (1963).
19. D. Aures and W. G. Clark, *Anal. Biochem.*, *9*, 35 (1964).
20. T. Teorell and E. Stenhagen, *Biochem. Z.*, *299*, 416 (1938).
21. F. H. Woeller, *Anal. Biochem.*, *2*, 508 (1961).
22. D. Aures, W. D. Davidson, and R. Hakanson, *Eur. J. Pharmacol.*, in press.
23. R. J. Levine and D. E. Watts, *Biochem. Pharmacol.*, *15*, 841 (1966).
24. G. A. Bray, *Anal. Biochem.*, *1*, 279 (1960).
25. R. D. Smith and C. F. Code, *Mayo Clin. Proc.*, *42*, 105 (1967).
26. R. W. Schayer, Z. Rothschild, and P. Bizony, *Amer. J. Physiol.*, *196*, 295 (1959).
27. I. F. Skidmore and M. W. Whitehouse, *Biochem. Pharmacol.*, *15*, 1965 (1966).
28. A. R. Somerville, *Biochem. Pharmacol.*, *13*, 1681 (1964).
29. M. A. Reilly and R. W. Schayer, *Brit. J. Pharmacol.*, *34*, 551 (1968).
30. W. Lorenz and E. Werle, *Z. Physiol. Chem.*, *348*, 319 (1967).
31. W. Lorenz, C. Pfleger, and E. Werle, *Arch. Exp. Pathol. Pharmakol.*, *258*, 150 (1967).
32. A. Burkhalter, *Biochem. Pharmacol.*, *11*, 315 (1962).
33. P. A. Shore, A. Burkhalter, and V. H. Cohn, *J. Pharmacol. Exp. Thera.*, *127*, 182 (1959).
34. Y. S. Kim and D. Glick, *J. Histochem. Cytochem.*, *15*, 347 (1967).
35. D. von Redlich and D. Glick, *Anal. Biochem.*, *10*, 459 (1965).
36. D. von Redlich and D. Glick, *Anal. Biochem.*, *29*, 167 (1969).
37. B. Grahn and E. Rosengren, *Brit. J. Pharmacol.*, *33*, 472 (1968).
38. D. V. Maudsley, A. G. Radwan and G. B. West, *Brit. J. Pharmacol, Chemother,* *31*, 313 (1967).

The Chemical Estimation of Catecholamines and Their Metabolites in Body Fluids and Tissue Extracts

H. WEIL-MALHERBE, *Section on Neurochemistry, Division of Special Mental Health Research, National Institute of Mental Health, St. Elizabeths Hospital, Washington, D.C. 20032*

I. INTRODUCTION

The elucidation of the biosynthesis and metabolism of the catecholamines in the mammalian organism was accompanied by a better understanding of the processes connected with their storage, release and re-uptake which take place in central and peripheral synaptic areas. These discoveries have shed light on the hitherto obscure mechanism of action of many drugs and have led to new insights and to productive hypotheses concerning the nature of diseases, such as manic-depressive psychosis, Parkinsonism, and essential hypertension.

These advances would not have been possible without sensitive and specific methods of analysis, not only of the catecholamines themselves but also of their major metabolites. In this Chapter some of these methods will be described, with the exception of those employing gas chromatography which will be dealt with in a separate chapter.

119

The numerous metabolites of the catecholamines are products or derivatives of products of two enzymes, monoamine oxidase (MAO) and catechol-O-methyltransferase (COMT). Methylation may precede deamination or vice versa. COMT transforms catechols into products methylated predominantly at the 3-hydroxyl group whereas MAO oxidizes amines to aldehydes and ammonia. By subsequent enzymic reactions, the aldehyde is either oxidized to the acid or reduced to the alcohol. The metabolites can thus be divided into three groups: deaminated but not O-methylated (Fig. 1, second row); O-methylated but not deaminated (Fig. 1, third row); and both deaminated and O-methylated (Fig. 1. fourth row).

In addition, catecholamines and their basic and neutral metabolites may be conjugated with sulfuric or glucuronic acids in position 4. Conjugation occurs mainly in the liver, but conjugating enzymes and conjugated metabolites were also found in the brain (1–3). Conjugated catecholamines have been demonstrated in the blood (4) and cerebrospinal fluid (5). Human cerebrospinal fluid has been shown to contain the sulfate of 3-methoxy-4-hydroxyphenylglycol (6). Conjugation with sulfuric acid seems to be the predominant reaction in man, whereas the rat seems to form mainly glucuronides, except for the conjugate of 3-methoxy-4-hydroxyphenylglycol which, in the rat as in man, is largely the sulfate.

There is still some doubt whether any of the acidic metabolites are excreted as conjugates and if so, what conjugates they form. In any case the conjugated fraction of the acidic metabolites, if it exists, is small (7–9).

II. THE ESTIMATION OF CATECHOLAMINES

1. Free (Unconjugated) Fraction

A. EXTRACTION FROM TISSUES AND BODY FLUIDS

The first steps in the estimation of catecholamines and their metabolites from biological material aim at the removal of interfering matter and the concentration of the substances of interest. Since many of them have an ionic charge in a certain pH range, fractionation into acidic, basic and neutral groups by solvent extraction or ion exchange chromatography is useful and sometimes sufficient. The ability of aluminum oxide ("alumina") to bind substances with two vicinal phenolic hydroxyl groups, i.e., catechol and catechol derivatives, from weakly alkaline solution has been of particular importance in this field. Although the adsorption of

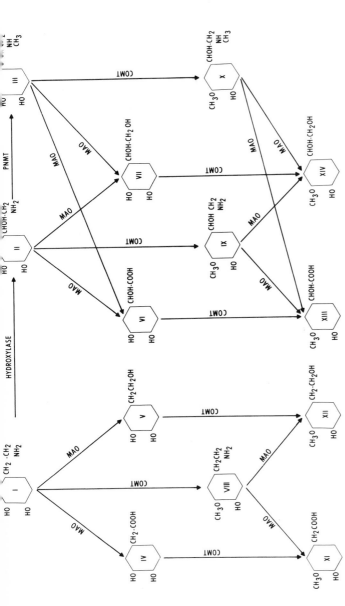

Fig. 1. Catecholamines and metabolites. First row (catecholamines): (I) 3,4-Dihydroxyphenylethylamine, dopamine; (II) 3,4-Dihydroxyphenyl-β-ethanolamine, norepinephrine; (III) N-Methyl-3,4-dihydroxyphenyl-β-ethanolamine, epinephrine. Second row (deaminated metabolites): (IV) 3,4-Dihydroxyphenylacetic acid, dopac; (V) 3,4-Dihydroxyphenylethanol; (VI) 3,4-Dihydroxymandelic acid; (VII) 3,4-Dihydroxyphenylglycol. Third row (O-methylated metabolites): (VIII) 3-Methoxy-4-hydroxyphenylethylamine, 3-methoxytyramine; (IX) 3-Methoxy-4-hydroxyphenyl-β-ethanolamine, normetanephrine; (X) N-methyl-3-methoxy-4-hydroxyphenyl-β-ethanolamine, metanephrine; Fourth row (deaminated, O-methylated metabolites): (XI) 3-Methoxy-4-hydroxyphenylacetic acid, homovanillic acid, HVA; (XII) 3-Methoxy-4-hydroxyphenylethanol; (XIII) 3-Methoxy-4-hydroxymandelic acid, vanillylmandelic acid, VMA; (XIV) 3-Methoxy-4-hydroxyphenylglycol, MHPG. Hydroxylase = Dopamine β-hydroxylase. PNMT = Phenylethanolamine N-methyl transferase. MAO = Monoamine oxidase. COMT = Catechol-O-methyl transferase.

catechol compounds by alumina is not completely specific and is shared, for instance, by acidic amino acids (10,11), it allows the separation of the catecholamines from their methylated metabolites as well as from other biological amines, such as tyramine and serotonin. Moreover, it is possible to separate catecholamines from the more firmly adsorbed acidic catechol compounds by fractional elution with acids of different strength.

The methods that have been devised for the purpose of isolating and concentrating catecholamines from tissue extracts and body fluids are based on solvent extractions, adsorption procedures, or combinations of both. Solvent extraction is generally restricted to the analysis of tissue extracts, although adsorption techniques are also suitable and, in the writer's opinion, preferable. The analysis of body fluids is usually performed by adsorption techniques.

Extracts suitable for adsorption or extraction procedures may be obtained by homogenizing a weighed sample of tissue in 0.4N perchloric acid. Excess perchloric acid may be removed from the filtrate as the little-soluble potassium salt. This is important when the electrolyte concentration has to be kept down to assure efficient retention by cation exchange resins, but the presence of perchlorate ions in the usual concentration does not interfere with the adsorption efficiency of alumina. Perchloric acid has no oxidizing effects under these conditions Certain deproteinizing agents have disadvantages: trichloroacetic acid may contain impurities which increase the blank and metaphosphoric acid interferes with the fluorimetric estimation of catecholamines in alumina eluates by the method of Weil-Malherbe and Bigelow to be described below. On the other hand, precipitation with tungstic acid has been found useful in our laboratory for the removal of proteins and mucoids from extracts treated with "glusulase" (see p. 146).

In some solvent extraction methods the tissue is homogenized in dilute HCl (12,13), acidified methanol (14), acidified ethanol (15), acidified acetone (16) or acidified n-butanol (17–19). Catecholamines may be adsorbed from protein-containing body fluids without previous deproteinization by adsorption on either alumina (20) or a strongly acidic cation exchange resin (21).

a. **Solvent Extraction Methods.** It is not feasible to extract catecholamines in their un-ionized form from alkaline solution owing to their extreme sensitivity to atmospheric oxygen in alkaline media. Their extraction in the cationic form requires a sufficiently polar solvent such as n-butanol. It does not seem to be generally known that the extraction of the cations of catecholamines with n-butanol was introduced by Weil-

Malherbe and Bone (22) and subsequently used by Montagu (14) in the writer's laboratory. These authors re-extracted the catecholamines from the solvent into dilute HCl and further purified them by adsorption on a column of alumina. Shore and Olin (12) added heptane to the butanol extract to facilitate the transfer of the amines to the aqueous phase, but they omitted the adsorption step. This procedure is rather unspecific and leads to high blanks, thus limiting the sensitivity of the assay. For these reasons Chang (17) and Ansell and Beeson (19) have now re-introduced the adsorption step.

Instead of heptane the use of cyclohexane (23) or isooctane (19) has been recommended. These solvents have lower blanks and may be used without previous washing. Anton and Sayre (24) reported improved recoveries when the tissue was homogenized in a solution containing 10 mg EDTA and 12 mg p-chloromercuriphenylsulfonic acid/ml. The latter reagent acts by combining with tissue sulfhydryl groups.

The solvent extraction methods are claimed to be especially useful for the combined estimation of serotonin and catecholamines. This may indeed be the case where the purification achieved by butanol extraction alone, without a subsequent adsorption step, is deemed adequate for the fluorimetric estimation of catecholamines. However, if further purification is desired, be it by adsorption on alumina (17,19) or on a cation exchange resin (16), solvent extraction appears to the writer to be redundant since the same degree of purification can be obtained by applying the adsorption procedures directly to a perchloric acid extract. Such extracts are also quite suitable for the combined estimation of serotonin and catecholamines as shown by Wada and McGeer (25). These authors adsorbed the catecholamines on alumina and extracted serotonin from the column effluent with n-butanol.

b. Adsorption on Alumina. As already mentioned, the adsorption of catecholamines on alumina is fairly specific and achieves a high degree of purification. The procedure is easy to perform, does not consume much time and has high recoveries. It should be part of every method for the estimation of catecholamines in biological material.

Alumina was first used by Lund (20) in the place of aluminum hydroxide previously employed (26,27). Weil-Malherbe and Bone (28) found that they obtained lower blanks, better recoveries, and faster rates of flow if they treated the commercial product with hot $2N$ hydrochloric acid, then washed free of acid by repeated decantation and finally dried at 300°C. This temperature is optimal for the preparation of highly adsorbent alumina from aluminum hydroxide (29). A drying temperature of 200°C gives a product that is only slightly less efficient than one dried at 300°C and may be substituted if facilities for heating at 300°C are not available. The adsorbing capacity of alumina dried at 100°C is, however, distinctly inferior to that of samples dried at 300 or 200°C (30). The alumina thus prepared is acid. It is neutralized before use by brief suspension in 0.1M tris buffer (pH 8.4).

As first established by Lund (20) and since confirmed by many authors, the adsorption of catecholamines on alumina is optimal at a pH of 8.4 or above. In such a solution catecholamines are liable to be oxidized by atmospheric oxygen, especially in the presence of traces of heavy metal catalysts. They can be protected from autoxidation to some extent by the addition of reducing and/or chelating compounds. Such additions are, however, not without danger since traces may be present in the eluate and interfere with the fluorimetric estimation. Fortunately, many tissue extracts, as well as plasma and urine, contain sufficient protective elements to prevent oxidation for a limited period. The addition of Na$_2$-EDTA at a final concentration of 0.01M and sodium bisulfite at a final concentration of 0.05% before pH adjustment was found to be safe in our procedure. To guard against overtitration, the extract is further mixed with 0.05–0.1 vol of 1M tris buffer (pH 8.4).

For the adsorption procedure the solution, after adjustment to pH 8.4, is either intimately mixed with the requisite amount of suspended alumina or passed through a column of the adsorbent. The column procedure has theoretical and practical advantages over the batch procedure since in the former a succession of "theoretical plates" leads to the establishment of successive equilibria with smaller and smaller concentrations of unadsorbed molecules. The adsorbed material separates on the column into sharp bands which allows separation from impurities and elution in a small volume. Moreover, washing and eluting can be carried out more simply, more efficiently, in less time, and with better recoveries on a column than by the repeated centrifugations or decantations required in the batch procedure. On the other hand, the delay between the pH adjustment and the time the last drop of the solution has entered the column may be sufficient to cause serious losses of catecholamines through autoxidation in some deproteinized tissue extracts. Advantages of both procedures have been combined in the compromise proposed by Crout (31) who, following Euler and Orwén (32), carries out the pH adjustment in the presence of the required amount of alumina. The mixture is stirred vigorously during the pH adjustment and for 7 min afterwards. It is then poured into an adsorption tube to form a column which is washed and eluted in the usual way. The catecholamines present are adsorbed as the pH is raised; once adsorbed they are no longer exposed to autoxidation (32). Crout used a magnetic stirrer for the pH adjustment.

In our experience magnetic stirring leads to excessive grinding of the brittle alumina particles, resulting in clogged columns and seepage of suspended alumina from the column. Fragmentation of alumina can, however, be avoided if a motor-driven stirring rod is used, a procedure also recommended by Anton and Sayre (33).

It has been found in our laboratory that recoveries are just as good when tissue extracts, containing EDTA and sodium bisulfite as antioxidants, are rapidly adjusted to pH 8.4 and immediately mixed with neutralized alumina as when the pH adjustment is performed in the presence of alumina. This procedure saves time if a series of samples are to be analyzed since they can be shaken, but usually not stirred, simultaneously.

A convenient way to monitor recoveries during the extraction and purification stages is by the addition of a radioactive tracer. The radioactivity in the purified solution serves as a measure for the recovery of the unlabeled substance.

Procedure. Mix 200 g of alumina ("for chromatographic adsorption analysis," British Drug Houses, U. S. distributor Gallard-Schlesinger Chemical Mfg. Corp., Carle Place, Long Island, N. Y.) with 1 liter of $2N$ HCl and heat to boiling for 20 min under vigorous stirring (caution: inefficient stirring may cause violent bumping). Filter on a sintered-glass funnel and wash with 1 liter of hot $2N$ HCl followed by ample water. Transfer the powder to a large beaker, stir it up repeatedly in about 1 liter of water, and decant the supernatant when the heavier particles have settled to the bottom. Filter again by suction and dry in a muffle oven at 300° for 2 hr. Cool in desiccator and store in closed bottle.

Suspend 0.5 g of alumina (0.7 g for the analysis of urine) in 10 ml 0.1M tris (hydroxymethyl) amino-methane (henceforth abbreviated "tris") buffer (pH 8.4) in a glass-stoppered tube, add 0.2 ml of $1N$ NaOH and invert gently and repeatedly. Allow the alumina to settle and remove the supernatant solution by aspiration through a piece of glass tubing of 2 mm i.d, with its lower end bent upward

Prepare a tissue extract by homogenizing 1–2 g in 5–10 vol of ice-cold 0.4M HClO$_4$ (approximately 2.6% w/v). Centrifuge at low temperature. Decant the supernatant solution and, if necessary, dilute to 10 ml with water. Add 0.1 vol of 1M tris buffer (pH 8.4), 0.05 vol of 10% Na$_2$-EDTA, and 0.01 vol of 5% sodium metabisulfite (prepared fresh). Titrate to pH 8.4, with $5N$ NaOH up to about pH 7.5, then with more dilute NaOH. Mix immediately with 0.5 g neutralized alumina in a glass-stoppered tube. Shake mechanically for 5 min. Remove the supernatant solution by aspiration.

Prepare an adsorption tube (consisting of a 50-ml spherical or cylindrical reservoir attached to a stem of 5-mm bore with a constriction about 15 cm below the reservoir [cf. 34]) by placing a plug of glass wool over the constriction, closing the lower end, and filling the stem with water. Wash the alumina into the adsorption tube through a funnel and allow the column to drain as the alumina settles. Excess water may be removed with a pipet, but of course the meniscus must always be maintained above the adsorbent. Wash the column 3 times by filling the stem with water. Finally elute with 3 ml 0.2N acetic acid, followed by 3 ml water.

For the analysis of urine use 25 ml and 0.7 g of alumina. Save the supernatant from the alumina adsorption for estimations of the conjugated fraction or metabolites if required. Use 5 ml 0.2N acetic acid followed by 5 ml of water for eluting the column.

c. **Adsorption on Cation Exchange Resin.** Cation exchange resins have been used for the extraction of catecholamines from tissue extracts or body fluids instead of alumina or for the further purification of alumina eluates, a procedure especially valuable in the analysis of urine (*vide infra*).

Methods differ in the choice of resin. Some authors, e.g., Bertler et al. (35) and Häggendal (36), prefer strongly acidic resins of the sulfonic acid type; others, e.g., Bergström and Hansson (37) and Kirshner and Goodall (38), prefer weakly acidic resins of the carboxylic acid type. Carboxylic resins have the advantage of greater ease of elution but are less efficient in the retention of amines and more sensitive to electrolyte concentration. Elution from sulfonic resins requires mineral acid at a strength of 1–2N in quantities dependent on the degree of crosslinking (the higher the degree of crosslinkage the more acid is needed). In the method of Bertler et al. (35) columns of 0.2 g of Dowex 50 are treated with 8 ml of 1.0N HCl for the elution of epinephrine and norepinephrine and with a further 8 ml of 2.0N HCl for the elution of dopamine. These high concentrations of HCl are not indifferent and may cause destruction of catecholamines (36,39). Moreover, these eluates often contain undesirable pigments, especially when urine is analyzed, as well as impurities dissolved from the resin. On the other hand, catecholamines may be eluted from carboxylic resins in a few milliliters of M acetic acid or in larger volumes of 0.4M ammonium acetate buffer at pH 5.0 (38,40).

If a cation exchange resin is to function, its active group has to be in anionic form. For a resin of the sulfonic acid type this is the case at a pH > 2, whereas for a carboxylic type resin the pH has to be about 6 or above. In view of their competitive effects, it is advisable to reduce the concentration of electrolytes to a minimum when cation exchange procedures are used. The removal of excess perchloric acid from tissue extracts as the slightly soluble potassium salt is therefore recommended. A preliminary adsorption on, and elution from, alumina may be used to separate catecholamines from media with a high electrolyte concentration, such as urine. Nothing is gained by diluting the sample (41).

The equilibrium of the exchange reaction is approached slowly, limiting the rate of flow through the column. Carboxylic resins shrink in acid medium; this may lead to a sudden undue acceleration of flow during the elution stage. It is advisable therefore to stop the flow completely for about 20–30 min after the resin bed has been filled with eluant, so as to ensure equilibration between the resin and the eluant.

Both types of resin have been used to separate catecholamines from each other as well as from the O-methylated amines (36,38,40). Techniques such as these are useful for the fractionation of labeled metabolites. An interesting method for the separation of catecholamines from their 3-O-methylated metabolites, described by Mattok et al. (42), makes use of an Amberlite CG 50 column buffered at pH 6.5. Elution with 4% boric acid removes the catecholamines due to the formation of strongly acidic catechol-borate complexes. The methylated amines are then eluted with $2N$ H_2SO_4. O. S. Steinsland (personal communication) uses Dowex 50, X4 for the adsorption of catecholamines from heparinized plasma at pH 6.5 by stirring, then washes the resin into a column and elutes catecholamines with sodium borate buffer at pH 9.1. Complex formation stabilizes the catecholamines and protects them from oxidation at this alkaline pH (43). A third ingenious use of the borate complexes is due to Wright (44). His method for the isolation of catecholamines from urine consists in passing acidified urine through a column of the anion exchange resin Dowex 2 which removes strong organic and mineral acids. Boric acid is then added to the filtrate which is passed through a second column containing the same resin in the borate form. The strongly acid complexes of boric acid and catecholamines are retained and eventually eluted with $0.2N$ HCl.

Example: *Purification of an alumina eluate prepared from urine by passage through a column of Amberlite CG 50, Type 2 (200–400 mesh).* Before use, prepare the resin by repeated recycling (at least 3 times) through the H+- and Na+-forms, suspending it alternately in N HCl and N NaOH for periods of 30 min. Wash copiously with distilled water between the additions of acid and alkali. In the process dust and fine particles are removed by decantation. Suspend the resin in the Na+-form in $1M$ sodium acetate buffer (pH 6.0). Titrate to pH 6.0 with $10M$ acetic acid until the pH remains stable for 30 min. Store under acetate buffer at 3°C.

Before use filter a portion of the resin on a sintered glass funnel and wash with water. Suck as free of adhering moisture as possible and weigh. Place 0.5 g in the reservoir of an adsorption tube. Fill the stem (5 mm i.d.) with water and allow the resin to settle through the water. Drain the column until the meniscus reaches the resin bed.

Add 0.1 vol of 1% Na_2-EDTA to the eluate from the alumina column, titrate to pH 6.0, and pass over the column at a rate of about 0.5 ml/min. Wash with 20 ml water and elute with 10 ml of $1.0M$ acetic acid. After about 3 ml of eluate have been collected stop the flow for 30 min before continuing the elution.

B. THE FLUORIMETRIC ESTIMATION OF EPINEPHRINE AND NOREPINEPHRINE BY THE TRIHYDROXYINDOLE METHOD

The history and chemistry of the trihydroxyindole (THI) method have been reviewed (45,46) and the interested reader is referred to these accounts. Briefly, epinephrine is oxidized to adrenochrome and the latter is rearranged to fluorescent N-methyl-3,5,6-trihydroxyindole (adrenolutin) in alkaline solution in which it is very unstable unless protected from oxidation by a suitable reducing agent. An analogous product, noradrenolutin, is presumably formed from norepinephrine.

In the classical form of the THI method ascorbic acid is used as reducing agent. It is well recognized, however, that ascorbic acid may itself form fluorescent oxidation products leading to high and unstable blanks. Various additions have been recommended designed to stabilize ascorbic acid, such as ethylenediamine (47), β-thiopropionic acid (48), or sodium borohydride (49).

Häggendal (21) replaced ascorbic acid by 2,3-dimercaptopropanol and obtained low and stable blanks. In the method here described (50) dimercaptopropanol has been replaced by mercaptoethanol because the former reagent, or at least the samples at our disposal, proved incompatible with cupric ion added as catalyst for the oxidation of epinephrine.

The stabilization of the fluorescent lutins in alkaline medium by sulfhydryl compounds is less efficient than that by ascorbic acid, the decay of adrenolutin fluorescence being faster than that of noradrenolutin fluorescence. This difference in stability has been utilized for the differentiation of epinephrine and norepinephrine in several automated versions of the THI-method: epinephrine and norepinephrine are measured together in an alkaline medium stabilized by ascorbic acid whereas in another alkaline sample, stabilized by thioglycollic acid (51), thiopropionic acid (52), or diethyldithiocarbamate (53), norepinephrine is estimated separately after the adrenolutin fluorescence has faded. However, although fading less rapidly, the fluorescence formed from norepinephrine is still sufficiently unstable to require the rigid timing afforded only by a fully mechanized procedure.

The stability of the fluorescence formed from both epinephrine and norepinephrine in the presence of a sulfhydryl compound is greatly enhanced by reacidification of the alkaline solution. Reacidification had previously been proposed by authors using ascorbic acid as reducing agent (17,54) but does not offer any clear advantage under those conditions. More recently, Laverty and Taylor (55) studied the optimal conditions for the development of fluorescence from a series of catecholamines and related compounds. They found that the oxidation products formed from epinephrine and norepinephrine by iodine could be sufficiently stabilized with a solution of sodium sulfite if followed by acidification. Oxidation with ferricyanide, in our experience, yields more fluorescence from epinephrine and norepinephrine than oxidation with iodine and, at the same time, reduces interference from dopamine. With ferricyanide as the oxidizing agent, omission of mercaptoethanol results in losses of intensity and stability of fluorescence.

The activation and emission spectra of adrenolutin are sufficiently different from those of noradrenolutin to permit a differentiation of epinephrine and norepinephrine by an optical method. Although this method works well with mixtures of the two amines free from contaminants, extended experience with it (56) has convinced us of its limitations when the mixtures are more complex. An alternative method of differentiation is based on the fact that epinephrine is oxidized at a pH of 3 much more readily than norepinephrine (57,58). It was found (50) that the oxidation of epinephrine by ferricyanide at pH 3 was dramatically catalyzed by the addition of cupric ions at a final concentration of $3 \times 10^{-4}M$ whereas zinc salts, previously recommended as catalysts (59), had no effect. The reason why the effect of Cu^{2+} has not been noticed before is explained by the fact that addition of ascorbic acid results in the formation of a colored complex and quenching of fluorescence. Although a first sample of dimercaptoethanol was used successfully as a stabilizer in the place of ascorbic acid, subsequent samples turned yellow in the presence of Cu^{2+} and again quenched all fluorescence. Eventually mercaptoethanol proved to be an entirely satisfactory substitute.

Copper ions catalyze the oxidant effect of ferricyanide not only on epinephrine but also on norepinephrine although the oxidation of norepinephrine only becomes significant when the pH is raised to 3.3–3.5. At a pH of 2.85 the oxidation of norepinephrine is negligible whereas that of epinephrine is still sufficiently rapid. This level of pH is therefore chosen for the differential oxidation of epinephrine in the presence of norepinephrine. The stability of the fluorescence, like that formed from norepinephrine at pH 6, is enhanced by acidification of the alkaline solution to pH 4.5–5.0. At this point the presence of the copper salt leads to the appearance of a white flocculent precipitate which is removed by brief centrifugation.

In another solution oxidized at pH 6.0 the fluorescence formed from epinephrine decays to a large extent during the time required for the alkaline tautomerization of the norepinephrine derivative. The result is that only a small correction has to be applied to allow for the presence of epinephrine and, where the amounts of epinephrine are small relative to those of norepinephrine, as in many tissue extracts, it may be omitted.

Whereas the activation maxima of the lutins are not appreciably affected by acidification, the fluorescence maxima are shifted to shorter wavelengths when the pH is changed from a strongly alkaline value to about 5.

The time intervals which are given in the following directions to allow for the oxidation and tautomerization reactions apply to an ambient temperature of $21 \pm 2°C$. Suitable modifications may be necessary if the temperature digresses appreciably from this range.

Procedure

Reagents. Potassium ferricyanide, 0.25%. Keep refrigerated; renew after 1 month.

Cupric acetate, $Cu(C_2H_3O_2) \cdot H_2O$, 0.2% (0.01M).

β-Mercaptoethanol, 1% (v/v) in 20% sodium sulfite (anhydr). Prepare fresh.

5N- and 10N NaOH-mercaptoethanol reagents: mix equal volumes of 5 or 10N NaOH and 1% β-mercaptoethanol in 20% Na_2SO_3 immediately before use.

Formic acid, 10M: 435 ml of 88% acid per liter.

Formic acid, 1M.

Acetic acid, 10M: 575 ml of glacial acetic acid per liter.

Epinephrine and norepinephrine standards: prepare dilute standards containing 1 µg of base per ml in 0.01N HCl from more concentrated stock solutions. Store at 3° for 1 week.

Use double glass-distilled water for the preparation of solutions. In the author's laboratory the second distillation is carried out over alkaline potassium permanganate. Triple glass-distilled water, with the second distillation performed over EDTA (1 g/liter), is used by Anton and Sayre (33).

Four 1-ml aliquots are required for the estimation of epinephrine, the same amount for the estimation of norepinephrine and three 1-ml aliquots for the estimation of dopamine (*vide infra*). When all three amines are to be determined, dilute the eluate to 13 ml. For fewer estimations dilute to a suitable smaller volume.

Estimation of epinephrine. Withdraw 4.5 ml eluate and adjust to a pH of 2.85 by the addition of 10 or $1M$ formic acid. An eluate from a column of Amberlite CG 50, containing $1M$ acetic acid, requires 0.14–0.15 ml of $10M$ formic acid. An eluate from an alumina column, containing $0.1M$ acetic acid, requires a similar amount of $1M$ formic acid. Dilute to 5.4 ml and measure out four aliquots (*a*, *b*, *c*, and *d*) of 1.2 ml, corresponding to 1.0 ml of eluate. Add 0.1 ml of epinephrine standard (1 µg/ml) to tube *b*, 0.1 ml of norepinephrine standard (1 µg/ml) to tube *c*, and 0.1 ml of $0.01N$ HCl to tubes *a* and *d*. To tubes *a*, *b*, and *c* add 0.1 ml of $0.01M$ cupric acetate and 0.1 ml of 0.25% ferricyanide, wait 5 min, then add 0.3 ml of $10N$ NaOH-mercaptoethanol reagent; wait 4 min, then add 0.3 ml of $10M$ acetic acid. Shake well after each addition. To the blank, tube *d*, add the reagents in the reverse order, i.e., acetic acid, followed by NaOH-mercaptoethanol reagent, ferricyanide, and cupric acetate, without waiting periods between additions. Centrifuge the tubes for about 5 min at 600*g*. Read the fluorescence in the spectrophotofluorometer at an excitation wavelength of 415 mµ and an emission wavelength of 500 mµ (4 mm slits). Read consecutively the four samples belonging to one group in the order, tubes *d*, *a*, *c*, and *b*. *Estimation of norepinephrine.* Titrate a 7.5 ml aliquot of the eluate to pH 6.0 and dilute to 9 ml. Withdraw four aliquots *a*, *b*, *c*, and *d* of 1.2 ml (equivalent to 1 ml of eluate). Put the remainder of the neutralized eluate aside (at 0°) for the estimation of dopamine. As before, add 0.1 ml of epinephrine and norepinephrine standards (1 µg/ml) to tubes *b* and *c*, respectively, and 0.1 ml of $0.01N$ HCl to tubes *a* and *d*. To tubes *a*, *b*, and *c* add 0.1 ml of 0.25% ferricyanide and, after 5 min, 0.3 ml of $5N$ NaOH-mercaptoethanol reagent. After a further 4 min, add 0.2 ml of $10M$ acetic acid. Shake well after each addition. To the blank, tube *d*, add the reagents in the reverse order starting with acetic acid and without waiting periods between additions. Read the fluorescence at 395 and 475 mµ for excitation and emission wavelengths, respectively (4 mm slits). Read standards and blank immediately after the corresponding sample.

Calculations

Epinephrine. In our experience the reading for the internal norepinephrine standard does not exceed 5% of that for the internal epinephrine standard after oxidation at pH 2.85. For most purposes, therefore, the error due to the presence of norepinephrine may be neglected. Corrections are possible and will be discussed below. The epinephrine content of the sample is obtained by subtracting the reading for the blank (tube d) from that of tube a and expressing the difference in terms of μg epinephrine by comparison with the internal standard, given by the difference of readings of tubes b and a. If the volume of the elute is V ml

$$\mu g\ E\ \text{in eluate} = 0.1V(a - d)/(b - a)$$

Norepinephrine. The amount of norepinephrine is obtained from the reading of tube a, corrected first by the blank (tube d) and second by the contribution of the epinephrine present. If oxidation at pH 2.85 has shown the presence of E μg of epinephrine per ml of eluate, if the volume of eluate is V ml and if b and c represent the fluorescence readings of the internal epinephrine and norepinephrine standards, respectively, then

$$\mu g\ NE\ \text{in eluate} = 0.1V[(a - d) - (b - a)E/0.1]/(c - a)$$

For eluates prepared from urine and tissues other than adrenal medulla, the correction $(b - a)E/0.1$ is small and can often be neglected. In this case the norepinephrine concentration determined by oxidation at pH 6.0 may be used to calculate an approximate correction of the epinephrine concentration measured at pH 2.85, if desired. If the concentrations of epinephrine and norepinephrine are less disparate, a more rigorous correction can be calculated with the aid of simultaneous equations based on the measured intensities of fluorescence of the internal standards at the two pH's. An example of such a calculation, applied to the optical differentiation of epinephrine and norepinephrine, is given below (p. 135).

C. THE FLUORIMETRIC ESTIMATION OF DOPAMINE

Carlsson and Waldeck (60) introduced a method in which dopamine is oxidized by iodine at pH 6. The oxidation product is protected by sodium sulfite, isomerized in alkaline solution presumably to a dihydroxyindole, and stabilized by re-acidification. This method has since been modified by many authors (for a review, see Ref. 30). The principal reason for dissatisfaction was the fact that the fluorescence continued to rise slowly for many hours in acid medium. Anton and Sayre (61) replaced iodine by periodate as the oxidant and claimed that this method yields a high and stable fluorescence immediately, but this claim could not be confirmed in the author's laboratory (30). Irradiation with uv light (60) or heating at 45°C (62) have been recommended as means of speeding up the development of fluorescence. Chang (17) found that heating the samples at 100°C for 5 min was effective and, more recently,

Laverty and Taylor (55) obtained the best results by heating at 100°C for 40 min. The method here described is that of Laverty and Taylor, with the inclusion of an internal standard and a nonoxidized blank. Results obtained by this method were in good agreement with those obtained by a method previously described (30).

The wavelength maxima for excitation and emission are at approximately 335 and 380 mμ. They are sufficiently far removed from the maxima of the fluorescent products of epinephrine and norepinephrine to eliminate mutual interference under the usual conditions. However, when one of the catecholamine is present in disproportionately high amounts (e.g., dopamine after an infusion of dopa) its interference in the estimation of other catecholamines must be considered. In such a case fractionation by column or paper chromatography may be necessary.

Procedure

Reagents. Phosphate buffer, 1M, pH 7.0: dissolve 17.4 g of K_2HPO_4 in about 50 ml of water, adjust the pH to 7.0 with 5N HCl, and make the volume up to 100 ml.

Iodine, 0.02N: 1.27 g of I_2 and 25 g of KI in 500 ml.

Alkaline sulfite: dissolve 1.25 g anhydr. Na_2SO_3 and 1 g Na_2-EDTA in 100 ml 2.5N NaOH. Prepare fresh.

Acetic acid, glacial.

Dopamine standard: 1 μg of base per ml in 0.01N HCl.

Withdraw three 1 ml samples (*a, b,* and *c*) from the eluate neutralized to pH 6.0 and add 0.2 ml of phosphate buffer. Add 0.1 ml of standard to *b*, 0.1 ml of 0.01N HCl to *a* and *c*. To tubes *a* and *b* add 0.05 ml 0.02N iodine solution and, after 3 min, 0.25 ml alkaline sulfite solution. Wait 5 min, then add 0.1 ml glacial acetic acid. Shake after each addition. To the blank (tube *c*) add the reagents in the reverse order (acetic acid, alkaline sulfite, iodine). Heat in a boiling water bath for 40 min in lightly stoppered tubes. Read in the spectrophotofluorometer at an activation wavelength of 335 mμ and an emission wavelength of 380 mμ.

Calculations

$$\mu g \text{ dopamine per aliquot} = 0.1(a - c)/(b - a)$$

where *a, b,* and *c* are the fluorescence readings obtained with the respective samples.

D. THE ESTIMATION OF CATECHOLAMINES IN PLASMA

a. **By the Ethylenediamine Condensation Method.** Owing to the limited amounts of plasma normally available and the extremely low concentration of catecholamines in peripheral venous plasma, their analysis poses special problems. Splitting the eluate into a series of fractions for the separate estimation of epinephrine and norepinephrine and for controlling each of these estimations by internal standards and blanks ought to be avoided to prevent a sacrifice of sensitivity. In the ethylenediamine (ED) condensation method (34), the fluorescent product is extracted into isobutanol. Scattering and quenching impurities are thereby largely eliminated (cf. 63) and controls by external standards and an external blank are adequate. Since epinephrine and norepinephrine may be differentiated optically, the entire eluate is available for the reaction. After the specificity of the procedure was increased by the inclusion of a cation exchange step, the results have been comparable with those obtained by the THI method. Dopamine, if present in plasma, would be included in the estimate obtained by the modified ED-method; however, all the evidence at present available points to the absence of detectable amounts of dopamine from normal plasma (22,61, 64–66).

In previous descriptions of the method clotting was prevented by an anticoagulant solution containing 1% EDTA and 2% sodium thiosulfate at pH 7.4, mixed with blood in the proportion of 1:4 (v/v). Essentially the same results are obtained with plasma prepared from heparinized blood. The dilution of plasma with an equal volume of $0.2N$ sodium acetate was also found to be unnecessary. The batch adsorption procedure described above for tissue extracts and urine could probably also be used for plasma, but this has not yet been checked and the column procedure will therefore be described.

Procedure

Reagents. *Aluminum oxide and Amberlite CG 50*, Type 2: prepared as described (p. 127).

Ethylenediamine: redistilled in an all-glass apparatus. The middle fraction is collected and stored in a dark bottle.

Isobutanol: reflux over NaOH, 10 g/liter, for 3 hr and rectify through a fractionating column in an all-glass apparatus. Store in a dark, glass-stoppered bottle at 3°C.

Na_2-EDTA, 1%.

Acetic acid, 0.2 and $1.0M$.

Standards containing 1 μg of epinephrine or norepinephrine (free bases) per ml in $0.01N$ HCl.

Suspend 0.5 g of alumina in 10 ml $0.1M$ tris buffer (pH 8.4) and wash into an adsorption tube whose stem (5 mm i.d.) has been filled with water. When the alumina has settled, drain the tube of excess fluid. Adjust a measured quantity of plasma to pH 8.4 by the cautious addition of a few drops of 0.5N sodium carbonate and pass through the column. Wash with 10 ml water. Elute with 3 ml 0.2M acetic acid, followed by 3 ml of water. Mix the eluate with 0.6 ml 1% EDTA and titrate to pH 6.0. Pass through a column of Amberlite CG 50 (0.4 g, moist weight). Wash with 20 ml of water. Elute with 3.5 ml of 1M acetic acid. Reject the first 0.3–0.4 ml of the effluent, collect the next 1 ml of the eluate, then stop the flow for 30 min. Finish the elution at a slow rate of flow. Remove the last drops of eluate by applying positive pressure to the top of the tube.

Mix the eluate, preferably in a "low-actinic" glass-stoppered tube, with 0.5 ml ED, shake vigorously, and heat at 55° for 25 min. Cool and saturate with NaCl (about 1 g). Extract with 1.8 ml isobutanol by shaking for 4 min. Transfer the organic phase to the fluorimeter cuvette and read the fluorescence at 510 mμ (reading "b") and at 580 mμ (reading "y") with activation at 420 mμ.

Carry a reagent blank through the adsorption, elution, and condensation procedures ("column blank") and correct the plasma readings by subtracting the column blank. Prepare standards of 0.05 μg epinephrine and 0.05 μg norepinephrine in 3 ml 1M acetic acid, together with a reagent blank of 3 ml 1M acetic acid and condense them with ED. Extract with isobutanol and correct the readings of the standards by subtracting the reagent blank.

Calculations

The fluorescence of the norepinephrine derivative at 510 mμ is about twice as great as that of the epinephrine compound. At 580 mμ the ratio is reversed. The following 2 equations may be set up:

$$E + N/m = y$$

and

$$E + N/n = b$$

where m = ratio of fluorescence of E/N standards of 580 mμ, n = ratio of fluorescence of E/N standards at 510 mμ, E = amount of epinephrine in sample, N = amount of norepinephrine in sample, y = amount of apparent epinephrine in sample indicated by the reading at 580 mμ, and b = amount of apparent epinephrine in sample indicated by the reading at 510 mμ. Hence

$$N = mn(b - y)/(m - n)$$

and

$$A = y - N/m$$

Recovery experiments have shown that the method of optical differentiation is reasonably reliable in plasma eluates.

An automated version of the ED method has been described by Viktora et al. (53).

b. The Estimation of Plasma Catecholamines by the Trihydroxyindole Method. Numerous modifications of the THI-method have been described for the estimation of catecholamines in plasma (33,54,57,67–69). The following is the method developed by Häggendal (21).

Procedure

Collect 18 ml blood into 2 ml 1% EDTA–0.9% NaCl solution. Pass the separated plasma over a column, 35 mm long, 2.7 mm i.d., of Dowex 50 W-X8 which has been treated with the following solutions: (1) 15 ml 2N NaOH containing 1% EDTA, (2) 40 ml water, (3) 20 ml 2N HCl, (4) 50 ml water, (5) 15 ml 0.1M phosphate buffer (pH 6.5) containing 0.1% EDTA, and (6) 5 ml water containing 0.1% EDTA. After the plasma has run through the column at a rate of 10 ml/hr, wash the column with 40 ml water containing 0.1% EDTA, then with 15 ml 0.1M phosphate buffer (pH 6.5) containing 0.1% EDTA, and finally with 10 ml water, at a rate of 15–20 ml/hr. Elute with 1.0N HCl and discard the eluate until its pH is below 2.5. Then collect 2.3 ml. Withdraw 0.6 ml for a nonoxidized blank (sample d). Adjust the pH of the remainder to 6.5 with 5N K$_2$CO$_3$. Divide the neutralized sample into 3 aliquots of 0.70 ml (samples a, b, and c). Add 0.10 ml of 0.1M phosphate buffer (pH 6.5) and 0.02 ml 0.0025% CuCl$_2$·2H$_2$O to samples a, b, and c. To sample b, the internal standard, add 0.01 μg epinephrine or, alternatingly, norepinephrine. Oxidize a, b, and c with 0.05 ml 0.25% K$_3$Fe(CN)$_6$. After 3 min stop the oxidation in a and b with 0.125ml of a solution containing 0.5 ml 2,3-dimercaptopropanol* and 50 g Na$_2$SO$_3$·7H$_2$O in 100 ml, followed immediately by 0.2 ml of 10N NaOH. In tube c, the faded blank, add 0.2 ml of 10N NaOH after completed oxidation and, after an interval of 10 min, add the dimercaptopropanol reagent. Mix sample d, the nonoxidized blank, with a freshly prepared mixture of 0.1 ml phosphate buffer and 0.125 ml of dimercaptopropanol reagent. Then add the solutions of CuCl$_2$, K$_3$Fe(CN)$_6$, and 10N NaOH as before and finally 0.1 ml 5N K$_2$CO$_3$.

After about 10 min read the samples at 400/515 and 450/515 mμ (activating/fluorescent wavelengths).

---

* Klensch (70) recommends the addition of cysteine instead of 2,3-dimercaptopropanol.

Häggendal obtained higher readings with a nonoxidized than with a faded blank and suggested using the mean value of the two blanks as a correction for the sample reading. External standards, containing 0.01 μg of epinephrine or norepinephrine, and a reagent blank are also prepared by replacing the sample with 0.70 ml water.

The results are calculated with the use of simultaneous equations as outlined in the preceding example (p. 135).

Gerst et al. (49) recommend the use of a constant temperature bath which fits into the well of the fluorometer since they found rigid control of sample temperature essential when measuring low fluorescence intensities.

E. OTHER METHODS FOR THE ESTIMATION OF CATECHOLAMINES

a. **Aminochrome-Bisulfite Addition Compounds.** Oesterling and Tse (71) prepared the aminochrome-bisulfite addition compounds of the catecholamines by oxidation with iodine at pH 5.0 for 15 min, followed by the addition of sodium bisulfite, and measured the increase of absorption at 355 mμ. Mattok et al. (42) adapted this method to the differential estimation of epinephrine, norepinephrine, and dopamine by varying the times of oxidation and development.

b. **Enzymic Labeling.** Methods have recently been described which are based on the enzymic incorporation of a labeled methyl group into the catecholamine molecule. If a mixture of the catecholamine-containing solution and the labeled methyl donor, S-adenosyl-L-methionine-methyl-^{14}C (^{14}C-SAM), is incubated with the enzyme phenylethanolamine N-methyltransferase (PNMT), methylation of suitable substrates occurs in the amine group; if it is incubated with the enzyme catechol-O-methyltransferase (COMT), methylation occurs at the 3-hydroxyl position. Such methods have a high degree of specificity and sensitivity and are therefore of great potential value.

(1) The PNMT Method for the Estimation of Norepinephrine. This method has been developed by Saelens et al. (72) for the estimation of norepinephrine in small areas of brain (as little as 10 mg). The tissue is homogenized in about 3 vol of a solution containing 0.2N HClO$_4$, 1mM pargyline, 1mM pyrogallol, and 1mM Na$_2$-EDTA. After 1 hr at 0°C an excess of magnesium carbonate is added and the mixture is centrifuged. Fifteen microliters of the supernatant is incubated for 1 hr at 37°C with 15 μl of a mixture of PNMT from rabbit adrenals* and ^{14}C-SAM. A 20-

* Preparations of PNMT are now available commercially from Gallard-Schlesinger Chemical Mfg. Corp., Biochemicals Department, Carle Place, L. I., New York.

138   H. WEIL-MALHERBE

μl portion is spotted on paper, carrier epinephrine added, and ascending chromatography carried out in butanol-1N HCl, 4:1. The spots are stained by alternate exposure to vapors of iodine and ammonia, cut out, and counted in a scintillation spectrometer. In a later modification (73) cation exchange paper (Whatman Chromedia P 81) is used for ascending chromatography in 0.2M ammonium acetate (pH 6)-isopropanol, 2:1. The authors claim that, owing to the much higher affinity of the enzyme for norepinephrine than for other amines, physiological concentrations of dopamine, epinephrine, metanephrine, and normetanephrine do not interfere. In that case, the tedious paper chromatography might perhaps be dispensed with and replaced by a simple extraction (74,75).

The method is specific for norepinephrine and is therefore not applicable where the estimation of other catecholamines is required.

(2) COMT Methods. Nikodijevic et al. (76) estimated norepinephrine in about 20 mg of heart tissue by incubating the tissue extract with COMT from rat liver and ^{14}C-SAM at pH 9.1 in a final volume of 0.5 ml at 37°. After an incubation of 20 min, carrier normetanephrine was added, the pH adjusted to 10, the reaction mixture first washed with benzene, and then extracted with toluene-isoamyl alcohol (3:7). A portion of the extract was evaporated and counted by liquid scintillation spectrometry.

Since COMT has a rather broad specificity, other catecholamines, i.e., dopamine and epinephrine, may interfere.

A similar method was used by Engelman et al. (77) to measure total catecholamines (norepinephrine + epinephrine) in plasma. A volume of plasma, about 10 ml, was diluted with 1–2 vol of water and passed through a 6 × 75 mm column of Amberlite IRC-50 at pH 6.0. After washing, the column was eluted with 8 ml 0.2N HCl, the first 2 ml being discarded. The eluates were lyophilized, redissolved, and incubated with ^{14}C-SAM and COMT at pH 8.5, 37°C, for 1 hr. The methylated amines formed + 100 μg carrier normetanephrine were passed through a column of Amberlite IRC-50 at pH 5.5–6.0 and eluted with 4N ammonia. They were then converted to vanillin by oxidation with sodium periodate. Vanillin was extracted with toluene at pH 6.5, returned into 1M K$_2$CO$_3$ solution and, after acidification, again extracted into toluene. The final extract was counted in a scintillation spectrometer. The recovery of norepinephrine was monitored by the addition of a trace of ^{3}H-DL-norepinephrine (approximately 0.2 ng) which was also added to controls in the absence and presence of COMT. Interference from dopamine is stated to be less than 2%. In its present form the method does not differentiate between epinephrine and norepinephrine.*

* Note added in Proof: In a more recent modification (120) metanephrine and normetanephrine formed in the reaction are separated by thin-layer chromatography

2. Hydrolysis of Conjugated Catecholamines

In many recent studies on the urinary excretion of catecholamines the investigators are content to measure only the unconjugated fraction of the catecholamines, assuming that this is a superior, or at least a sufficient, index of stress response and sympathetic activity. This view may probably be traced to the observation that ingested catecholamines are excreted mainly in conjugated form. This does, however, not justify the reverse conclusion that the conjugated fraction is of exogenous origin and of little physiological significance. The recent interest in the estimation of the major metabolites of the catecholamines is largely due to an acknowledgment of the fact that the turnover of endogenous catecholamines can only be gauged by establishing as complete a record as possible of the excretion of catecholamines and their metabolites. Surely the conjugated fraction of the catecholamines cannot be overlooked in such a balance sheet.

The reluctance of many analysts to determine the conjugated fraction of urinary catecholamines may be explained by the added complications inherent in acid hydrolysis. The alumina eluates of urines after acid hydrolysis are often yellow, their blanks are high, quenching is serious, and the specificity of the measurements is decreased. It is therefore essential to purify them further by a cation exchange adsorption-elution cycle. As has been shown previously (30), this procedure, while not significantly affecting recoveries, reduces the estimates for epinephrine, norepinephrine, and dopamine to 36, 43, and 58%, respectively, of the estimates given by the alumina eluate.

The usual method for the hydrolysis of catecholamine conjugates is heating at 100°C for 20–30 min at a pH of 0 (32) or of 1.0–1.5 (78). As a precaution, a stream of nitrogen may be passed over the liquid during the heating (56). Although this procedure appears to be satisfactory for the hydrolysis of the easily hydrolyzable sulfates, it does not completely split the more stable glucuronides. Since only a small proportion of the conjugates of catecholamines and catecholamine metabolites are excreted as glucuronides in man (7,79), it may be neglected. Another objection to acid hydrolysis is its destructive effect which increases with time of heating and increasing acidity and which is particularly serious in the case of some metabolites such as 3-methoxy-4-hydroxymandelic acid (VMA) and 3-methoxy-4-hydroxyphenylglycol (MHPG).

Enzymic hydrolysis is a much milder procedure. The enzyme preparation usually employed ("glusulase") is a crude extract of Helix pomatia containing both sulfatase and β-glucuronidase activity and therefore suitable for hydrolyzing both types of conjugates. It entails, however, certain problems which have so far restricted its use to the analysis of some catecholamine metabolites: (1) Glusulase is strongly inhibited by the sulfates and phosphates present in urine. They are precipitated, according to LaBrosse et al. (80), by barium ions at a pH of 11.5—a procedure which involves the destruction of free catecholamines. (2) The prolonged incubation at 37°C necessary for quantitative hydrolysis is a further cause of poor recoveries of free catecholamines. (3) Glusulase contains mucins which are not precipitated by trichloroacetic or perchloric acids and are therefore likely to gum up the adsorption columns and to cause emulsions during extractions.

Experiments which are now in progress in our laboratory (with Dr. E. R. B. Smith) are designed to overcome these difficulties and to produce a method suitable for the enzymic hydrolysis of both sulfates and glucu-ronides of catecholamines and catecholamine metabolites. It has been possible to remove inhibitory anions in a slightly acid medium by the use of an anion exchange resin, but attempts to prevent losses of free catechol-amines during the incubation have so far met with only partial success. The addition of ascorbic acid, EDTA, borate, bisulfite, or dithionite pro-vided little or no protection, even in an atmosphere of nitrogen; on the other hand, the addition of mercaptoethanol produced a substantial improvement in recoveries.

III. THE ESTIMATION OF THE 3-O-METHYLATED AMINES

The 3-O-methylated amines, metanephrine, normetanephrine, and 3-methoxy-p-tyramine, are oxidized and tautomerized to fluorescent compounds similar to those formed from catecholamines under conditions similar to those used in the tri- (or di-) hydroxyindole reaction. The method for the estimation of metanephrine and normetanephrine here described (41) incorporates the modifications of the THI-method devel-oped in our laboratory (50) except that iodine, rather than ferricyanide, and a higher level of pH are used for the simultaneous oxidation of the two amines. In the otherwise similar methods of Häggendal (81) and of Laverty and Taylor (55) for the estimation of normetanephrine, an alkaline sulfite solution containing EDTA is added for tautomerization and stabilization. We found that substitution of mercaptoethanol for EDTA considerably increased the fluorescence intensity. The method has so far only been applied to the analysis of human and rat urine but there is no reason why—mutatis mutandis—it should not be used for the analysis of tissue extracts.

It is the usual practice to determine the conjugated and unconjugated fractions of the O-methylated amines together, after acid or enzymic hydrolysis. Epinephrine and norepinephrine interfere with the estima-

tion and have to be removed. Whether or not their simultaneous estimation is desired, the initial steps of the two procedures, including the
hydrolysis of urine and the treatment with alumina at pH 8.4, are identical. The O-methylated amines contained in the supernatant after the
alumina treatment (or the effluent if a column is used) and the first 5-ml
of washings are adsorbed on a column of Amberlite CG-50 at pH 6. In
contrast to sulfonic acid type resins which yield colored eluates, purification on a carboxylic type resin is satisfactory, but retention of the amines
is limited by two factors, the size of the column and the salt content of the
urine. It has been found empirically that a 7 × 120 mm column can
cope with a volume of acid-hydrolyzed urine containing 15 mg of creatinine or 25 ml, whichever is less. In most cases this amount of urine has an
apparent electrolyte content (expressed in terms of NaCl conductivity)
well below 7.2 meq which approximately marks the limit for the capacity
of the column when adsorbing methylated amines from urine. In the
case of tissue extracts or urine hydrolyzed with glusulase where perchloric
acid has been added for deproteinization, the electrolyte content may be
reduced by titrating to pH 8.4 with potassium hydroxide and removing
the precipitate of $KClO_4$ before the alumina treatment. Before passing
the solution through the resin column, its salt content may be checked
with a conductivity bridge.

Procedure

Reagents

Some of the reagents are the same as those described for the estimation
of epinephrine and norepinephrine. In addition, the following are
required:

Tris (hydroxymethyl) aminomethane, $1M$: 121 g/liter. Store at 3°C.
Potassium ferricyanide, 0.5%. Store at 3°C for 1 week.
Iodine, 0.005N: prepare daily from 0.1N iodine.
Standards containing 1 μg metanephrine or normetanephrine (free
bases) per ml of 0.01N HCl.
Prepare a bed, 7 × 120 mm, of Amberlite CG-50, type 2, treated as
described (p. 127), by pouring a resin suspension into a column filled with
water. Allow the resin to settle, then aspirate and discard excess resin
and buffer with a pipet, thus eliminating dust and fine particles. Wash
with 10 ml water.
Hydrolyse 25 ml of urine by heating with acid or by incubation with
glusulase and treat with alumina at pH 8.4 as previously described.
From the supernatant of the alumina treatment withdraw a portion
containing 15 mg of creatinine or 25 ml whichever is less, adjust its pH to
6.2, and pass over the Amberlite column at a rate of 0.3–0.5 ml/min.

Wash the column with 20 ml of water.   Elute with $1.0M$ formic acid at a rate of 0.25 ml/min.   Discard the first 3 ml of eluate.*   Collect the following 10 ml.

Estimation of metanephrine.   Withdraw 4.5 ml from the eluate, titrate to pH 2.7–2.8 with $10M$ formic acid, and dilute to 5.4 ml.   Measure out four portions (a, b, c, and d) of 1.2 ml, corresponding to 1.0 ml of the eluate. Add 0.1 ml (0.1 μg) of metanephrine standard to tube b, 0.1 ml (0.1 μg) of normetanephrine standard to tube c, and 0.1 ml of $0.01N$ HCl to tubes a and d.   To tubes a, b, and c add 0.1 ml 0.2% cupric acetate followed by 0.1 ml 0.5% $K_3Fe(CN)_6$.   Leave for 10 min, shaking the tube occasionally. Add 0.5 ml of $10N$ NaOH-mercaptoethanol reagent.   After another 3 min add 0.2 ml $10M$ acetic acid.   Shake after each addition.   To the blank, tube d, add the reagents in reverse order, without waiting periods: acetic acid, alkaline mercaptoethanol, ferricyanide, and cupric acetate. Remove the precipitate appearing after acidification by centrifugation at $700g$ for 5 min.   Read the samples in the spectrophotofluorometer at excitation/emission wavelengths of 415/500 mμ, with 4 mm slits, in the order a, b, c and d for each eluate.

Estimation of normetanephrine.   Add 0.4 ml of $1M$ tris base to 4.5 ml of the eluate and titrate to pH 7.4–7.6 with $5N$, finally with $0.5N$ NaOH. Dilute to 6.3 ml.   Withdraw four 1.4-ml portions a, b, c, and d, corresponding to 1.0 ml of eluate.   Add internal standards and $0.01N$ HCl as in the estimation of metanephrine.   Add 0.1 ml $0.005N$ iodine to tubes a, b, and c.   Shake immediately and once more during the oxidation interval.   After 2 min add 0.2 ml $10N$ NaOH-mercaptoethanol reagent. After a further 2.5 min add 0.2 ml $10M$ acetic acid.   To the blank, tube d, add the reagents in the reverse order.   After 30–40 min read the samples at excitation/emission wavelengths of 390/475 mμ.

The calculations are analogous to those described for epinephrine and norepinephrine.

Estimation of 3-methoxytyramine.   The eluate from the Amberlite CG 50 column also contains 3-methoxytyramine, the 3-O-methylated metabolite of dopamine.   Laverty and Taylor give the following directives for its estimation.

Dilute 0.5 ml of the neutralized eluate to 1.1 ml with $0.1M$ borate buffer of pH 8.8, which should be the final pH of the mixture.   Add 0.05 ml of $0.02N$ iodine solution and shake vigorously.   After 3 min add 0.25 ml of alkaline sulfite solution (2.5% $Na_2SO_3 \cdot 7H_2O$, 1% Na_2-EDTA in $2.5N$ NaOH).   After a further 5 min add 0.25 ml glacial acetic acid. Heat for 40 min at 100°C.   Read at 320/375 mμ (excitation/emission

---

* This volume may have to be modified according to the dead space of the column used.

wavelengths). Run a sample with added internal standard and a reversed blank in parallel with the test sample.

IV. THE ESTIMATION OF 3-METHOXY-4-HYDROXY-MANDELIC ACID (VANILLYLMANDELIC ACID, VMA)

The most convenient methods for the routine estimation of VMA are based on its oxidation to vanillin, introduced by Sandler and Ruthven (82). Ferricyanide (83), periodate (84), an alkaline copper reagent (85), and catalytic oxidation under pressure (82) have been used to bring about this conversion. Of these, the periodate method is probably most widely used, but we have found it to be subject to inhibition by urinary constituents present in some samples of urine. The method described is a modification of the method of Sunderman et al. (83). It uses ferricyanide as the oxidizing agent which, in our experience, has proved more reliable than periodate. The color reaction with indole has been replaced by absorptiometry at 360 mμ according to Pisano et al. (84), and two internal standards are run together with each sample. For the sake of simplicity the internal standards are only added after the extraction of the urine with ethyl acetate, but control experiments have shown that this is permissible since losses up to this point are minimal. An unoxidized blank allows for the presence of vanillin from dietary sources.

The methods based on the formation of vanillin from VMA are subject to interference mainly from two natural sources, unconjugated MHPG and p-hydroxymandelic acid. Unconjugated MHPG seems to be absent or is at most a small proportion of the total MHPG and is therefore of little consequence. Moreover, if the ethyl acetate extract containing VMA is re-extracted with phosphate buffer of pH 7.0, rather than the carbonate solution used in the original method, MHPG would be expected to remain largely in the organic phase. Alternatively, MHPG may be removed by an extraction at pH 6, prior to the extraction of VMA at pH 1 (86,87).

p-Hydroxymandelic acid is oxidized in the assay to p-hydroxybenzaldehyde, with an absorption peak at 330 mμ, somewhat lower than the absorption peak for vanillin at 348 mμ. By taking the reading at 360 mμ rather than at the maximum of 348 mμ, the interference from p-hydroxybenzaldehyde is diminished or eliminated at the expense of a small loss in sensitivity. However, this correction is valid only if the amount of p-hydroxymandelic acid present in urine does not exceed 50–100% of the VMA content. This condition is usually met in human urine as well as in the urine of cats, mice, and chimpanzees (88). Dogs, however, excrete 30–60 times more p-hydroxymandelic acid and 10 times less VMA, per mg creatinine, than man (88). A similar ratio seems to occur in rat urine. Attempts to differentiate *p*-hydroxybenzaldehyde from vanillin optically by the method of simultaneous equations were not successful in extracts of rat urine, owing to the preponderance of *p*-hydroxybenzaldehyde and the proximity of the two absorption curves (L. B. Bigelow, personal communication). It is advisable to record complete absorption spectra of the vanillin-containing extract, together with internal and external standards, between 300 and 450 mμ; the quantity of *p*-hydroxybenzaldehyde present may then be roughly assessed by the shift of the vanillin absorption peak to shorter wavelengths. Normally, an extract, with or without added standard, has the same absorption as the unoxidized blank at wavelengths above 410 mμ. If the curve does not return to the base line in this region, it should alert the investigator to the presence of interfering material or to faulty techniques. A reading at 450 mμ may be substituted if the recording of the complete spectrum is not feasible.

Other procedures for the estimation of VMA are based on paper chromatography (89), thin-layer chromatography (90–93), electrophoresis on paper (94,97), or on cellulose acetate (98). A column chromatographic procedure suitable for research purposes has been developed by Weise et al. (99). An extract from urine is fractionated on a column of Dowex 1 by 1*M* ammonium formate (pH 8.0). The band of VMA is located with the aid of a radioactive tracer, 7-^{3}H-VMA, isolated from the urine of subjects who had received an infusion of 7-^{3}H-DL-epinephrine.

Another isotope technique has recently been proposed by O'Gorman (100); he adsorbs VMA onto silica gel from a dichloromethane extract. VMA is then acetylated with tritiated acetic anhydride in methyl cyanide and the radioactivity of the isolated product determined. The method is suitable for the estimation of VMA in serum.

Procedure

Reagents. *Phosphate buffer*, 0.5*M* (pH 7.0). Dissolve 68.0 g KH_2PO_4 in about 800 ml, adjust the pH to 7.0 with 5*N* NaOH, and dilute to 1 liter.

Potassium phosphate, dibasic, 4*M*. Dissolve 139.3 g of K_2HPO_4 to 200 ml.

Potassium ferricyanide, 0.6%. Keep refrigerated. Renew after 1 month.

Zinc sulfate, $ZnSO_4·7H_2O$, 1.2%.

Ammonia, 0.5*N* solution, containing 6% sodium chloride.

Ethyl acetate (reagent grade).

Toluene (reagent grade).

Florisil (Floridin Co., Pittsburgh, Pa.).

Standard, containing 100 μg/ml in 0.1*N* HCl. Keep refrigerated. Renew after 3 months.

Shake 50 ml of urine, 5 ml of $10N$ H_2SO_4, and 4 g of Florisil for 10 min and centrifuge. Saturate 44 ml of the supernatant ($= 40$ ml of urine) with NaCl and extract first with 88, then with 44 ml of ethyl acetate by shaking for 10 min. Re-extract the combined extracts with 10 ml $0.5M$ phosphate buffer (pH 7.0). Separate the aqueous layer and wash the ethyl acetate layer with 2 ml water. Measure the volume of the aqueous solution, including the washings (p ml). Transfer to a small beaker and add slowly, with stirring, 5 ml of conc. HCl at 0°. Measure out 4 aliquots (a, b, c, and d) of 3 ml. Add 0.1 ml of standard ($= 10$ μg VMA) to b, 0.2 ml of standard ($= 20$ μg VMA) to c, and 0.6 ml of zinc sulfate solution to a, b, c, and d. To a, b, and c add 0.6 ml 0.6% $K_3Fe(CN)_6$; to d add 0.6 ml of water. Incubate for 2 hr at 37°C, protected from bright light. Cool, add 0.5 ml $4M$ K_2HPO_4, and adjust the pH to 7.0 with $5N$ NaOH. Extract with 5 vol (about 25 ml) toluene by shaking for 10 min. Re-extract in 3 ml of $0.5N$ ammonia-6% NaCl reagent (5 min shaking) and separate the aqueous layer containing vanillin. Measure the optical density of samples a, b, and c at 360 and 450 mμ against sample d. Alternatively, record the absorption spectra between 300 and 450 mμ.

Calculation

The corrected reading of the 20 μg internal standard is usually within $\pm 15\%$ of twice the corrected reading of the 10 μg standard. A result is calculated based on the 10 μg standard and another one based on the 20 μg standard, and the mean of the two is used as the concentration of the unknown sample.

Since four portions of 3 ml were taken out of a total of $p + 5$ ml, the result is multiplied by the factor $(p + 5)/12$ to obtain the concentration of VMA in 10 ml of urine.

V. THE ESTIMATION OF 3-METHOXY-4-HYDROXYPHENYLGLYCOL (MHPG)

Like VMA, MHPG may be measured after oxidation to vanillin. A method based on this reaction and applicable to human urine has been described by Ruthven and Sandler (101). According to these authors, urine is hydrolyzed enzymically and passed over a column of Amberlite CG-50 at pH 6.2–6.3 to remove metanephrine and normetanephrine. The effluent of the column containing MHPG is extracted with ethyl acetate, the extract evaporated, and the residue oxidized with periodate. This method has been improved by Dr. L. B. Bigelow in this laboratory to the extent that the recoveries of MHPG which averaged 12.3% in the original method are now about 80%. It has been ascertained that the losses occur mainly during the evaporation of the ethyl acetate extract.

This step has therefore been eliminated. The ethyl acetate extract is shaken with ammonia in the presence of periodate to effect simultaneous oxidation and extraction of the vanillin formed into the ammonia layer. The passage of urine through the Amberlite column was also found to be unnecessary since the O-methylated amines are not extracted into ethyl acetate at pH 6. Troublesome emulsification which previously accompanied the solvent extraction, mainly owing to the presence of components of glusulase, was prevented by precipitation with tungstic acid. The method here described has the advantage over the gas-chromatographic estimation of MHPG that it includes unconjugated MHPG which, along with interfering material, is removed by solvent extraction before gas chromatography. Inhibitory anions have to be removed before glusulase addition by barium precipitation at pH 11.5 rather than by anion exchange resin which seems to involve losses of MHPG.

Procedure for the estimation of total MHPG in urine

Reagents. $BaCl_2$, saturated solution.

Glusulase, β-glucuronidase Type H2 from Helix pomatia, Sigma Chemical Company, St. Louis, Mo.

Sodium tungstate, 40%.

Perchloric acid, 60%, w/v.

Sodium phosphate, basic, Na_2HPO_4, $0.5M$.

Florisil (Floridin Co., Pittsburgh, Pa.).

Ammonia solution, $4N$ and $1N$.

Sodium periodate, 5%, prepared fresh.

Sodium metabisulfite, 20%, prepared fresh.

Phosphoric acid, approximately $7.2M$: 500 ml of 85% H_3PO_4 per liter.

Ethyl acetate, toluene (reagent grade).

Standard, containing 200 μg of MHPG per ml of abs ethanol. Stable for >6 months at 3°C.

To 30 ml urine add 3 ml of saturated $BaCl_2$ and $5N$ NaOH to pH 11.5. Centrifuge and acidify the supernatant solution to pH 5.0–5.5. Add 1 ml glusulase and incubate at 37° for 16 hr. Cool in ice and add 1 ml of 40% sodium tungstate and 60% $HClO_4$ to a pH of 1.0–1.1. Leave in ice for 20 min, then centrifuge at about 10,000 g for 10 min. Adjust the pH of the supernatant solution to 6.2–6.5 by the addition of 2 ml $0.5M$ Na_2HPO_4 and $5N$, then $1N$ NaOH. Add 3 g Florisil and a saturating amount of NaCl (10–12 g) and shake first with 70 ml, then with 50 ml ethyl acetate for 5 min each time. Combine the extracts. Bring the volume up to 120 ml with ethyl acetate and divide into 3 equal portions, a, b, and c. Add 0.1 ml of MHPG standard to b, 4 ml $4N$ ammonia to a, b, and c and 0.1 ml of 5% $NaIO_4$ to a and b. Shake for 20 min, then add

0.5 ml of 20% sodium metabisulfite to a and b. To c, the blank, add a premixed sample of 0.5 ml $Na_2S_2O_5$ and 0.1 ml $NaIO_4$. Shake again for 5 min, centrifuge, and transfer the aqueous phase to a 10-ml beaker. After $\frac{1}{2}$ hr (to allow ethyl acetate to evaporate) adjust to pH 7–7.5 first with $7.2M$ H_3PO_4, later with more dilute H_3PO_4. Decant into 50 ml glass-stoppered extraction tubes and add 1.5 ml of water. Add about 0.5 g of Florisil, saturate with NaCl, and extract for 10 min with 35 ml toluene. Centrifuge. Transfer the toluene phase to another extraction tube containing 1.5 ml $1N$ ammonia. Shake for 5 min. Place the aqueous extract in a cuvet and record the absorption spectrum of the sample and standard against the blank, between 300 and 450 mμ. Use the absorption at 348 mμ to calculate the concentration of MHPG in the sample.

It is important that the absorption curve should return to the baseline, or close to it, at 450 mμ, indicating the absence of interfering absorption.

VI. HOMOVANILLIC ACID, 3,4-DIHYDROXYPHENYLACETIC ACID AND 3,4-DIHYDROXYMANDELIC ACID

Methods for the determination of these metabolites of the catecholamines (see Fig. 1, Nos. 4, 6, 11) are discussed but are not described in detail.

1. Homovanillic Acid (HVA, 3-Methoxy-4-hydroxyphenylacetic Acid)

Anden, Roos, and Werdinius (102) found that HVA is oxidized to a fluorescent compound by treatment with ferricyanide in ammoniacal solution for 4 min, when the reaction is stopped by the addition of cysteine. The reaction product has activation and fluorescence maxima at 320 and 420 mμ and has been identified as 2,2'-dihydroxy-3,3'-dimethoxybiphenyl-5,5'-diacetic acid (103; Fig. 2). The same compound is formed from HVA in the presence of H_2O_2 and peroxidase (104). Gjess-

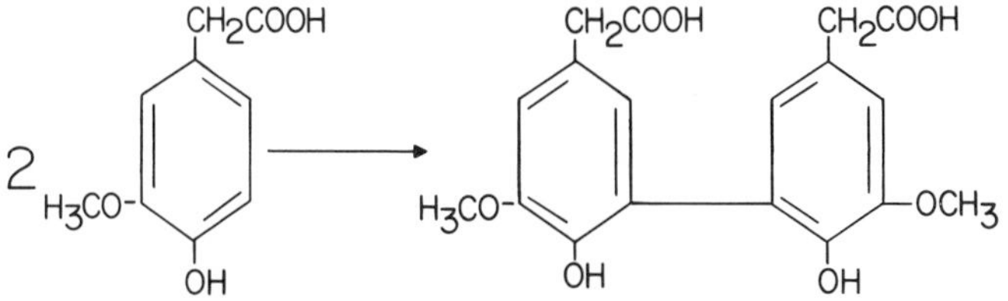

Fig. 2. Formation of 2,2'-dihydroxy-3,3'-dimethoxybiphenyl-5,5'-diacetic acid from homovanillic acid.

ing et al. (105) have studied the specificity of the reaction and concluded that the structure required was a 3-methoxy-4-hydroxyphenyl compound with a side chain containing at least two carbon atoms, the first of which was a CH_2-group, i.e.,

$$3\text{-}CH_3O\text{---}4\text{-}OH\text{---}C_6H_3\text{---}CH_2\text{---}C\equiv$$

A positive reaction is therefore also given by vanilalanine and vanillactic acid, substances that may occur in urine (106). Anden et al. (102) applied their method to deproteinized brain extracts extracted with ether. Sato (107) found that solvent extraction alone provided insufficient purification for the estimation of HVA in urine where blanks were excessive. He succeeded in obtaining satisfactory extracts by adsorbing urinary HVA on a 0.5×3 cm column of Dowex 1, X4, eluting with $1.5M$ NaCl and extracting the eluate with a mixture of toluene-chloroform, 4:1. Weil-Malherbe and Van Buren (108) obtained good results with Sato's method but found it an advantage to use 2 ml of urine instead of 1 ml.

2. 3,4-Dihydroxyphenylacetic Acid (Dopac)

The most convenient and sensitive way of estimating dopac is by condensation with ethylenediamine to a fluorescent product (109). Methods based on this principle were used for the assay of dopac in brain (110,111) and in urine (108). Since dopac has the catechol structure, a useful first step in its extraction and purification is by adsorption on alumina at pH 8.4. It is, however, more firmly bound than the catecholamines and requires strong acid for elution (109,112). We have had little success in numerous attempts to further purify the alumina eluate containing dopac by anion exchange (56) owing to low recoveries and have therefore extracted the acidified eluate from the alumina adsorption with ethyl acetate before condensation with ethylenediamine (108). The assay as applied to urine may include other 3,4-dihydroxyphenylic acids, such as protocatechuic and 3,4-dihydroxymandelic acids, but these acids are present in human urine in much lower concentrations than dopac, if at all (113,114).

According to von Euler et al. (66), a part of dopac present in human urine and in bovine plasma exists in an acid-hydrolyzable conjugated form. We have been unable to obtain evidence for the presence of conjugated dopac in human urine by acid hydrolysis, in spite of good recoveries.

3. 3,4-Dihydroxymandelic Acid (DHMA)

DHMA is a very unstable compound which seems to be even more sensitive to oxygen than other catechols. Its oxidation to protocatechuic aldehyde has been used for its assay in urine (115,116), but interference by dopac made these methods of doubtful validity (113). Conditions were eventually found which eliminated the interference of dopac (114). The method is based on the observation (117) that aromatic aldehydes in methanol solution are nonfluorescent but become fluorescent on acidification due to formation of acetals. It consists of the following steps: adsorption on alumina and elution with $1N$ H_2SO_4 (like dopac, DHMA is adsorbed more firmly than the catecholamines); passage through a column of Dowex 50, X4; extraction with ethyl acetate; oxidation to protocatechuic aldehyde by shaking the extract with ammonia solution; extraction of the aldehyde with ethyl acetate; evaporation and fluorimetry of the residue in methanol solution before and after acidification.

Recently, Sato and DeQuattro (118) described an enzymic assay for urinary DHMA in which the acid is isotopically labeled by incubation with ^{14}C-SAM and COMT. The ^{14}C-VMA formed is oxidized to ^{14}C-vanillin by periodate; ^{14}C-vanillin is extracted and counted. Owing to the limited specificity of COMT, any dopac present is methylated to HVA and since HVA interferes to some extent in the periodate method of VMA estimation (119), the prior removal of dopac is essential. Although previous attempts in our laboratory to separate DHMA and dopac by solvent—including ether—extraction were not promising, Sato and DeQuattro report the successful removal of dopac by ether extraction with only slight losses of DHMA. If this point is confirmed, their method is the method of choice as it is certainly much more sensitive than our previous method. Sato and DeQuattro have even used it to determine the concentration of DHMA in plasma and tissues. Incidentally, their figure for the average 24 hr-excretion in man, 90–100 μg, is in excellent agreement with our own results.

References

1. N. R. Clendenon and N. Allen, *Fed. Proc.*, *26*, 707 (1967).
2. H. Hidaka, T. Nagatsu, K. Takeya, S. Matsumoto, and K. Yagi, *J. Pharmacol. Exp. Ther.*, *166*, 272 (1969).
3. S. M. Schanberg, J. J. Schildkraut, G. R. Breese, and I. J. Kopin, *Biochem. Pharmacol.*, *17*, 247 (1968).
4. J. Häggendal, *Acta Physiol. Scand.*, *59*, 255 (1963).
5. J. Häggendal, *Acta Physiol. Scand. suppl.*, *277*, 84 (1966).
6. S. M. Schanberg, G. R. Breese, J. J. Schildkraut, E. K. Gordon, and I. J. Kopin, *Biochem. Pharmacol.*, *17*, 2006 (1968).
7. McC. Goodall and L. Rosen, *J. Clin. Invest.*, *42*, 1578 (1963).
8. McC. Goodall and H. Alton, *Biochem. Pharmacol.*, *17*, 905 (1968).
9. J. Alton and McC. Goodall, *Biochem. Pharmacol.*, *17*, 2163 (1968).
10. T. Wieland, *Z. Physiol. Chem.*, *273*, 24 (1942).
11. F. Turba and M. Richter, *Ber. Deutsch. Chem. Ges.*, *75*, 340 (1942).
12. P. A. Shore and J. S. Olin, *J. Pharmacol. Exp. Ther.*, *122*, 295 (1958).

13. J. A. R. Mead and K. F. Finger, *Biochem. Pharmacol.*, *6*, 52 (1961).
14. K. A. Montagu, *Biochem. J.*, *63*, 559 (1956).
15. T. B. B. Crawford and A. S. Outschoorn, *Brit. J. Pharmacol. Chemother.*, *6*, 8 (1951).
16. R̃. M. Fleming, W. G. Clark, E. D. Fenster, and F. C. Towne, *Anal. Chem.*, *37*, 692 (1951).
17. C. C. Chang, *Int. J. Neuropharmacol.*, *3*, 643 (1964).
18. R. P. Maickel, R. H. Cox, Jr., J. Saillant, and F. P. Miller, *Int. J. Neuropharmacol.*, *7*, 275 (1968).
19. G. B. Ansell and M. F. Beeson, *Anal. Biochem.*, *23*, 196 (1968).
20. A. Lund, *Acta Pharmacol.*, *5*, 231 (1949).
21. J. Häggendal, *Acta Physiol. scand.*, *59*, 242 (1963).
22. H. Weil-Malherbe and A. D. Bone, *Biochem. J.*, *58*, 132 (1954).
23. D. C. Wise, *Anal. Biochem.*, *18*, 94 (1967).
24. A. H. Anton and D. F. Sayre, *Eur. J. Pharmacol.*, *4*, 435 (1968).
25. J. A. Wada and E. G. McGeer, *Arch. Neurol.*, *14*, 129 (1966).
26. F. H. Shaw, *Biochem. J.*, *32*, 19 (1938).
27. U. S. von Euler, *Arch. Int. Pharmacodyn. Ther.*, *77*, 477 (1948).
28. H. Weil-Malherbe and A. D. Bone, *Biochem. J.*, *51*, 311 (1952).
29. A. S. Russell and C. N. Cochran, *Ind. Eng. Chem.*, *42*, 1336 (1950).
30. H. Weil-Malherbe, in *Methods of Biochemical Analysis*, Vol. 16, D. Glick, Ed., Wiley (Interscience), New York, 1968, pp. 293–326.
31. J. R. Crout, in *Standard Methods of Clinical Chemistry*, Vol. 3, D. Seligson, Ed., Academic Press, New York, 1961, pp. 62–80.
32. U. S. von Euler and I. Orwén, *Acta Physiol. Scand. Suppl.*, *118*, 1 (1955).
33. A. H. Anton and D. F. Sayre, *J. Pharmacol. Exp. Ther.*, *138*, 360 (1962).
34. H. Weil-Malherbe, in *Methods in Medical Research*, Vol. 9, J. H. Quastel, Ed., Yearbook Pub., Chicago, 1961, pp. 130–146.
35. A. Bertler, A. Carlsson, and E. Rosengren, *Acta Physiol. Scand.*, *44*, 273 (1958).
36. J. Häggendal, *Scand. J. Clin. Lab. Invest.*, *14*, 537 (1962).
37. S. Bergström and G. Hansson, *Acta Physiol. Scand.*, *22*, 87 (1951).
38. N. Kirshner and McC. Goodall, *J. Biol. Chem.*, *226*, 207 (1957).
39. L. M. Gunne, *Acta Physiol. Scand. Suppl.*, *204*, 5 (1963).
40. D. T. Masuoka, W. Drell, H. F. Schott, A. F. Alcaraz, and E. C. James, *Anal. Biochem.*, *5*, 426 (1963).
41. L. B. Bigelow and H. Weil-Malherbe, *Anal. Biochem.*, *26*, 92 (1968).
42. G. L. Mattok, D. L. Wilson, and R. A. Heacock, *Clin. Chim. Acta*, *14*, 99 (1966).
43. E. M. Trautner and M. Messer, *Nature*, *169*, 31 (1952).
44. J. T. Wright, *Lancet*, *1958-II*, 1155.
45. U. S. von Euler, *Pharmacol. Rev.*, *11*, 262 (1959).
46. J. Häggendal, *Pharmacol. Rev.*, *18*, 325 (1966).
47. U. S. von Euler and F. Lishajko, *Acta Physiol. Scand.*, *45*, 122 (1959); *51*, 348 (1961).
48. J. F. Palmer, *West Indian Med. J.*, *13*, 38 (1964).
49. E. C. Gerst, O. S. Steinsland, and W. W. Walcott, *Clin. Chem.*, *12*, 659 (1966).
50. H. Weil-Malherbe and L. B. Bigelow, *Anal. Biochem.*, *22*, 321 (1968).
51. R. J. Merrills, *Anal. Biochem.*, *6*, 272 (1963).
52. L. E. Martin and C. Harrison, *Anal. Biochem.*, *23*, 529 (1968).
53. J. K. Viktora, A. Baukal, and F. W. Wolff, *Anal. Biochem.*, *23*, 513 (1968).
54. H. L. Price and M. L. Price, *J. Lab. Clin. Med.*, *50*, 769 (1957).

THE CHEMICAL ESTIMATION OF CATECHOLAMINES 151

55. R. Laverty and K. M. Taylor, *Anal. Biochem.*, *22*, 269 (1968).
56. H. Weil-Malherbe, *Z. Klin. Chem.*, *2*, 161 (1964).
57. A. Lund, *Acta Pharmacol.*, *6*, 137 (1950).
58. U. S. von Euler and U. Hamberg, *Acta Physiol. Scand.*, *19*, 74 (1949).
59. U. S. von Euler and I. Floding, *Acta Physiol. Scand. suppl.*, *118*, 45 (1955).
60. A. Carlsson and B. Waldeck, *Acta Physiol. Scand.*, *44*, 293 (1958).
61. A. H. Anton and D. F. Sayre, *J. Pharmacol. Exp. Ther.*, *145*, 326 (1964).
62. S. Udenfriend, *Fluorescence Assay in Biology and Medicine*, Academic Press, New York, 1962, p. 137.
63. J. R. Crout, *Pharmacol. Rev.*, *11*, 296 (1959).
64. H. Weil-Malherbe and A. D. Bone, *Biochem. J.*, *67*, 65 (1957).
65. W. M. Manger, K. G. Wakim, and J. L. Bollman, *Chemical Quantitation of Epinephrine and Norepinephrine in Plasma*, Thomas, Springfield, Ill., 1959, p. 56.
66. U. S. von Euler, I. Floding, and F. Lishajko, *Acta Soc. Med. Upsalien.*, *64*, 217 (1959).
67. G. Cohen and M. Goldenberg, *J. Neurochem.*, *2*, 58, 71 (1957).
68. R. A. Miller and B. G. Benfey, *Brit. J. Anaesth.*, *30*, 159 (1958).
69. A. Vendsalu, *Acta Physiol. Scand. Suppl.*, *173*, (1960).
70. H. Klensch, *Z. Kreislaufforsch*, *54*, 771 (1965).
71. M. J. Oesterling and R. L. Tse, *Amer. J. Med. Technol.*, *27*, 112 (1961).
72. J. K. Saelens, M. S. Schoen, and G. B. Kovacsics, *Biochem. Pharmacol.*, *16*, 1043 (1967).
73. G. B. Kovacsics and J. K. Saelens, *Arch. Int. Pharmacodyn. Ther.*, *174*, 481 (1968).
74. J. Axelrod, *J. Biol. Chem.*, *237*, 1657 (1962).
75. L. A. Pohorecky, M. Zigmond, H. Karten, and R. J. Wurtman, *J. Pharmacol. Exp. Ther.*, *165*, 190 (1969).
76. B. V. Nikodijevic, J. W. Daly, S. Senoh, B. Witkop, and C. R. Creveling, *Fed. Proc.*, *27*, 602 (1968).
77. K. Engelman, B. Portnoy, and W. Lovenberg, *Amer. J. Med. Sci.*, *255*, 259 (1968).
78. H. Weil-Malherbe and A. D. Bone, *J. Clin. Path.*, *10*, 138 (1957).
79. E. R. B. Smith and H. Weil-Malherbe, *J. Lab. Clin. Med.*, *60*, 212 (1962).
80. E. H. LaBrosse, J. Axelrod, I. J. Kopin, and S. S. Kety, *J. Clin. Invest.*, *40*, 253 (1961).
81. J. Häggendal, *Acta Physiol. Scand.*, *56*, 258 (1962).
82. M. Sandler and C. R. J. Ruthven, *Biochem. J.*, *80*, 78 (1961).
83. F. W. Sunderman, Jr., P. D. Cleveland, N. C. Law, and F. W. Sunderman, *Amer. J. Clin. Pathol.*, *34*, 293 (1960).
84. J. J. Pisano, J. Crout, and D. Abraham, *Clin. Chim. Acta*, *7*, 285 (1962).
85. H. Weil-Malherbe, *Anal. Biochem.*, *7*, 485 (1964).
86. L. R. Gjessing, *Scand. J. Clin. Lab. Invest.*, *15*, 463 (1963).
87. J. D. Sapira, *Clin. Chim. Acta*, *20*, 139 (1968).
88. L. R. Gjessing, M. Maeda, O. Borud, and E. J. Vellan, *Scand. J. Clin. Lab. Invest.*, *18*, 638 (1966).
89. M. D. Armstrong, K. N. F. Shaw, and P. E. Wall, *J. Biol. Chem.*, *218*, 293 (1956).
90. E. Schmid and N. Henning, *Klin. Wochenschr.*, *41*, 566 (1963).
91. J. P. O'Neal, J. W. Traubert, and S. Meites, *Clin. Chem.*, *12*, 441 (1966).
92. R. F. McGregor, M. Khan, D. Marrack, and M. P. Sullivan, *Amer. J. Clin. Pathol.*, *46*, 163 (1966).

152   H. WEIL-MALHERBE

93. J. M. C. Gutteridge, *Clin. Chim. Acta*, *21*, 211 (1968).
94. W. von Studnitz, *Scand. J. Clin. Lab. Invest. Suppl. 48*, (1960).
95. A. Randrup, *Scand. J. Clin. Lab. Invest.*, *14*, 262 (1962).
96. D. Klein and J. M. Chernaik, *Clin. Chem.*, *7*, 257 (1961).
97. F. Eichhorn and A. Rutenberg, *Clin. Chem.*, *9*, 615 (1963).
98. T. J. Butler, *Clin. Chem.*, *13*, 488 (1967).
99. V. K. Weise, R. K. McDonald, and E. H. LaBrosse, *Clin. Chim. Acta*, *6*, 79 (1961).
100. L. P. O'Gorman, *Clin. Chim. Acta*, *23*, 247 (1969).
101. C. R. J. Ruthven and M. Sandler, *Clin. Chim. Acta*, *12*, 318 (1965).
102. N.-E. Anden, B.-E. Roos, and B. Werdinius, *Life Sci.* (Oxford), *2*, 448 (1963).
103. H. Corrodi and B. Werdinius, *Acta Chem. scand.*, *19*, 1854 (1965).
104. G. G. Guilbault, *Anal. Chem.*, *39*, 271 (1967).
105. L. R. Gjessing, E. J. Vellan, B. Werdinius, and H. Corrodi, *Acta Chem. Scand.*, *21*, 820 (1967).
106. L. R. Gjessing, *Scand. J. Clin. Lab. Invest.*, *15*, 649 (1963).
107. T. L. Sato, *J. Lab. Clin. Med. 66*, 517 (1965).
108. H. Weil-Malherbe and J. M. Van Buren, *J. Lab. Clin. Med.*, *74*, 305 (1969).
109. H. Weil-Malherbe and A. D. Bone, *J. Neurochem.*, *4*, 251 (1959).
110. E. Rosengren, *Acta Physiol. scand.*, *49*, 370 (1960).
111. N.-E. Anden, B.-E. Roos, and B. Werdinius, *Life Sci.* (Oxford), *2*, 319 (1963).
112. W. Drell, *Fed. Proc.*, *16*, 174 (1957).
113. H. Weil-Malherbe and E. R. B. Smith, *Pharmacol. Rev.*, *18*, 331 (1966).
114. H. Weil-Malherbe, *J. Lab. Clin. Med.*, *69*, 1025 (1967).
115. H. Miyake, H. Yoshida, and R. Imaizumi, *Japan. J. Pharmacol.*, *12*, 79 (1962).
116. V. DeQuattro, D. Wybenga, W. von Studnitz, and S. Brunjes, *J. Lab. Clin. Med.*, *63*, 864 (1964).
117. E. P. Crowell and C. J. Varsel, *Anal. Chem.*, *35*, 189 (1963).
118. T. Sato and V. DeQuattro, *Fed. Proc.*, *28*, 543 (1969).
119. B. Parrad and C. Bohuon, *Ann. Biol. Clin.*, *26*, 635 (1968).
120. K. Engelman and B. Portnoy, *Circul. Res.*, *26*, 53 (1970).

Assay of Enzymes of Catecholamine
Biosynthesis and Metabolism

C. R. Creveling and J. W. Daly,
*Laboratory of Chemistry, National Institute of Arthritis and Metabolic Diseases,
National Institutes of Health, Bethesda, Maryland 20014*

I. INTRODUCTION*

Biosynthesis of the physiologically important catecholamines, dopamine, norepinephrine, and epinephrine, is generally thought to proceed by the following enzymatic reactions: (*1*) Hydroxylation of tyrosine to form DOPA, catalyzed by tyrosine hydroxylase; (*2*) decarboxylation of DOPA to the amine, dopamine, catalyzed by dopa decarboxylase; (*3*) Hydroxylation of dopamine to form norepinephrine, catalyzed by dopamine-β-hydroxylase; (*4*) *N*-methylation of norepinephrine to form epinephrine, catalyzed by phenethanolamine-*N*-methyltransferase.

---

* The following abbreviations are used in this Chapter: TH, tyrosine hydroxylase; DC, dopa decarboxylase; DβH, dopamine-β-hydroxylase; PNMT, phenethanolamine-*N*-methyltransferase; COMT, catechol-*O*-methyltransferase; MAO, monoamine oxidase; DOPA, 3,4-dihydroxyphenylalanine; DA, 3,4-dihydroxyphenethylamine; NE, norepinephrine; E, epinephrine; DMPH$_4$, 6,7-dimethyltetrahydropteridine; μm, μmoles.

Metabolism of catecholamines proceeds by two pathways: O-methyla-tion, catalyzed by catechol-O-methyltransferase, and oxidative deamina-tion, catalyzed by monoamine oxidase. Monoamine oxidase catalyzes the oxidative deamination of both catecholamines and their O-methylated metabolities, giving rise to the corresponding aldehydes which are subse-quently either enzymatically reduced to the corresponding alcohol or oxidized to the corresponding acid. The catecholic acids and alcohols are further metabolized by catechol-O-methyltransferase. The biosyn-thesis and metabolism of catecholamines is the subject of a recent compre-hensive review (1).

From the variety of assays for these enzymes that have been developed, an attempt has been made to choose and present a method(s) for the assay of each of these enzymes which combines reliability, simplicity, and sensitivity, and which can be applied to both purified and crude enzyme preparations. As far as possible, the conditions for measuring enzyme activity have been optimized with respect to substrate, cofactors, and protein concentration. Reaction times have been chosen which will give nearly linear reaction rates. For TH and DC, assays based on tyrosine and DOPA, respectively have been presented. In the case of DβH, PNMT, COMT, and MAO, advantage has been taken of the rather broad substrate specificity exhibited by these enzymes in presenting assays based, respectively, on tyramine, phenethanolamine, 3,4-di-hydroxypropriophenone, and tyramine. Literature references to alter-nate assays and selected references to various applications have been cited.

II. TYROSINE HYDROXYLASE

The enzyme TH initiates and appears to be the rate limiting step in the biosynthesis of catecholamines (2). In addition to L-tyrosine, hydroxyla-tion of α-methyltyrosine (3) and L-phenylalanine (4,5) have been reported. D-Tyrosine is not a substrate for TH (6). TH is widely distributed in sympathetically innervated tissues and in the central nervous system. The subcellular distribution of TH is still under investigation, but in brain and perhaps in adrenal the enzyme appears to be associated with subcellular particles from which it is partially solubilized during homo-genization. TH has been purified from adrenal medulla (6,7). It requires a reduced pteridine cofactor for optimal activity (6,8,9). The naturally occurring cofactor is probably dihydrobiopterin, but a synthetic compound, 6,7-dimethyltetrahydropteridine (DMPH$_4$), along with an excess of a reducing agent, 2-mercaptoethanol, proves quite satisfactory for use in TH assays. The properties of TH (6,7,10–12), circadian rhythm of pineal TH (13), the effect of various physical and chemical treatments on TH levels (14–19), subcellular distribution of TH (6,7,15, 20–23), levels of TH activity in various tissues (6,24), brain (6,25–27), salivary gland (18), and inhibition using TH from either adrenal medulla (3,6,7,28–38), heart (28,34) or brain (39–42) are among the investigations of this enzyme that have been reported.   Certain inhibitors of TH have been shown to be competitive with substrate (3,5,6,7,32,41), or with cofactor $DMPH_4$ (3,7,10,28,30,31,33,34), or noncompetitive with either substrate or cofactor (29,35–37).   In addition, differences in K_i values for an inhibitor have been noted in investigations with either soluble or particulate TH from bovine adrenal medulla (32).   Thus, for studies on the inhibition of TH, cofactor, and substrate levels, and source of enzyme are important variables.   Inhibition by excess cofactor has been reported (9).   Stimulation of TH by Fe^{++} has been noted under certain conditions (6,7,12).

Methods for the assay of TH activity are based on the conversion of tyrosine to DOPA.   Two radiometric procedures have found acceptance. In the first method, L-tyrosine-3,5-^{3}H is converted by the action of TH to L-DOPA-3-^{3}H (43).   One tritium is released into the medium as water. Separation of the tritiated water from the starting material and L-DOPA-3-^{3}H is easily accomplished using a microcolumn of Dowex-50 (H^+ form). The amino acids are retained on the resin and the effluent, tritiated water is collected and measured by liquid scintillation techniques.   This method has been widely used for the assay of TH activity in both crude and purified tissue preparations.   A similar assay has been independently developed for study of tyrosinase (44–47).   For comparison of the properties of TH and tyrosinase, see Ref. 6.

In an alternate radiometric method, L-tyrosine-^{14}C is used as substrate and the product, L-DOPA-^{14}C, is separated by adsorption and elution from alumina (6,12).   This method, while more time consuming, has found some application.   A recent modification of this assay involves fluorometric measurement of L-DOPA after separation by both column chromatography and alumina absorption (48).   As the L-tyrosine-^{14}C method offers no distinct advantages except perhaps slightly lower blank values and is more time consuming, only the L-tyrosine-3,5-^{3}H method is reported.   Both radiometric methods have the disadvantage that endogenous tyrosine, normally present at approximately $10^{-4}M$ in crude tissue, dilutes the L-tyrosine-3,5-^{3}H employed in the assay.   Corrections based on fluorimetric determination of endogenous tyrosine (3,49) in crude preparations may be made but are probably not critical except where tissue dilution is minimal.

1. Reagents

2-Mercaptoethanol, DMPH₄, and L-*tyrosine-3,5-³H* (5–25 mCi/μM) are commercially available.

The $DMPH_4$ must be stored at $-20°C$ under anhydrous conditions. A solution containing $DMPH_4$ ($10^{-2}M$) and 2-mercaptoethanol ($1M$) in $0.1M$ sodium phosphate buffer (pH 7.4) is prepared and kept in ice for each set of determinations. The solution is stable for several hours.

L-*Tyrosine-3,5-³H*(5–25 mCi/μM) is purified in the following manner: L-tyrosine-3,5-H³ (~ 1 μM) in 2 ml of $0.1N$ HCl is applied to a small column (0.5×3 cm) of Dowex-50 (H^+). The column is washed with 100 ml of water and the tyrosine then eluted with 20 ml of $3N$ HCl. The acid eluent is lyophilized and the residue dissolved in 2.0 ml $0.01N$ HCl. Aliquots of this solution can then be stored until use at $-20°C$ for several months. For each set of determinations an aliquot in $0.01N$ HCl is evaporated to dryness in a stream of nitrogen and then dissolved in an appropriate volume of $2 \times 10^{-3}M$ L-tyrosine so as to contain 5–20 μCi/ml.

The decarboxylase inhibitor, p-*bromo*-m-*hydroxybenzyloxyamine* (NSD 1055, Smith Nephews, Ltd., Great Britain) (cf. 50), is prepared as a $0.01M$ solution in water and stored at $+5°$.

Sodium acetate buffer (pH 6.0) is $1.0M$.

2. Method

Tissue is chilled on ice, weighed, minced, diluted with 2 volumes (v/w) of cold $0.32M$ sucrose, and homogenized with a motor-driven Teflon or glass pestle. A tissue press has been used successfully with heart and adrenal tissue (28,36).

The reaction mixture is prepared in a conical tube (13 ml) immersed in ice with the following components added in this sequence: 0.2 ml of $1M$ acetate buffer (pH 6.0); 0.1 ml of $0.1M$ phosphate buffer (pH 7.4) containing 1 μM of $DMPH_4$ and 100 μM of mercaptoethanol; 0.1 ml of solution containing 0.2 μM of L-tyrosine and 0.5–2 μCi of L-tyrosine-3,5-³H; sufficient water for the *final* volume of the incubation medium to be 1.0 ml; and the TH preparation. $FeSO_4$(0.5μM) may enhance TH activity.

For crude tissues, 0.01 ml of the solution containing 0.1 μM of p-bromo-m-hydroxybenzyloxyamine is included in the mixture. Linear rates of TH activity with respect to tissue are obtained with crude tissue homogenates representing up to approximately 100 mg of tissue from adrenal medulla, heart, or brain. With purified adrenal TH (6), 0.1 to 1 mg of protein may be used. The reaction is nearly linear with respect to time for 20 min. The levels of tyrosine ($2 \times 10^{-4}M$) and cofactor, $DMPH_4$ ($10^{-3}M$), are saturating and the pH of 6.0 is near optimal.

The reaction is started by the addition of TH preparation, incubated at 37° with gentle shaking in air for 15 min, stopped by the addition of 50% trichloroacetic acid (0.05 ml), centrifuged, and the supernatant passed over a small column (1 × 0.3 cm) of Dowex-50 (H^+ form). The incubation tube is washed with 1 ml of distilled water, recentrifuged, and the supernatant used to wash the column. When more precision is required, an aliquot (0.50 ml) of the original supernatant can be transferred to the column followed by distilled water (2 ml). The effluent and wash (2 ml) from the column are collected in a counting vial, 10 ml of Brays phosphor solution added, and the tritiated water determined by liquid scintillation counting and corrected for counting efficiency. Either incubations with TH preparations that have been heated at 95° for 3 min or zero time incubations are used as controls.

III. DOPA DECARBOXYLASE

The enzyme DC was the first of enzymes concerned with catecholamine biosynthesis to be discovered and has been extensively studied (for recent reviews, see Refs. 1 and 51). The enzyme appears to have a broad substrate specificity (52–55). However, only DOPA and 5-hydroxytryptophan of the possible physiological substrates undergo decarboxylation at significant rates (56–60). In view of its apparent substrate specificity, DC has also been referred to as 5-hydroxytryptophan decarboxylase and as aromatic L-amino acid decarboxylase. No evidence for the existence of separate DC and 5-hydroxytryptophan decarboxylase enzymes has been obtained as yet. DC is widely distributed in mammalian tissues with especially high activity found in kidney. Most studies on the subcellular localization of DC have indicated that the enzyme is a soluble component of cytoplasm. However, certain investigations of DC activity (23,61,62) suggest that DC may also be associated with some particulate component of the cell. DC has been purified from kidney (53,58,59,63) and adrenal (55). The activity of DC with DOPA as substrate is stimulated by the addition of small amounts of pyridoxal phosphate. This stimulation is not well understood and is dependent on the state of the enzyme, pH, substrate, and other factors (59,64). For studies on inhibition of DC, the order of addition of cofactor, inhibitor, and substrate can greatly influence the results (65). The properties of DC (53,56,58), effects of various physical and chemical treatments on DC levels in various tissues (76,77), brain (25,26,61,78–83), pineal (84–87), gastric mucosa (88), liver (59), adrenal (55,56), spleen (89), skin (90), tumors (56,91–93), heart (94), spinal cord (67,89), sympathetic nerves (67,68), and inhibition using DC from either kidney (42,50,65,95–101) or brain (83,102,103) are among the many published studies on this enzyme. Assays of DC activity have been based on either DOPA or 5-hydroxy-tryptophan as substrate. Methods based on 5-hydroxytryptophan are covered in the chapter by Lovenberg and Engleman. Assays based on DOPA are linear for much shorter periods of time than those based on 5-hydroxytryptophan because of the lability of this compound. The earliest method for the assay of DC was based on manometric measurement of CO_2 formation (104,105) and has been widely used for studies involving preparations of DC with high activities. Sensitive radioisotopic methods based on evolution of $^{14}CO_2$ from carboxyl-labeled amino acid have also been developed (60,83,106). In addition to such techniques, DC activity has been measured by separation of the product, DA, followed by its fluorometric determination. The DA has been separated by butanol extraction (78), by chromatography on Permutite (76), and by ion exchange chromatography (53,77). Radioisotopic adaptations of these procedures have now been reported using DL-DOPA-2-^{14}C. The assay based on butanol extraction of DA-2-^{14}C is quite convenient (107) but appears to result in higher blanks than the method presented below, which is based on separation of DA-^{14}C from DOPA-^{14}C on an IRC-50 column. The use of Dowex-50 column chromatography (108), paper chromatography (20), paper electrophoresis (109), and thin-layer chromatography (110) for the separation of DA-^{14}C from DOPA-^{14}C has been employed as the basis of assays for DC. An interesting method based on the differential extraction of a trinitrophenyl derivative of DA or serotonin has been reported (111). None of the methods yet available for the assay of DC is completely satisfactory, and more sensitive and simpler methods will probably be developed.

1. Reagents

DL-*DOPA-2-^{14}C* (1–5 μCi/μM), *DA HCl* and *pyridoxal phosphate* are commercially available.

The DL-*DOPA-2-^{14}C* should be purified as follows: Approximately 50 μM of DL-DOPA-2-^{14}C (1–5 μCi/μM) is dissolved in 1 ml of 0.01M phosphate buffer (pH 6.5) and transferred to a small column (0.6 × 3 cm) of IRC-50 (Na$^+$ form, buffered at pH 6.5) and washed with 3 ml of water. The effluent containing approximately 50 μM of DL-DOPA-2-^{14}C is lyophilized and then made up in 0.01M HCl with sufficient L-DOPA added, so that 1.0 ml contains 10 μM of L-DOPA and 1–5 μCi of DL-DOPA-2-^{14}C. The solution is divided into portions and stored at $-20°$ until use (Method 1).

The *pyridoxal phosphate* is prepared as a $10^{-3}M$ solution in water and stored at 5°.

The monoamine oxidase inhibitor, *pargyline* (N-benzyl-N-methylprop-argylamine, Abbott Laboratories Inc.) (112) is prepared as a $0.01M$ solution in water and stored at 5°. The monoamine oxidase inhibitor tranylcypromine (*trans*-2-phenylcyclopropylamine, Smith, Kline and French) (113) in a $0.02M$ solution in water may be substituted for pargyline.

Potassium phosphate buffer (pH 7.0) is $0.5M$.

L-*DOPA* is prepared frequently as a $10^{-2}M$ solution in $0.01M$ HCl and stored at −20° (Method 2).

A $10^{-3}M$ solution of DA HCl in $0.01M$ HCl is prepared frequently and stored at −20° for use as carrier in Method 1.

2. Methods

Method 1. Crude tissue extracts may be prepared in a manner similar to that described for TH. The reaction mixture is prepared in conical tubes (13 ml) immersed in ice with the following components added in this sequence: 0.2 ml of $0.5M$ phosphate buffer (pH 7.0); 0.1 ml solution containing 0.1 μM pyridoxal phosphate; sufficient distilled water for the *final* volume of the incubation mixture to equal 1.0 ml; the DC preparation; and 0.1 ml of $0.01N$ HCl containing 1 μM of L-DOPA and 0.1–0.5 μCi of DL-DOPA-2-¹⁴C.

Preincubation with pyridoxal phosphate before addition of substrate affords only slightly higher activities. However, in the case of inhibition studies, the effect of preincubations and the order of addition of substrate, cofactor, and inhibitor may be very significant (65). For crude tissues, 0.01 ml of the solution containing 0.1 or 0.2 μM, respectively, of pargyline or tranylcypromine is included in the mixture. Linear rates of DC activity with respect to tissue are obtained with crude tissue homogenates representing up to approximately 100 mg of tissue from kidney, adrenal, heart, and brain. The quantity of tissue required is, of course, dependent on the level of DC activity. With enzyme purified through the second ammonium sulfate precipitation of the procedure of Clark et al. (63), 0.05 to 0.5 mg of protein may be used. The reaction with DC is nearly linear with respect to time for 20 min. DOPA (10^{-3}) and pyridoxal phosphate (10^{-4}) concentrations are saturating and the pH of 7.0 is near optimal for DC with L-DOPA as substrate.

The reaction is started with the addition of DL-DOPA-2-¹⁴C and is incubated at 37° for 15 min and stopped by the addition of 0.1 ml of 50% trichloroacetic acid. After addition of 0.1 ml of $10^{-3}M$ DA HCl as carrier, the mixture is centrifuged. Ascorbic acid (10 mg) is added to the supernatant solution and the pH adjusted to 6.5 with 2 ml of $0.5M$ potassium phosphate buffer (pH 6.5) and sufficient $3N$ NH₄OH. The buffered solution is made up to 5 ml and transferred to a small (0.6 × 3 cm) IRC-50 column (Na⁺ form, buffered at pH 6.5). The column is washed with 5 ml water to remove traces of DOPA. The DA is then eluted with 10 ml of $1M$ acetic acid and the radioactivity in a 2-ml aliquot measured in 10 ml of Bray's phosphor by liquid scintillation techniques. Appropriate corrections for counting efficiency must be made. The recovery of carrier DA by this technique is 80 to 90%. Individual corrections can be determined by fluorometric assay of the dopamine present in the acetic acid eluate (excitation 280 mμ, emission 330 mμ). Either incubations with DC preparations that have been heated at 95° for 3 min or zero time incubations are used as controls. Corrections for use of DL-DOPA-2-¹⁴C must be made since only L-DOPA is a substrate for DC.

Method 2. When employing active tissue extracts or purified enzyme, the sensitivity of the radiometric assay is not required and the following fluorometric assay based on the procedure in Method 1 can be substituted. This nonradiometric procedure is less costly and is the same as Method 1 except that 0.1 ml of 0.01N HCl containing 1 μM of L-DOPA is used instead of DL-DOPA-¹⁴C. No DA is added as carrier before chromatography on IRC-50. The DA is eluted from the column with 5 ml of 0.5N HCl and DA is measured fluorometrically (excitation 280 mμ, emission 330 mμ). A standard curve for the recovery of DA from 0.01 to 0.1 μM is determined.

IV. DOPAMINE-β-HYDROXYLASE

The third step in the biosynthesis of NE is catalyzed by the enzyme DβH (114). DβH has a broad substrate specificity and catalyzes the β-hydroxylation of many phenethylamines (115–121) with the formation of their respective β-phenethanolamines. DβH is present in high concentration in the chromaffin granules of the adrenal medulla (15,21,122–124) and is present in most sympathetically innervated tissues and various areas of the CNS. The enzyme contains copper (125–127) and has been isolated in pure form from beef adrenal medulla (126). A variety of studies have been reported on the mechanism of the reaction (115,116, 126–135), on DβH activity in ligated sympathetic nerves (15,136), in pheochromocytoma and neuroblastoma tissue (137,138), and in heart (139), on the concommitant release of catecholamines and DβH from adrenal medulla (16,122,140,141), on association of DβH with granules in the sympathetic nerves (15,152), on the effect of immunosympathectomy (73), and on inhibitors of DβH (117,119–121,143–153). Recent interest has been focused on the presence of endogenous inhibitors (122,145, 154–156) of DβH. Antibodies specific for DβH have been prepared using

DβH from adrenal medulla (157). An immunohistochemical technique, which provided evidence for the presence of DβH in sympathetic nerves, has been reported (158).

A number of methods for the assay of DβH have been reported, which while satisfactory for the measurement of DβH activity in purified preparations cannot be applied directly with crude tissue preparations. These methods are based on measurements of the rate of hydroxylation of various substrates including DA, epinine, tyramine, α-methyltyramine, and phenethylamine. Assay procedures for purified DβH utilizing DA (128) or epinine (116) as substrates are based upon fluorometric measurement of the enzymatic products, NE and E, by the trihydroxyindole method (159). DβH activity has also been assayed using DA-^{14}C as substrate followed by the addition of carrier NE and isolation of NE-^{14}C by column (132) or paper chromatography (128,131) or a combination of both techniques (123). In another alternate procedure, DA-^{14}C and NE-^{14}C were acetylated (119,127) and the derivatives separated by paper chromatography. Another assay method employed DA-^{14}C as substrate (128). Periodate oxidation of the NE-^{14}C to formaldehyde-^{14}C was followed by isolation of the dimedon derivative and radiometric assay. All of these methods are extremely time consuming.

An assay based on the stoichiometric formation of tritiated water and NE-β-^{3}H following hydroxylation of DA-β,β-^{3}H (131,160) has been reported (138). Tritiated water was separated by column chromatography and measured radiometrically in a manner similar to that described for TH (Section II).

All of these methods, where DA has been employed as the substrate, are subject to various limiting factors. Thus, DA is readily oxidized both to aminochromes and to 6-hydroxydopamine, a compound which is isographic with NE (131,160). In addition, DA and NE are substrates for MAO and COMT, so that in crude tissue preparations, inhibition of both catabolic enzymes is necessary in order to prevent rapid metabolism of both the substrate DA and the product NE.

Several assay methods for DβH have now been reported based on the hydroxylation of tyramine or α-methyltyramine. In the original procedure, tyramine and its hydroxylated product, octopamine, were isolated by ion exchange chromatography followed by periodate cleavage of the octopamine to form p-hydroxybenzaldehyde which has a characteristic λ_{max} at 330 mμ in base (161). The method has been modified by the use of tyramine-^{3}H (G) (126). With this modification, the p-hydroxy-benzaldehyde-^{3}H formed by periodate cleavage is extracted into toluene and measured radiometrically. This modification increases the sensitivity of the method and permits the measurement of p-hydroxybenzaldehyde in the presence of endogenous or exogenous compounds which would interfere with an assay based on ultraviolet absorption. For the measurement of DβH activity in crude tissue, with tyramine-^{3}H as substrate, the presence of a MAO inhibitor such as pargyline or tranylcypromine is required. The use of α-methyltyramine-^{3}H, a compound which is not a substrate for MAO, would overcome this problem. Nonradioactive α-methyltyramine (paredrine) has been used with the periodate procedure to measure DβH activity (162). It appears likely, based on studies with amphetamines (121), α-methyl-m-tyramines, and α-methyldopamines (118), that only the D-isomer of α-methyltyramine is a substrate for DβH. Thus appropriate corrections must be made when DL-α-methyltyramine is used as substrate. However, this specificity would permit the determination of a "true blank" reaction with L-α-methyltyramine-[3]H. A similar experimental design has been suggested for the measurement of tyrosine hydroxylase using D- and L-tyrosine (6).

An assay procedure similar to that described for DA-β,β-^{3}H has been reported using tyramine-β,β-^{3}H as substrate. As with DA-β,β-^{3}H, the stoichiometric release of tritium into tritiated water was measured (163, 164).

Phenethylamine-^{14}C has been used as a substrate to measure DβH activity in purified preparations (126). The product, phenethanolamine-^{14}C, was isolated by paper chromatography and measured radiometrically.

The periodate cleavage method for the measurement of octopamine-^{3}H formation, from tyramine-^{3}H, appears to be the method of choice for the assay of DβH at the present time. Briefly, this procedure involves incubation of tyramine-^{3}H with the DβH preparation in the presence of a MAO inhibitor followed by periodate cleavage of the octopamine-^{3}H formed, and radiometric measurement of extracted p-hydroxybenzaldehyde. The reaction mixture must contain ascorbate and oxygen as coreactants in addition to the substrate. To obtain maximal reaction rates, fumarate and catalase are necessary. The amount of catalase, which serves to protect the copper protein, DβH, from H_2O_2, must be ascertained by titration of added catalase versus DβH activity (115). The optimum pH for the reaction is between 5.5 and 7.5 depending upon the enzyme preparation.

Unfortunately, the determination of DβH activity in tissues is difficult due to inhibitory substance(s) either present in tissue or formed during cellular disruption. Thiol reagents such as N-ethylmaleimide, p-hydroxymercuribenzoate, and cupric ions are capable of partially reversing the effect of the endogenous inhibitor(s) (154–156).

Consistently during purification of DβH, an accompanying increase in the number of units of DβH has been observed (128) which may be related to disappearance of endogenous inhibitory substances. Such partial purification procedures have been utilized as a method for the measurement of DβH activity in pairs of adrenal glands from guinea pig and rabbit (162). The DβH activity was determined both by chromatographic isolation of NE-^{14}C formed from DA-^{14}C and by the periodate ultraviolet absorption method using DL-α-methyltyramine as substrate. A different approach to the measurement of adrenal DβH activity in crude preparations has been the addition of cupric ion, p-hydroxymercuribenzoate or N-ethylmaleimide to reverse the effect of endogenous inhibitor(s). Optimal concentrations of these reagents for stimulation of DβH activity vary with the DβH preparation, and must be determined in each case. Maximal stimulations were delineated by reaction mixtures containing from 0.5 to 5 $\times$ 10^{-5}M cupric chloride (16,122,140,156) or by 5 to 500 $\times$ 10^{-5}M p-hydroxymercuribenzoate (122,140). Optimal stimulation of DβH activity by N-ethylmaleimide was observed at a concentration of 2.5 $\times$ 10^{-2}M with adrenal medulla preparations (154, 165). N-Ethylmaleimide, in contrast to cupric ions and p-hydroxymercuribenzoate, does not inhibit purified DβH when present at greater than optimal concentrations. The authors, therefore, suggest the presence of N-ethylmaleimide (4 $\times$ 10^{-2}M) when DβH activity is measured in crude preparations of adrenal medulla. Preincubation of tissue with N-ethylmaleimide may, however, be necessary (154).

Although the presence of DβH activity has been demonstrated in many tissues other than adrenal medulla (15,136–139,142,166–168), no satisfactory technique has been developed for the determination of the level of DβH in tissues other than the adrenal medulla. The measurement of DβH activity was not possible in brain homogenates in the presence of N-ethylmaleimide (162) despite the ability of N-ethylmaleimide to reverse the inhibition of purified DβH caused by crude homogenates of brain (154,155).

Further investigations are obviously necessary before DβH activity can be measured in crude tissue preparations with any degree of reliability. There are, at the present time, a number of publications on the levels and localization of DβH in the brain (25,161,168). It is the view of the authors that such published values must be regarded as, at best, semiquantitative in nature. The method presented below is satisfactory for the assay of DβH in the absence of endogenous inhibitor(s).

1. Reagents

Tyramine-^3H (G) (1–5 mCi/μM) is commercially available and if necessary it can be purified by passing the tyramine-^{3}H solution over a small column (3 $\times$ 0.3 cm) of Dowex-50 (H$^+$ form), washing the column with 50 ml water and eluting the tyramine-^{3}H with 20 ml of $2N$ HCl followed by lyophilization. The tyramine-^{3}H can then be dissolved in $0.01M$ HCl with sufficient tyramine so that 0.1 ml contains 10 μM tyramine and 1–5 μCi of tyramine-^{3}H (G). Aliquots are stored at $-20°$ (Method 1).

For preparation of stock solutions of *MAO inhibitors, pargyline,* and *tranylcypromine,* see Section III, DC Reagents.

Sodium fumarate is prepared as a $0.1M$ aqueous solution. If fumaric acid is used, the pH of the stock solution should be adjusted to 6.2 with $0.1N$ sodium hydroxide. Fumarate solutions should be stored at $-20°$.

Sodium ascorbate is prepared as a $0.1M$ aqueous solution. If ascorbic acid is used, the pH should be adjusted to 6.2 with $0.1N$ sodium hydroxide. Ascorbate solutions should be prepared fresh.

Catalase (Worthington crystalline catalase, beef liver or its equivalent) is prepared as an aqueous suspension containing approximately 4000 units/ml. Catalase suspensions are stable if kept refrigerated but are rapidly inactivated if frozen and thawed. A quantity of 400 units per incubation is suggested, but the optimum catalase concentration must be determined for each DβH preparation (113).

Potassium phosphate buffer (pH 6.2) is $0.5M$.

Tyramine HCl is prepared as a $0.1M$ solution and stored at 5°C (Method 2).

Sodium periodate is prepared as a 2% solution in water. This solution will remain stable for several weeks if stored at 5°.

Sodium metabisulfite is prepared fresh as a 10% aqueous solution.

2. Methods

Method 1. A partially purified DβH preparation is necessary for the direct application of this method. The preparative procedure of Levin et al. (128) carried through the calcium phosphate gel step gives a satisfactory preparation for a variety of studies. For exploratory investigation of crude tissue preparations, the procedures described for TH (Section II) but with 10 vol of $0.32M$ sucrose are suggested.

Reaction mixtures are prepared in conical tubes (13 ml) immersed in ice with the following components added in this sequence: 0.2 ml of $0.5M$ phosphate buffer (pH 6.2); 0.1 ml of a solution containing 10 μM of sodium fumarate (pH 6.2); 0.025 ml of solution containing 2.5 μM of sodium ascorbate (pH 6.2); 0.1 ml of $0.01M$ HCl containing 10 μM tyramine and 1–5 μCi tyramine-^{3}H(G); sufficient distilled water so that the *final* volume of the incubation mixture will be 1.0 ml; sufficient catalase suspension (400 units or the quantity necessary for maximal stimulation); and the DβH preparation.

For crude tissue preparations, 0.025 ml of a solution containing 0.25 or 0.5 μM, respectively, of pargyline or tranylcypromine, is included in the mixture. When purified DβH (128) is used, 0.5 mg of protein or less is sufficient. The reaction is nearly linear for 40 min. The concentration of tyramine (10^{-2}) is saturating and the other cofactors and pH near optimal.

The reaction is started by the addition of DβH and is incubated with gentle shaking at 37° in air for 20 min. The reaction is stopped by the addition of 0.5 ml of 10% perchloric acid and the mixture centrifuged to sediment protein. An aliquot of the supernatant fluid (1 ml) is transferred to a screw cap tube, to which 12N NH$_4$OH (0.25 ml) and 2% sodium periodate (0.3 ml) are added. After 4 min at room temperature, 10% sodium metabisulfite (0.3 ml) is added to the mixture followed by 5N HCl (1 ml). The mixture is then extracted with 10 ml toluene. The mixture is clarified by centrifugation and the radioactivity in a 5-ml aliquot of the toluene extract is determined in a toluene based phosphor by liquid scintillation techniques. Corrections for counting efficiency should be applied. Incubations with DβH preparations that have been heated at 95° for 3 min are used as controls. For studies with crude tissue preparations, the use of either cupric chloride, p-hydroxymercuribenzoate, or N-ethylmaleimide to prevent the inhibition of DβH by endogenous substances may be tried (see above, and Refs. 122,140,145, 154,156, and 165).

Method 2. For studies in which the sensitivity of the radiometric assay is not necessary and in which no substances are present which might interfere with measurement of the ultraviolet absorption of p-hydroxybenzaldehyde at 330 mμ, the following modification is satisfactory. Nonisotopic tyramine (10 μM) is used instead of tyramine-^{3}H. After termination of the reaction with perchloric acid, the supernatant is applied to a small column (0.6 $\times$ 3 cm) of Dowex-50 (H$^+$ form) and the column washed with water (10 ml). The amines are eluted with 3N NH$_4$OH (3 ml) and subjected to periodate cleavage by the addition of 0.3 ml of 2% sodium periodate. After 4 min at room temperature, the oxidation is stopped by the addition of 0.30ml of 10% sodium metabisulfite, and the optical density at 330 mμ is measured directly. The optical density is corrected for the value obtained by carrying out the procedure with heat denatured enzyme. Standard quantities of commercially available octopamine are carried through the entire procedure for reference. Suggested octopamine standards are 0.05, 0.1, and 0.15 μM.

V. PHENETHANOLAMINE-N-METHYLTRANSFERASE

The conversion of NE to E is catalyzed by the enzyme PNMT (169). This enzyme exhibits broad substrate specificity (170,171), and not only NE, but a variety of phenethanolamines including E (172), normetanephrine, metanephrine, octopamine, and phenethanolamine serve as substrates. Methylation of other classes of amines, such as dopamine, serotonin, tryptamine, and histamine, does not occur (170). Both D- and L-isomers of the phenethanolamines are substrates for PNMT. L-Norepinephrine is, however, a better substrate (170,181) than D-norepinephrine. PNMT is a soluble enzyme and is found principally in the adrenal medulla although low level activity has been demonstrated in heart (170) and brain (170,173–175). Partial purification of PNMT from monkey (170), rabbit (176,177), and beef adrenals (178) has been reported. The enzyme requires S-adenosyl-L-methionine as the methyl donor, shows no metal requirement and is inhibited by sulfhydryl reagents such as p-chloromercuribenzoate. The purified enzyme has been used as the basis for the assay of endogenous NE (177,179) and octopamine (180). Partial purification of PNMT from adrenal glands through the first ammonium sulfate fractionation followed by dialysis provides a preparation suitable for investigations of substrates (177,181) and inhibitors (178,181,182). In mammals the synthesis of adrenal PNMT is stimulated by endogenous glucocorticoids released from the adrenal cortex by ACTH (183,184). Adrenal medullary TH, COMT, and MAO are uneffected by changes in glucocorticoid level. This demonstration of a relationship between the glucocorticoid-ACTH axis and the level of PNMT and thus the capacity for E biosynthesis has resulted in several investigations of the nature and significance of this apparent control mechanism (171,182,185–187).

Two procedures have been used for the assay of PNMT activity. Both are based on radiometric measurement of product formation utilizing S-adenosyl-L-methionine-methyl-^{14}C as the methyl donor. In one method, NE is used as substrate and after incubation the remaining S-adenosyl-L-methionine-methyl-^{14}C is precipitated as the reineckate salt, sedimented by centrifugation, and the supernatant containing the product, E-^{14}C, measured radiometrically (176). In the second more widely used method, normetanephrine (184), phenethanolamine (171, 175,185), or norepinephrine (170) is used as substrate and the radioactive N-methylated product is separated from the remaining S-adenosyl-L-methionine-methyl-^{14}C by extraction into an organic solvent and measured radiometrically. The sensitivity of the assay is enhanced and the blank is lower when phenethanolamine is used as the substrate for PNMT.

When PNMT activity is measured in soluble fraction of crude tissue preparations, separation from MAO has been accomplished by centrifugation. In studies with crude preparations containing mitochondria or microsomes, MAO inhibitors must be present to prevent the rapid oxidation of phenethanolamines. Caution must be used in the selection of the MAO inhibitor since certain inhibitors, such as tranylcypromine and pargyline, also inhibit PNMT (178). The assay of PNMT in crude preparations may also be complicated by the presence of endogenous substrates of PNMT and by the presence of other methyltransferases. A "blank" reaction mixture in which the substrate, phenethanolamine, is omitted must be employed to detect the presence of other substrates in the enzyme preparation. In the case of high "blank" values, the endogenous substrates can be removed by dialysis against $0.001M$ phosphate buffer (pH 7.6). PNMT is stable to dialysis at 4°C (170). The other complicating factor is the possible presence of COMT and/or a nonspecific N-methyl-transferase (188) in the PNMT preparation. Removal of catechols by dialysis circumvents any interference due to COMT activity. The presence of "nonspecific N-methyltransferase" which catalyzes the N-methylation of serotonin, normetanephrine, metanephrine, tyramine, dopamine, etc., would still, however, interfere with the measurement of PNMT activity. The presence of the nonspecific N-methyltransferase in crude preparations may be estimated with another substrate such as phenethylamine. Assays for the nonspecific N-methyltransferase, which also appears only in the soluble protein fraction, are the same as for PNMT (see Method) except that an amine such as phenethylamine is used as substrate (188) instead of phenethanolamine. High activity for this nonspecific N-methyltransferase has been reported in lung, with moderate activity in adrenal gland and kidney and low activity in spleen, heart, and brain (188).

1. Reagents

Phenethanolamine is commercially available and is prepared as a $0.03M$ solution in $0.01M$ hydrochloric acid and stored at 5°C.

S-*Adenosyl*-L-*methionine* is commercially available as the iodide or chloride and may be used without purification to dilute S-adenosyl-L-methionine-methyl-[14]C.

S-*Adenosyl*-L-*methionine-methyl-[14]C* (30–50 μCi/μM) is commercially available and is usually of sufficient purity for use without additional purification. If necessary, the S-adenosyl-L-methionine-methyl-^{14}C can be purified by ascending paper chromatography on Whatman No. 1 paper in butanol:acetic acid:water (60:15:25) and elution in $0.1N$ HCl

(176). S-Adenosyl-L-methionine is prepared in dilute acid solution or in 0.01M phosphate buffer (pH 5.0) at a concentration of 2 × 10^{-3}M so that 1 ml contains 5–20 μCi of S-adenosyl-L-methionine-methyl-^{14}C. The solution is stored at −20°.*

Tris buffer, pH 8.5 is 0.5M.

2. Method

Extracts of crude tissue are prepared as described for TH except that 10 volumes of 1.15% KCl are used as the homogenizing medium, followed by centrifugation for 60 min at 100,000g. For studies with purified PNMT, the procedure described by Axelrod (170) or by Saelens (177) through the first ammonium sulfate fractionation followed by dialysis against 0.001M phosphate buffer (pH 7.6) may be used.

The reaction mixture is prepared in either 15 ml glass stopper conical tubes or screw cap culture tubes immersed in ice with the following components added in this sequence: 0.1 ml of 0.5M Tris buffer, pH 8.5; 0.05 ml of 0.01N HCl solution containing 1.5 μM of DL-phenethanolamine; sufficient water so that the final volume of the incubation mixture will be 0.5 ml; the PNMT preparation; 0.01 ml of a solution containing 0.02 μM S-adenosyl-L-methionine; and 0.05–0.2 μCi of S-adenosyl-L-methionine-methyl-^{14}C.

The reaction is started with the addition of S-adenosyl-L-methionine-methyl-^{14}C and incubated at 37° for 30 min. In studies with soluble supernatant fraction 0.05 to 0.2 ml of supernatant equivalent to 5–20 mg of tissue may be used. With purified enzyme 100 μg of protein is satisfactory. The reaction is nearly linear for 60 min (170). The DL-phenethanolamine and S-adenosyl-L-methionine concentrations are saturating and the pH of 8.5 is optimal.

The reaction is stopped by the addition of 0.5 ml of 0.5M sodium borate buffer (pH 10) (189) and extracted with 6 ml of a toluene-isoamyl alcohol mixture (30:1), followed by centrifugation. An aliquot (4 ml) of the organic phase is transferred to a scintillation counting vial containing 10 ml of a toluene based phosphor solution. Radioactivity is deter-

---

* On occasion it may be necessary to neutralize the commercial S-adenosyl-L-methionine-methyl-^{14}C solutions. An acidic solution of S-adenosyl-methionine may be neutralized by the following procedure: small amounts of dry Dowex 2x-8 (pH 5–5.5, HCO$_3^-$) resin are cautiously added to the acidic solution until the pH is approximately 5 to 6. The pH should never exceed pH 7. The suspension is centrifuged and the supernatant decanted. The resin is prepared as follows: Dowex 2x-8, 50–100 mesh, Cl$^-$ form, is washed exhaustively with distilled water, 0.5M and 1M NaHCO$_3$, and finally with water until the pH of the wash is between pH 5.0 and 5.5. After drying the resin is ready for use.

mined by liquid scintillation spectrometry with appropriate corrections for counting efficiency. Recovery of N-methyl-phenethanolamine by this procedure is $>95\%$. A reaction carried out in the absence of phenethanolamine serves as a blank.

VI. CATECHOL-O-METHYLTRANSFERASE

Catecholamines are metabolized *in vivo* by two alternate pathways; O-methylation by COMT (190) and oxidative deamination by MAO. COMT catalyzes the O-methylation of a large variety of catechols (191–202) which include such potential physiological substrates as DOPA, DA, NE, E, 3,4-dihydroxyphenylacetic acid, 3,4-dihydroxymandelic acid, 3,4-dihydroxyphenylglycol, 3,4-dihydroxyphenylethanol, 3,4-dihydroxy-benzoic acid, and N-acetyldopamine. The enzyme exhibits little stereospecificity. Thus, D- and L-E are O-methylated at similar rates (191). In the case of substituted catechols, both isomeric O-methylated products are formed. However, with the above mentioned physiological substrates, the *meta*-O-methylated isomer predominates (195,196,198, 203). COMT is a soluble enzyme found in most animal tissues (191). A portion of COMT activity is, however, associated with particulate matter in brain (204) and liver (205). The ratio of soluble and particulate COMT activity varies greatly with species (205). COMT has been purified from rat liver (191,192) and requires S-adenosyl-L-methionine as the methyl donor and a divalent cation. Magnesium ions have commonly been used, but other cations can be substituted (191,206). The purified enzyme has been used for the assay of plasma, urinary, and tissue levels of NE (207,208). The levels of COMT activity in various tissues (191,209–211), in brain (25,175,209,210), uterus (211,212), skeletal muscle (213), ocular tissue (214), skin (211,215,216), neuroblastoma (217), pineal and pituitary glands (218), and adipose tissue (219), the effect of age (220), denervation (214,221,222), estrus (211), congestive heart failure (223), immunosympathectomy (73), and hypophysectomy (224) on COMT levels, the effect of adrenergic blocking agents on extrane-uronal COMT activity (225), and the inhibition of COMT (226–237) are among the studies reported on this enzyme.

The first procedure described for the assay of COMT activity involved the use of NE or E as substrate, followed by extraction and fluorometric measurement of the product, (nor)metanephrine (191,241). Similar methods based on the colorimetric determination of normetanephrine (229,231) or on ultraviolet spectroscopy after periodate cleavage of metanephrine (237) have been reported. Another fluorometric assay has been described (238) based on the substrate, 3,4-dihydroxyphenylacetic acid.   In this case, the O-methylated product was separated from the catecholic substrate by alumina chromatography.   An extremely simple colorimetric assay using pyrocatecholphthalein as substrate has been reported (192,239).

The introduction of radioassay techniques has increased the sensitivity and greatly simplified the measurement of COMT.   Two approaches for radiometric assay have been used.   In the first technique, S-adenosyl-L-methionine-methyl-^{14}C (225) or -methyl-^{3}H (213) is used as the methyl donor.   After enzymatic transfer of the isotopically labeled methyl group to the substrate, the O-methylated product is separated and measured radiometrically.   This approach has been used with E or NE as substrate, followed by extraction of the radioactive product (nor)-metanephrine into a toluene-isoamyl alcohol mixture and measurement of the radioactivity in the organic phase by liquid scintillation techniques. A more sensitive assay for COMT based on solvent extraction has been reported (25,204,240) using 3,4-dihydroxybenzoic acid as the substrate. An earlier assay based on this same substrate used paper chromatography to separate the product(s) (235).   In principle, almost any catechol could be employed as a substrate and of the catechols which have been studied in our laboratory as substrates for COMT, 3,4-dihydroxy-propiophenone is one of the most active and stable and is recommended by the authors.   Reactions carried out in the absence of substrate are used to determine the blank.   This type of assay cannot be used to measure inhibition by other catechols and pyrogallols which, as substrates, would also form extractable radioactive products.

In the second approach, which has also been widely used, the substrate has been DL-E-^{3}H or DL-NE-^{3}H.   The product, (nor)metanephrine-^{3}H, is extracted into toluene-isoamyl alcohol and assayed radiometrically (209,241).   This technique must be used with caution with crude tissue preparations, since other metabolic pathways are available which could result in the formation of tritiated products which would also extract into the organic phase.   A reaction carried out without addition of S-adenosyl-L-methionine should be used as a blank.   Inhibition of MAO with either pargyline or tranylcypromine is advised when E-^{3}H or NE-^{3}H is used as substrate with crude tissue preparations.   This technique may be used in studies of the inhibition of COMT by catechols and pyrogallols, since the O-methylated products formed from these competitive substrates will not be radioactive.

Inhibition of COMT activity in brain homogenates by pyrogallol has been reported, with a method that utilized norepinephrine-^{14}C as substrate, followed by isolation of all metabolic products by a combination of thin-layer chromatography and electrophoresis (242).

Regardless of the type of assay employed for COMT, the final concentrations of substrate, magnesium ion, and S-adenosyl-L-methionine should be such that maximal reaction rates are obtained. Marked inhibition of COMT activity is produced by magnesium ion concentrations greater than 2.0 mM. The optimum concentration of magnesium needed for maximal activity in crude tissue preparations is dependent upon the type of tissue and the extent of dilution in the homogenizing media. This variability is probably due to differences in endogenous cation concentration in tissue. Most assays for COMT reported in the literature have employed nonsaturating concentrations ($\ll 10^{-3}M$) of S-adenosyl-L-methionine. A concentration of S-adenosyl-L-methionine of $1 \times 10^{-3}M$ is required for saturation of COMT.

1. Reagents

S-*Adenosyl*-L-*methionine-methyl*-^{14}C (see Section IV, PNMT, Reagents) is stored as a $0.01M$ solution which contains 1–10 μCi of S-adenosyl-L-methionine-methyl-^{14}C per ml.

3,4-Dihydroxypropiophenone is available commercially and should be recrystallized from chloroform and stored as a $0.01M$ solution in $0.001N$ HCl (Method 1). 3,4-Dihydroxybenzoic acid, an alternate substrate (Method 1), is commercially available and can be used without further purification as a $0.01M$ solution in $0.001N$ HCl.

The $MgCl_2$ *solution* is $0.1M$.

The *potassium phosphate buffer* (pH 8) is $0.5M$.

DL-*Epinephrine*-7-^3H (5 μCi/μM) is available commercially and usually requires no further purification. DL-Epinephrine bitartrate should be prepared as a $0.01M$ solution in $0.01N$ HCl containing 4–20 μCi of DL-epinephrine-7-^{3}H per ml and can be stored at 4° (solutions of catecholamines should not be exposed to sunlight or other sources of uv irradiation) (Method 2).

S-*Adenosyl*-L-*methionine* is commercially available and is prepared as a $0.01M$ solution in $0.01N$ HCl (Method 2).

For *tranylcypromine* or *pargyline*, see Section II, TH, Reagents.

2. Methods

Method 1. The preparation of crude tissue is similar to that described for TH except that 1 to 4 vol of 1.15% KCl should be used as the homogenizing media. For most tissues (liver, kidney, heart) the COMT activity is most conveniently determined on the supernatant obtained by centrifugation of homogenates at 100,000g for 60 min. The activity of

COMT in brain is quite low and for satisfactory results the measurements are best made with crude homogenates. Enzyme purified by the procedure described by Axelrod and Tomchick (191) though the first dialysis is suitable for a variety of studies.

Reaction mixtures are prepared in 15 ml glass-stopper conical tubes or screw cap culture tubes immersed in ice with the following components added in this sequence: 0.1 ml of $0.5M$ sodium phosphate buffer (pH 8); 0.01 ml of $0.1M$ magnesium chloride; sufficient distilled water so that the final volume of the incubation mixture will be 0.5 ml; the COMT preparation; 0.05 ml of solution containing 0.5 μM of S-adenosyl-L-methionine and 0.05 to 0.5 μCi of S-adenosyl-L-methionine-methyl-^{14}C; and 0.1 ml of $0.01M$ 3,4-dihydroxypropiophenone or 0.1 ml of $0.01M$ 3,4-dihydroxybenzoic acid.

The reaction is started by the addition of substrate and is incubated for 20 min at 37°. For tissue preparations, the reaction is linear with protein for homogenates equivalent to 10–100 mg of tissue. With purified enzyme approximately 0.1 mg of protein is sufficient. The reaction is nearly linear for 40 min. The concentration of substrate ($2 \times 10^{-3}M$) and cofactor ($10^{-3}M$) are saturating and the concentration of magnesium ion and the pH of 8 are optimal.

The reaction is stopped by the addition of 0.1 ml of $1N$ HCl and the mixture extracted with 10 ml of toluene. The recovery of the O-methylated product is 98%. Following centrifugation, an aliquot (5 ml) of the organic phase is transferred to a scintillation counting vial, a toluene based phosphor solution (10 ml) added, and the radioactivity determined. Corrections are made for counting efficiency. The results are corrected for blank values obtained by carrying out the incubation without substrate.

Method 2. Reaction mixtures are prepared in 15 ml glass-stoppered conical tubes or screw cap culture tubes, immersed in ice with the following components added in this sequence: 0.1 ml of $0.5M$ phosphate buffer (pH 8); 0.01 ml of $0.1M$ magnesium chloride; sufficient water so that the final volume of the incubation mixture will be 0.5 ml; the COMT preparation; 0.01 ml of a solution containing 0.5 μM of S-adenosyl-L-methionine; and 0.05 ml of solution containing 0.5 μM of DL-epinephrine and 0.2–1 μCi of DL-epinephrine-7-[3]H.

The reaction is started by the addition of substrate and incubated for 20 min at 37°. For crude preparations, the use of tranylcypromine (0.2 μM) or pargyline (0.1 μM) (see Section II, DC, Reagents) is advised. For enzyme levels, see Method 1. The reaction is nearly linear for 20 min. The concentration of S-adenosyl-L-methionine (10^{-3}) and substrate (10^{-3}) are saturating and the pH with phosphate buffer and magnesium ion concentrations are optimal. Higher activities with norepinephrine as substrate for COMT have been found in tris buffer (pH 9.0) (208) than in the usual pH 7.9 phosphate buffer (209,241). The reaction is terminated by the addition of 0.5 ml of 0.5M sodium borate buffer (pH 10) (189). Carrier DL-metanephrine HCl (2.5 μM) is added and the mixture extracted with 5 ml of toluene-isoamyl alcohol (7:3). The extraction of metanephrine is 88 $\pm$ 5%. Following centrifugation, an aliquot (5 ml) of the organic phase is transferred to a scintillation counting vial, Bray's phosphor solution (10 ml) and methanol (1 ml) are added, and the radioactivity measured. Corrections are made for counting efficiency. The results are corrected for blank values obtained by carrying out the incubation in the absence of S-adenosylmethionine.

VII. MONOAMINE OXIDASE

Oxidative deamination catalyzed by mitrochondrial monoamine oxidase is the second principle pathway for the metabolism of catecholamines (1). This enzyme(s), referred to simply as MAO, acts on a variety of monoamines and occurs in nearly all tissues. MAO is predominately associated with mitochondria, but in certain tissues, namely heart and splenic nerve, microsomal preparations are reported to exhibit MAO activity (243). A number of recent studies (244–250) have been reported on the solubilization and purification of MAO. The presence of a number of monoamine oxidases with differing substrate profiles has been indicated by a variety of studies including partial separation by chromatographic (251,252) and electrophoretic (253,254) techniques. Mitochondrial preparations (255) are suitable for the study of MAO inhibition, but it should be noted that comparisons of the results obtained with the action of various inhibitors on crude and purified MAO preparations may reveal marked differences (256,257). Many of the studies cited earlier regarding other enzymes concerned with biosynthesis and metabolism of catecholamines have also dealt with MAO activities. These included studies on subcellular distribution (20,21,204) levels in various tissues (210,211, 212,214), in brain (25,79,80,210,258), and the effects of various parameters (136,211,213,220–224,259) on enzyme activities.

A great variety of procedures for the assay of MAO activity have been developed. The subject is considered in detail in the chapter by Kapeller-Adler. A method for MAO assay (80,260) based on the use of isotopically labeled tyramine as the substrate is presented here. The deaminated products, p-hydroxyphenylacetaldehyde, p-hydroxyphenylacetic acid, and p-hydroxyphenylethanol, are extracted from the reaction mixture and measured radiometrically. The p-hydroxyphenylacetic acid and p-hydroxyphenylethanol are secondary products arising by the action of alcohol and aldehyde dehydrogenase (1). This assay was chosen for its simplicity, sensitivity, and because tyramine most closely resembles the catecholamines in structure. The method has been employed in a number of investigations of MAO (80,122,252,260–264). Similar radiometric procedures for MAO assay (271) have been reported using such substrates as dopamine-^{14}C (80), serotonin-^{14}C (80), and tryptamine-^{14}C (265). The tryptamine-^{14}C assay has been widely used. Two spectrophotometric procedures permitting the continuous measurement of reaction rates have been described for MAO using m-iodobenzylamine (266) or kynuramine (267) as substrates. A modification of the latter method, based on a fluorometric assay of 4-hydroxyquinoline formed by oxidative cyclization of the enzyme product, has been reported (268,269). Another sensitive fluorometric assay is based on the use of tryptamine as the MAO substrate (270). Because of the possible occurrence of various MAOs with differing substrate activities, the use of a number of assays with different substrates (tyramine, serotonin, tryptamine, m-iodobenzylamine, kynuramine) is sometimes advisable.

1. Reagents

Tyramine-^{3}H (G) is commercially available. (See Section IV, DβH, Reagents.) It is prepared in 0.01N HCl as a 0.01M solution of tyramine containing 0.5–2 μCi of tyramine-^{3}H (G) per ml.

The *potassium phosphate buffer* (pH 8) is 0.5M.

2. Method

Crude tissue preparation is similar to that described for TH except that 2 to 5 vol of cold 1.15% KCl is recommended as the homogenizing media. Mitochondrial preparations from rat liver are suitable for many studies (255).

Reaction mixtures are prepared in 15 ml glass-stopper conical tubes or screw cap culture tubes immersed in ice with the following components added in this sequence: 0.1 ml of 0.5M phosphate buffer (pH 8); 0.1 ml of 0.01N HCl solution containing 1 μM of tyramine and 0.05–0.2 μCi of tyramine-^{3}H (G); sufficient distilled water so that the final volume of the incubation mixture will be 0.5 ml; and the MAO preparation.

The reaction is started by the addition of substrate and incubated for 30 min at 37° with shaking. Aliquots of crude tissue preparations corresponding to 5–100 mg of original tissue may be used. The concentra-

CATECHOLAMINE BIOSYNTHESIS AND METABOLISM   175

tion of tyramine $(2 \times 10^{-3}M)$ is saturating and the pH is optimal. Reactions with MAO are nearly linear for 40 min. The reaction is terminated by the addition of 0.2 ml of $2N$ HCl. The mixture is extracted with 3.0 ml of ethyl acetate, clarified by centrifugation, and an aliquot (1.0 ml) of the organic phase containing the deaminated products transferred to a scintillation counting vial, and radioactivity determined by liquid scintillation techniques in 10 ml of Bray's phosphor solution. The extraction efficiency is approximately 93%. The procedure carried out with MAO preparations heated at 95° for 3 min serves as a blank.

References

1. M. Sandler and C. R. J. Ruthven, in *Progress in Medicinal Chemistry*, Vol. 6, G. P. Ellis and G. B. West, Eds., Butterworths, London, 1969, p. 200.
2. S. Udenfriend, *Pharmacol. Rev.*, *18*, 43 (1966).
3. S. Udenfriend, P. Zaltzman-Nirenberg, and T. Nagatsu, *Biochem. Pharmacol.*, *14*, 837 (1965).
4. M. Ikeda, M. Levitt, and S. Udenfriend, *Biochem. Biophys. Res. Commun.*, *18*, 482 (1965).
5. M. Ikeda, M. Levitt, and S. Udenfriend, *Arch. Biochem. Biophys.*, *120*, 420 (1967).
6. T. Nagatsu, M. Levitt, and S. Udenfriend, *J. Biol. Chem.*, *239*, 2910 (1964).
7. B. Petrack, F. Sheppy, and V. Fetzer, *J. Biol. Chem.*, *243*, 743 (1968).
8. A. R. Brenneman and S. Kaufman, *Biochem. Biophys. Res. Commun.*, *17*, 177 (1964).
9. L. Ellenbogen, R. J. Taylor, Jr., and G. B. Brundage, *Biochem. Biophys. Res. Commun.*, *19*, 708 (1965).
10. M. Ikeda, L. A. Fahien, and S. Udenfriend, *J. Biol. Chem.*, *241*, 4452 (1966).
11. T. H. Joh, R. Kapit, and M. Goldstein, *Biochem. Biophys. Acta*, *171*, 378 (1969).
12. E. G. McGeer, S. Gibson, and P. L. McGeer, *Can. J. Biochem.*, *45*, 1557 (1967).
13. E. G. McGeer and P. L. McGeer, *Science*, *143*, 73 (1966).
14. H. Thoenen, R. A. Mueller, and J. Axelrod, *Nature*, *221*, 1264 (1969).
15. P. Laduron and F. Belpaire, *Nature*, *217*, 1155 (1968).
16. O. H. Viveros, L. Arqueros, R. J. Connett, and N. Kirshner, *Mol. Pharmacol.*, *5*, 69 (1969).
17. B. Burkard, K. F. Gey, and A. Pletscher, *Nature*, 213, 732 (1967).
18. G. C. Sodvall and I. J. Kopin, *Biochem. Pharmacol.*, *16*, 39 (1967).
19. W. Dairman, R. Gordon, S. Spector, A. Sjoerdsma, and S. Udenfriend, *Mol. Pharmacol.*, *4*, 457 (1968).
20. P. L. McGeer, S. P. Bagchi, and E. G. McGeer, *Life Sci.* (Oxford), *4*, 1859 (1965).
21. P. Laduron and F. Belpaire, *Biochem. Pharmacol.*, *17*, 1127 (1968).
22. J. M. Musacchio, *Biochem. Pharmacol.*, *17*, 1470 (1968).
23. L. Stjärne and F. Lishajko, *Biochem. Pharmacol.*, *16*, 1719 (1967).
24. T. Nagatsu, M. Levitt, and S. Udenfriend, *Biochem. Biophy. Res. Commun.*, *14*, 543 (1964).
25. W. H. Vogel, V. Orfei, and B. Century, *J. Pharmacol. Exp. Ther.*, 165, 196 (1969).

26. S. Udenfriend, in *Biochem. Pharmacol. Basal Ganglia Proc.*, 2nd Symp. Coll. Phys. Surg. Columbia Univ., E. Costra, L. J. Côté, and M. D. Yaur, Eds., Raven Press, New York, 1966, p. 101.

27. E. G. McGeer, S. Gibson, J. A. Wada, and P. L. McGeer, *Can. J. Biochem.*, *45*, 1943 (1967).

28. M. Levitt, J. W. Gibb, J. W. Daly, M. Lipton, and S. Udenfriend, *Biochem. Pharmacol.*, *16*, 1313 (1967).

29. S. Ayukawa, T. Takeuchi, M. Sezaki, T. Hara, H. Umczawa, and T. Nagatsu, *J. Antibiot.* (Tokyo), *21*, 350 (1968).

30. G. A. Johnson, E. G. Kim, P. A. Platz, and M. M. Mickelson, *Biochem. Pharmacol.* *17*, 403 (1968).

31. M. Goldstein, H. Gang, and B. Anagnoste, *Life Sci.* (Oxford), *6*, 1457 (1967).

32. P. A. Weinhold and V. B. Rethy, *Biochem. Pharmacol.*, *18*, 677 (1969).

33. W. S. Saari, J. Williams, S. F. Britcher, D. E. Wolf, and F. A. Kuehl, Jr., *J. Med. Chem.*, *10*, 1008 (1967).

34. D. K. Zhelyaskov, M. Levitt, and S. Udenfriend, *Mol. Pharmacol.*, *4*, 445 (1968).

35. R. J. Taylor, Jr., and L. Ellenbogen, *Life Sci.* (Oxford), *6*, 1463 (1967).

36. R. J. Taylor, Jr., C. S. Stubbs, Jr., and L. Ellenbogen, *Biochem. Pharmacol.*, *17*, 1779 (1968).

37. R. J. Taylor, Jr., C. S. Stubbs, Jr., and L. Ellenbogen, *Biochem. Pharmacol.*, *18*, 587 (1969).

38. B. N. Lutsky and N. Zenker, *J. Med. Chem.*, *11*, 1241 (1968).

39. S. P. Bagchi and P. L. McGeer, *Life Sci.* (Oxford), *3*, 1195 (1964).

40. E. G. McGeer and P. L. McGeer, *Can. J. Biochem.*, *45*, 115 (1967).

41. E. G. McGeer, P. L. McGeer, and D. A. Peters, *Life Sci.* (Oxford), *6*, 2221 (1967).

42. A. Parulkar, A. Burger, and D. Aures, *J. Med. Chem.*, *9*, 738 (1966).

43. T. Nagatsu, M. Levitt, and S. Udenfriend, *Anal. Biochem.*, *9*, 122 (1964).

44. S. H. Pomerantz, *Biochem. Biophys. Res. Commun.*, *16*, 188 (1964).

45. S. H. Pomerantz, *J. Biol. Chem.*, *241*, 161 (1966).

46. D. Gaudin and J. H. Fellman, *Biochim. Biophys. Acta*, *141*, 64 (1967).

47. S. H. Pomerantz, *Science*, *164*, 838 (1969).

48. T. Nagatsu and T. Yamamoto, *Experientia*, *24*, 1183 (1968).

49. T. P. Waalkes and S. Udenfriend, *J. Lab. Clin. Med.*, *50*, 733 (1957).

50. D. J. Drain, M. Horlington, R. Lazare, and G. A. Poulter, *Life Sci.* (Oxford), *1*, 93 (1962).

51. T. L. Sourkes, *Pharmacol. Rev.*, *18*, 53 (1966).

52. H. Weissbach, W. Lovenberg, and S. Udenfriend, *Biochem. Biophys. Res. Commun.*, *3*, 225 (1960).

53. W. Lovenberg, H. Weissbach, and S. Udenfriend, *J. Biol. Chem.*, *237*, 89 (1962).

54. R. Ferrini and A. Glaesser, *Biochem. Pharmacol.*, *13*, 798 (1964).

55. J. H. Fellman, *Enzymologia*, *20*, 366 (1959).

56. P. Hagen, *Brit. J. Pharmacol.*, *18*, 175 (1962).

57. E. Werle, A. Schauer, and H. W. Buehler, *Arch. Int. Pharmacodyn. Ther.*, *145*, 198 (1963).

58. E. Werle, and D. Aures, *Hoppe-Seyler's Z. Physiol. Chem.*, *316*, 45 (1959).

59. J. Awapara, R. T. Sandman, and C. Hanly, *Arch. Biochem. Biophys.*, *98*, 520 (1962).

60. J. Awapara, T. L. Perry, C. Hanly, and E. Peck, *Clin. Chim. Acta*, *10*, 286 (1964).

61. H. Langemann and H. Ackermann, *Helv. Physiol. Acta*, *19*, 399 (1961).

62. G. Rodriquez de Lores Arnaiz and E. de Robertis, *J. Neurochem.*, *11*, 213 (1964).

63. C. T. Clark, H. Weissbach, and S. Udenfriend, *J. Biol. Chem.*, *210*, 139 (1954).
64. H. F. Schott and W. G. Clark, *J. Biol. Chem.*, *196*, 449 (1952).
65. W. Lovenberg, J. Barchas, H. Weissbach, and S. Udenfriend, *Arch. Biochem. Biophys.*, *103*, 9 (1963).
66. R. Y. Moore, R. K. Bhatnagar, and A. Heller, *Int. J. Neuropharmacol.*, *5*, 287 (1966).
67. N. E. Anden, T. Magnusson, and E. Rosengren, *Acta Physiol. Scand.*, *64*, 127 (1965).
68. P. Holtz and E. Westerman, *Naunyn-Schmiedenbergs Arch. Pharmakol. Exp. Pathol.*, *227*, 538 (1956).
69. A. Dahlström and J. Jonason, *Eur. J. Pharmacol.*, *4*, 377 (1968).
70. C. Giordano, H. A. Samiy, J. Bloom, F. W. Haynes, and J. P. Merrill, *Experientia*, *17*, 558 (1961).
71. G. I. Klingman, *J. Pharmacol. Exp. Ther.*, *148*, 14 (1965).
72. C. Ernzerhoff, P. Holtz, and D. Palm, *Biochem. Pharmacol.*, *15*, 1880 (1966).
73. L. L. Iverson, J. Glowinski, and J. Axelrod, *J. Pharmacol Exp. Ther.*, *151*, 273 (1966).
74. W. P. Burkard, R. Pavlin, A. Pletscher, and K. F. Gey, *Int. J. Neuropharmacol.*, *1*, 233 (1962).
75. H. Blaschko, P. Hagen, and A. D. Welch, *J. Physiol.* (London), *129*, 27 (1955).
76. L. S. Dietrich, *J. Biol. Chem.*, *204*, 587 (1953).
77. V. E. Davis, and J. Awapara, *J. Biol. Chem.*, *235*, 124 (1960).
78. R. Kuntzman, P. A. Shore, D. Bogdanski, and B. B. Brodie, *J. Neurochem.*, *6*, 226 (1961).
79. D. F. Bogdanski, H. Weissbach, and S. Udenfriend, *J. Neurochem.*, *1*, 272 (1962).
80. R. E. McCaman, M. W. McCaman, J. M. Hunt, and M. S. Smith, *J. Neurochem.*, *12*, 15 (1965).
81. E. Robins, J. M. Robins, A. B. Croninger, S. G. Moses, S. J. Spencer, and R. W. Hudgens, *Biochem. Med.*, *1*, 240 (1967).
82. N. Kärki, R. Kuntzman, and B. B. Brodie, *J. Neurochem.*, *9*, 53 (1962).
83. R. S. De Ropp and A. Furst, *Brain Res.*, *2*, 323 (1966).
84. S. H. Snyder and J. Axelrod, *Biochem. Pharmacol.*, *13*, 805 (1964).
85. R. Håkanson and C. Owman, *J. Neurochem.*, *12*, 417 (1965).
86. R. Håkanson and C. Owman, *J. Neurochem.*, *13*, 597 (1966).
87. R. Håkanson, M. N. Lombard des Gouttes, and C. Owman, *Life Sci.*, (Oxford) *6*, 2577 (1967).
88. R. Håkanson and C. Owman, *Biochem. Pharmacol.*, *15*, 489 (1966).
89. N. E. Andén, T. Magnusson, and E. Rosengren, *Experientia*, *20*, 328 (1964).
90. R. Håkanson and H. Moller, *Acta Dermato-Venereol.*, *43*, 485 (1963).
91. R. Håkanson, H. Moller, and N. G. Stromby, *Experientia*, *21*, 265 (1965).
92. D. Mackey, J. F. Riley, and D. M. Shepard, *J. Pharm. Pharmacol.*, *13*, 257 (1961).
93. H. Langemann, A. Boner, and P. B. Mueller, *Schweiz. Med. Wochschr.*, *92*, 1621 (1962).
94. P. B. Mueller and H. Langemann, *Klin. Wochschr.*, *40*, 911 (1962).
95. C. C. Porter, L. S. Watson, D. C. Titus, J. A. Totaro, and S. S. Byer, *Biochem. Pharmacol.*, *11*, 1067 (1962).
96. T. L. Chrusciel, *Int. J. Neuropharmacol.*, *1*, 137 (1962).
97. M. Sletzinger, J. M. Chemerda, and F. W. Bollinger, *J. Med. Chem.*, *6*, 101 (1963).

98. P. Oehme, H. Rex, and E. Ackermann, *Acta Biol. Med.*, *12*, 234 (1964).

99. J. A. Buzard and P. D. Nytch, *J. Biol. Chem.*, *234*, 884 (1959).

100. W. P. Buckard, K. F. Gey, and A. Pletscher, *Arch. Biochem. Biophys.*, *107*, 187 (1964).

101. E. J. Glamkowski, G. Gal, M. Sletzinger, C. C. Porter, and L. S. Watson, *J. Med. Chem.*, *10*, 852 (1967).

102. B. B. Brodie, R. Kuntzman, C. W. Hirsch, and E. Costa, *Life Sci.*, (Oxford), *3*, 81 (1962).

103. J. H. Merritt and T. S. Sulkowski, *Biochem. Pharmacol.*, *16*, 369 (1967).

104. H. Blaschko, *J. Physiol.*, *101*, 337 (1942).

105. P. Holtz and E. Westermann, *Naunyn-Schmiedebergs Arch. Pharmakol. Exp. Pathol. Pharmakol*, *227*, 538 (1956).

106. D. A. Buyski, in *Isotopes in Experimental Pharmacology*, L. J. Roth, Ed., Univ. Chicago, Chicago, 1965, p. 275.

107. P. Laduron and F. Belpaire, *Anal. Biochem.*, *26*, 210 (1968).

108. R. Håkanson, *Acta Pharmacol. Toxicol.*, *24*, 217 (1966).

109. A. R. Somerville, *Biochem. Pharmacol.*, *13*, 1861 (1964).

110. D. Aures, R. Fleming, and R. Håkanson, *J. Chromatogr.*, *33*, 480 (1968).

111. C. Streffer, *Biochem. Biophys. Acta*, *139*, 193 (1967).

112. J. D. Taylor, A. A. Wykes, Y. C. Gladish, and W. B. Martin, *Nature*, *187*, 941 (1960).

113. A. R. Maass and M. J. Nimmo, *Nature*, *184*, 547 (1959).

114. S. Kaufman and S. Friedman, *Pharmacol. Rev.*, *11*, 350 (1965).

115. E. Y. Levin and S. Kaufman, *J. Biol. Chem.*, *236*, 2043 (1961).

116. W. F. Bridgers and S. Kaufman, *J. Biol. Chem.*, *237*, 526 (1962).

117. J. B. Van der Schoot and C. R. Creveling, *Advan. Drug Res.*, *2*, 47 (1964).

118. M. L. Torchiana, C. C. Porter, and C. A. Stone, *Arch. Int. Pharmacodyn. Ther.*, *174*, 118 (1968).

119. M. Goldstein and J. F. Contrera, *J. Biol. Chem.*, *237*, 1898 (1962).

120. C. R. Creveling, J. W. Daly, B. Witkop, and S. Udenfriend, *Biochim. Biophys. Acta*, *64*, 125 (1962).

121. M. Goldstein, M. R. McKereghan, and E. Lauber, *Biochim. Biophys. Acta*, *89*, 191 (1964).

122. O. H. Viveros, L. Arqueros, R. J. Connett, and N. Kirshner, *Mol. Pharmacol.*, *5*, 60 (1969).

123. F. Belpaire and P. Laduron, *Biochem. Pharmacol.*, *17*, 411 (1968).

124. M. Oka, K. Kajikawa, T. Ohuchi, H. Yoshida, and R. Imaizumi, *Life Sci.* (Oxford), *6*, 461 (1967).

125. S. Friedman and S. Kaufman, *J. Biol. Chem.*, *240*, PC 552 (1965).

126. S. Friedman and S. Kaufman, *J. Biol. Chem.*, *240*, 4763 (1965).

127. M. Goldstein, E. Lauber, and M. R. McKereghan, *J. Biol. Chem.*, *240*, 2066 (1965).

128. E. Y. Levin, B. Levenberg, and S. Kaufman, *J. Biol. Chem.*, *235*, 2080 (1960).

129. S. Kaufman, W. F. Bridgers, F. Eisenberg, and S. Friedman, *Biochem. Biophys. Res. Commun.*, *9*, 497 (1962).

130. S. Friedman and S. Kaufman, *J. Biol. Chem.*, *241*, 2256 (1966).

131. S. Senoh, C. R. Creveling, S. Udenfriend, and B. Witkop, *J. Amer. Chem. Soc.*, *81*, 6236 (1959).

132. J. W. Smith and N. Kirshner, *J. Biol. Chem.*, *237*, 1890 (1962).

133. M. Goldstein and T. H. Joh, *Mol. Pharmacol.*, *3*, 396 (1967).

134. M. Goldstein, T. H. Joh, and T. Q. Garvey III, *Biochem.*, *7*, 2724 (1968).
135. W. E. Blumberg, M. Goldstein, E. Lauber, and J. Peisbach, *Biochim. Biophys. Acta*, *99*, 187 (1965).
136. P. Laduron and F. Belpaire, *Life Sci.* (Oxford), *7*, 1 (1968).
137. C. Bohuon and F. Guerinot, *Clin. Chim. Acta*, *19*, 125 (1968).
138. M. Goldstein, N. Prochoroff, and S. Sirlin, *Experientia*, *21*, 592 (1965).
139. L. T. Potter and J. Axelrod, *J. Pharmacol. Exp. Ther.*, *142*, 299 (1963).
140. O. H. Viveros, L. Arqueros, and N. Kirshner, *Life Sci.* (Oxford), *7*, 609 (1968).
141. J. W. Smith and N. Kirshner, *Mol. Pharmacol.*, *3*, 52 (1967).
142. L. Stjärne, R. H. Roth, and F. Lishajko, *Biochem. Pharmacol.*, *16*, 1729 (1967).
143. C. R. Creveling, J. B. Van der Schoot, and S. Udenfriend, *Biochem. Biophys. Res. Commun.*, *8*, 215 (1962).
144. J. B. Van der Schoot, C. R. Creveling, T. Nagatsu, and S. Udenfriend, *J. Pharmacol. Exp. Ther.*, *141*, 74 (1963).
145. I. W. Chubb, B. N. Preston, and L. Austin, *Biochem. J.*, *111*, 243 (1969).
146. M. Goldstein and J. F. Contrera, *Experientia*, *18*, 332 (1964).
147. M. Goldstein and J. F. Contrera, *Experientia*, *18*, 334 (1964).
148. M. Goldstein, E. Lauber, and M. R. McKereghan, *Biochem. Pharmacol.*, *13*, 1103 (1964).
149. A. L. Green, *Biochim. Biophys. Acta*, *81*, 394 (1964).
150. M. Goldstein, B. Anagnoste, E. Lauber, and M. R. McKereghan, *Life Sci.* (Oxford), *3*, 763 (1964).
151. M. Goldstein and J. F. Contrera, *Nature*, *142*, 1081 (1961).
152. J. Augstein, S. M. Green, A. M. Monro, T. I. Wrigley, A. R. Katritzky, and G. J. T. Tiddy, *J. Med. Chem.*, *10*, 391 (1967).
153. T. Nagatsu, S. Ayukawa, and H. Umezawa, *J. Antibiot.* (Tokyo), *21*, 354 (1968).
154. T. Nagatsu, H. Kuzuya, and H. Hidaka, *Biochim. Biophys. Acta*, *139*, 319 (1967).
155. T. Nagatsu, in *Biological and Chemical Aspects of Oxygenases*, K. Bloch and O. Hayaishi, Eds., Maruzen, Tokyo, 1966, p. 273.
156. D. S. Duch, O. H. Viveros, and N. Kirshner, *Biochem. Pharmacol.*, *17*, 255 (1968).
157. J. W. Gibb, S. Spector, and S. Udenfriend, *Mol. Pharmacol.*, *3*, 473 (1967).
158. B. G. Livett, L. B. Geffen, and R. A. Rush, *Biochem. Pharmacol.*, *18*, 923 (1969).
159. U. S. Von Euler and I. Floding, *Acta Physiol. Scand. Suppl.*, *118*, 45 (1955).
160. S. Senoh, B. Witkop, C. R. Creveling, and S. Udenfriend, *J. Amer. Chem, Soc.*, *81*, 1768 (1959).
161. J. J. Pisano, C. R. Creveling, and S. Udenfriend, *Biochim. Biophys. Acta*, *43*. 566 (1960).
162. T. Nagatsu, J. B. Van der Schoot, M. Levitt, and S. Udenfried, *J. Biochem.* (Japan), *64*, 39 (1968).
163. S. Kaufman, W. F. Bridgers, and J. Baron, *Oxidation of Organic Compounds III*, (Advances in Chemistry Series, Vol. 77), Reinhold, New York, 1968, p. 172.
164. C. R. Creveling and J. W. Daly, *Pharmacologist*, *7*, 157 (1965).
165. H. Kuzuya and T. Nagatsu, *Enzymologia*, *36*, 31 (1969).
166. R. Håkanson and H. Moller, *Acta Dermato-Venereol.*, *43*, 548 (1963).
167. C. A. Chidsey, G. A. Kaiser, and B. Lehr, *J. Pharmacol. Exp. Ther.*, *144*, 393 (1964).
168. S. Udenfriend and C. R. Creveling, *J. Neurochem.*, *4*, 350 (1959).
169. J. Axelrod, *Pharmacol. Rev.*, *18*, 95 (1966).

170. J. Axelrod, *J. Biol. Chem.*, *237*, 1657 (1962).

171. R. J. Wurtman, J. Axelrod, E. S. Vesell, and G. T. Ross, *Endocrinology, 82,* 584 (1968).

172. J. Axelrod, *Biochim. Biophys. Acta*, *45*, 614 (1960).

173. P. L. McGeer and E. G. McGeer, *Biochem. Biophys. Res. Commun.*, *17*, 502 (1964).

174. R. D. Ciaranello, R. E. Barchas, G. S. Byers, D. W. Stemmle, and J. D. Barchas, *Nature, 221*, 368 (1969).

175. L. A. Pohorecky, M. Zigmond, H. Karten, and R. J. Wurtman, *J. Pharmacol. Exp. Ther.*, *165*, 190 (1969).

176. R. W. Fuller and J. M. Hunt, *Anal. Biochem.*, *16*, 349 (1966).

177. J. K. Saelens, M. S. Schoen, and G. B. Kovacsics, *Biochem. Pharmacol.*, *16,* 1043 (1967).

178. L. R. Krakoff and J. Axelrod, *Biochem. Pharmacol.*, *16*, 1384 (1967).

179. G. B. Kovacsics and J. K. Saelens, *Arch. Int. Pharmacodyn. Ther.*, *174*, 481 (1968).

180. P. Molinoff and J. Axelrod, *Science, 164*, 428 (1969).

181. R. W. Fuller and J. M. Hunt, *Biochem. Pharmacol.*, *14*, 1896 (1965).

182. R. W. Fuller and J. M. Hunt, *Life Sci.* (Oxford), *6*, 1107 (1967).

183. R. J. Wurtman and J. Axelrod, *Science, 153*, 1464 (1965).

184. R. J. Wurtman and J. Axelrod, *J. Biol. Chem.*, *241*, 2301 (1966).

185. R. J. Wurtman, E. Noble, and J. Axelrod, *Endocrinology, 80*, 825 (1967).

186. F. L. Margolis, J. Roffi, and A. Jost, *Science, 154*, 275 (1966).

187. R. W. Fuller and J. M. Hunt, *Nature, 214*, 190 (1967).

188. J. Axelrod, *J. Pharmacol. Exp. Ther.*, *132*, 28 (1962).

189. W. M. Clark and H. A. Lubs, *J. Bacteriol.*, *2*, 1 (1917).

190. J. Axelrod, *Rec. Progr. Hormone Res.*, *21*, 597 (1965).

191. J. Axelrod, and R. Tomchick, *J. Biol. Chem.*, *233*, 702 (1958).

192. P. J. Anderson and A. D'Iorio, *Biochem. Pharmacol.*, *17*, 1943 (1968).

193. M. S. Masri, A. M. Booth, and F. DeEds, *Biochim. Biophys. Acta, 65*, 495 (1962).

194. J. Daly, J. Axelrod, and B. Witkop, *Ann. N.Y. Acad. Sci.*, *96*, 37 (1962).

195. S. Senoh, J. W. Daly, J. Axelrod, and B. Witkop, *J. Amer. Chem. Soc.*, *81*, 6240 (1959).

196. J. W. Daly, J. Axelrod, and B. Witkop, *J. Biol. Chem.*, *235*, 1155 (1960).

197. A. M. Booth, M. S. Masri, D. J. Robbins, O. H. Emerson, F. T. Jones, and F. DeEds, *J. Biol. Chem.*, *234*, 3014 (1959).

198. E. Van Winkle and A. J. Friedhoff, *Life Sci.* (Oxford), *7*, 1135 (1968).

199. M. S. Masri, J. Robbins, O. H. Emerson, and F. DeEds, *Nature, 202*, 878 (1964).

200. J. W. Daly, J. K. Inscoe, and J. Axelrod, *J. Med. Chem.*, *8*, 153 (1965).

201. J. Axelrod and A. B. Lerner, *Biochim. Biophys. Acta, 71*, 650 (1963).

202. J. Fishman, M. Miyazaki, and I. Yoshizawa, *J. Amer. Chem. Soc.*, *89*, 7147 (1967).

203. F. A. Kuehl, Jr., M. Hichens, R. E. Ormond, M. A. P. Meisinger, P. H. Gale, V. J. Cirillo, and N. G. Brink, *Nature, 203*, 154 (1964).

204. M. Alberici, G. Rodriquez de Lores Arnaiz, and E. de Robertis, *Life Sci.* (Oxford), *4*, 1951 (1965).

205. J. K. Inscoe, J. W. Daly, and J. Axelrod, *Biochem. Pharmacol.*, *14*, 1257, (1965).

206. S. Senoh, Y. Tokuyama, and B. Witkop, *J. Amer. Chem. Soc.*, *84*, 1719 (1962).

207. K. Engelman, B. Portnoy, and W. Lovenberg, *Amer. J. Med. Sci.*, *255*, 259 (1968).

208. B. Nikodijevic, J. W. Daly, and C. R. Creveling, *Biochem. Pharmacol.*, *18*, 1577 (1969).
209. J. Axelrod, W. Albers, and C. D. Clemente, *J. Neurochem.*, *5*, 68 (1959).
210. J. R. Crout, C. R. Creveling, and S. Udenfriend, *J. Pharmacol. Exp. Ther.*, *132*, 269 (1961).
211. M. M. Salseduc, I. J. Jofre, and J. A. Izquierda, *Med. Pharmacol. Exp.*, *14*, 113 (1966).
212. R. J. Wurtman, J. Axelrod, and L. T. Potter, *J. Pharmacol. Exp. Ther.*, *144*, 150 (1964).
213. M. Assicot and C. Bohuon, *Nature*, *212*, 861 (1966).
214. S. Waltman and M. Sears, *Invest. Ophthalmol.*, *3*, 601 (1964).
215. R. Håkanson and H. Moller, *Acta Dermato-Venereol.*, *43*, 552 (1963).
216. J. Bamshad, A. B. Lerner, and J. S. McGuire, *J. Invest. Dermatol.*, *43*, 111 (1964).
217. E. H. LaBrosse and M. Karon, *Nature*, *196*, 1222 (1962).
218. J. Axelrod, P. D. MacLean, R. W. Albers, and H. Weissbach, in *Proc. 4th Int. Neurochem. Symp.*, Varenna, Italy, S. S. Kety and J. Elkes, Eds., Pergamon, London, 1960, p. 307.
219. K. Stock and E. O. Westermann, *J. Lipid Res.*, *4*, 297 (1963).
220. A. J. Prange, Jr., J. E. White, M. A. Lipton, and A. M. Kinkead, *Life Sci.* (Oxford), *6*, 581 (1967).
221. J. R. Crout and T. Cooper, *Nature*, *196*, 387 (1962).
222. L. T. Potter, T. Cooper, V. L. Willman, and D. E. Wolfe, *Circ. Res.*, *16*, 468 (1965).
223. L. R. Krakoff, R. A. Buccino, J. F. Spann, Jr., and J. DeChamplain, *Amer. J. Physiol.*, *215*, 549 (1968).
224. L. Landsberg, J. DeChamplain, and J. Axelrod, *J. Pharmacol. Exp. Ther.*, *165*, 102 (1969).
225. A. Eisenfeld, J. Axelrod, and L. Krakoff, *J. Pharmacol. Exp. Ther.*, *156*, 107 (1967).
226. Z. M. Bacq, L. Gosselin, A. Dresse, and J. Renson, *Science*, *130*, 453 (1959).
227. J. Axelrod and M. Laroche, *Science*, *130*, 800 (1959).
228. J. R. Crout, *Biochem. Pharmacol.*, *6*, 47 (1961).
229. J. V. Burba and M. F. Murnaghan, *Biochem. Pharmacol.*, *14*, 823 (1965).
230. E. T. Abbs, K. G. Broadley, and D. J. Roberts, *Biochem. Pharmacol.*, *16*, 279 (1967).
231. B. Belleau and J. Burba, *Biochim. Biophys. Acta*, *54*, 195 (1961).
232. C. Mavrides, K. Missala, and A. D'Iorio, *Can. J. Biochem. Physiol.*, *41*, 1581 (1963).
233. T. Sasaki, *Ann. Sankyo Res. Lab.*, *18*, 93 (1966).
234. S. B. Ross and Ö. Haljasmaa, *Acta Pharmacol. Toxicol.*, *21*, 205 (1964).
235. A. D'Iorio and C. Mavrides, *Can. J. Biochem. Physiol.*, *40*, 1454 (1962).
236. A. D'Iorio and C. Mavrides, *Can. J. Biochem. Physiol.*, *41*, 1779 (1963).
237. R. W. Gardier, G. L. Endahl, and W. Hamelberg, *Anesthesiology*, *28*, 677 (1967).
238. M. Assicot and C. Bohuon, *Life Sci.* (Oxford), *8*, 93 (1969).
239. P. J. Anderson and A. D'Iorio, *Can. J. Biochem.*, *44*, 347 (1966).
240. R. E. McCaman, *Life Sci.* (Oxford), *4*, 2353 (1965).
241. J. Axelrod, *Methods Enzymol.*, *5*, 748 (1962).
242. J. W. Schweitzer and A. J. Friedhoff, *Life Sci.* (Oxford), *8*, 173 (1969).
243. L. Stjärne, R. H. Roth, and N. J. Giarman, *Biochem. Pharmacol.*, *17*, 2008 (1968).

244. S. Nara, B. Gomes, and K. T. Yasunobu, *J. Biol. Chem.*, *241*, 2774 (1966).
245. V. G. Erwin and L. Hellerman, *J. Biol. Chem.*, *242*, 4230 (1967).
246. T. Nagatsu, *J. Biochem.* (Japan), *59*, 606 (1966).
247. S. R. Guha and K. Murti, *Biochem. Biophys. Res. Commun.*, *18*, 350 (1965).
248. L. M. Barbato and L. G. Abood, *Biochim. Biophys. Acta*, *67*, 531 (1963).
249. M. B. H. Youdim and T. L. Sourkes, *Can. J. Biochem.*, *44*, 1397 (1966).
250. W. Hardegg and E. Heilbron, *Biochim. Biophys. Acta*, *51*, 553 (1961).
251. V. Z. Gorkin, *Nature*, *200*, 77 (1963).
252. J. B. Ragland, *Biochem. Biophys, Res. Commun.*, *31*, 203 (1968).
253. H. C. Kim and D'Iorio, A., *Can. J. Biochem.*, *46*, 295 (1968).
254. M. B. H. Youdim and M. Sandler, *Biochem. J.*, *105*, 43P (1967).
255. D. Johnson and H. Lardy, *Methods Enzymol.*, *10*, 94 (1967).
256. A. Burger and Nara, S., *J. Med. Chem.*, *8*, 859 (1965).
257. L. Hellerman and V. G. Erwin, *J. Biol. Chem.*, *243*, 5234 (1968).
258. N. Weiner, *J. Neurochem.*, *6*, 79 (1960).
259. S. H. Snyder, J. Fischer, and J. Axelrod, *Biochem. Pharmacol.*, *14*, 363 (1965).
260. S. Otsuka and Y. Kobayashi, *Biochem. Pharmacol.*, *13*, 995 (1964).
261. W. M. McIsaac and V. Estevez, *Biochem. Pharmacol.*, *15*, 1625 (1966).
262. R. E. McCaman, G. Rodriques de Lores Arnaiz, and E. DeRobertis, *J. Neurochem.*, *12*, 927 (1965).
263. B. Jarrott and L. L. Iverson, *Biochem. Pharmacol.*, *17*, 1619 (1968).
264. C. B. Smith, *J. Pharmacol. Exp. Ther.*, *151*, 207 (1966).
265. R. J. Wurtman and J. Axelrod, *Biochem. Pharmacol.*, *12*, 1439 (1963).
266. V. Zeller, G. Ramachander, and E. A. Zeller, *J. Med. Chem.*, *8*, 440 (1965).
267. H. Weissbach, T. E. Smith, J. W. Daly, B. Witkop, and S. Udenfriend, *J. Biol. Chem.*, *235*, 1160 (1960).
268. M. Krajl, *Biochem. Pharmacol.*, *14*, 1683 (1965).
269. B. D. Drujan and J. M. Diaz Borges, *Fresenius' Z. Anal. Chem.*, *243*, 662 (1968).
270. W. Lovenberg, R. J. Levine, and A. Sjoerdsma, *J. Pharmacol. Exp. Ther.*, *135*, 7 (1962).
271. J. Southgate and G. G. S. Collins, *Biochem. Pharmacol.*, *18*, 2285 (1969).

Measurement of Choline Esters

Donald J. Jenden and L. B. Campbell, *UCLA School of Medicine and Brain Research Institute, Los Angeles, California 90024*

I. INTRODUCTION

Esters of choline are of widespread occurrence in nature and are major components of lipids. Several esters of small molecular size are of particular importance because of their powerful pharmacological properties and the presumptive role of some of them as neurohumors. Methods for the identification and measurement of these substances are necessary to develop an understanding of the cellular mechanisms involved in their synthesis, release, effect, destruction, and possible reuptake. Until recently the only methods with the required sensitivity were bioassays, the limitations of which are discussed below. In the past few years several chemical methods have been described which portend a growth of knowledge of cholinergic mechanisms comparable to that which followed the development of specific and sensitive chemical methods for the study of catecholamines.

Acetylcholine is the only choline ester which has been shown beyond reasonable doubt to have a transmitter function. It was originally isolated from horse spleen by Dale and Dudley (1) although it had previously been shown to occur in ergot (2) and has since been identified rigorously in several mammalian tissues, including sympathetic ganglia (3,4) and fresh rat brain (5). Synthesis of acetylcholine from isotopically labeled choline by brain slices and by brain *in vivo* has been demonstrated by several investigators (6–9). Together with propionylcholine and butyrylcholine, it has been identified in homogenates of tissue which have been allowed to autolyze (10–12). In view of the ability of many tissues to generate choline esters when appropriately incubated, it is not always possible to decide whether a given ester occurs in living tissue unless the tissue is frozen immediately after the animal is sacrificed, or choline acetyltransferase is inactivated by immediate homogenization in a suitable medium. Other choline esters of pharmacological interest which have been identified in animal tissues include propionylcholine (13), acrylylcholine (14), dimethylacrylylcholine (15), urocanylcholine (murexine) (16), and γ-aminobutyrylcholine (17,18). Several related substances may be present in tissue extracts, and are important because they interfere with the bioassay of choline esters or have similar effects on the test organs (12,19–21). There is ample reason to believe that the acetylcholine-like activity in normal mammalian brain, which has been the subject of intensive investigation by bioassay, is almost entirely due to acetylcholine itself (22–28); nevertheless, the work of Hosein and his collaborators (12,19–21) has served to point up the dangers of bioassay and to focus attention on the need for more specific methods of identification and measurement.

Several published reviews are excellent sources of information on the biological and chemical properties of acetylcholine and related substances, and of the methods available for their extraction, identification, and measurement (29–33). In this chapter the subject will not be reviewed comprehensively. Instead, recent advances dealing particularly with chemical methods of analysis will be summarized, and selected procedures will be described in sufficient detail to permit their immediate use. These methods have been chosen and evaluated to permit the reader to select the one most appropriate to a given problem. Each represents a different compromise between the conflicting objectives of simplicity, speed, sensitivity, specificity, and accuracy.

Analytical procedures for estimating the concentration of a compound in tissue may be divided into four stages: extraction, concentration, separation, and quantitative detection. These are not always discrete steps, nor are they all employed in every procedure. For example, in estimating the quantity of acetylcholine released into a perfusate, many investigators have applied bioassay directly without preliminary concentration and separation. In doing so they rely upon the specificity of the assay tissue to identify the active component, and on its sensitivity to obviate the need for concentration. Preliminary separation and purification by paper chromatography or paper electrophoresis has nevertheless been frequently used to remove interfering materials, and to differentiate between tissue components with similar effects on the assay tissue. When a chemical method of detection and measurement is employed, preliminary concentration and separation are usually essential to achieve reasonable specificity.

II. EXTRACTION PROCEDURE

In order to measure the total level of acetylcholine in a tissue it is necessary to use conditions in which the ester is completely released from the "bound" form, i.e., from subcellular organelles, to prevent its destruction by tissue cholinesterases or by chemical hydrolysis, and to prevent its continued synthesis by choline acetyltransferase. Since the acetylcholine content may change rapidly during or after excision of the tissue, it is desirable to remove and freeze the tissue as rapidly as possible. Many investigators freeze or rapidly chill the animal before dissection by immersion in liquid nitrogen or oxygen, or a solvent maintained at a comparable temperature (30,34,35). Although rapid freezing of the brain *in situ* might appear to offer a means of fixation which avoids losses resulting from decapitation and dissection, the freezing process is by no means instantaneous. Takahashi and Aprison (34) reported that the rat thalamus took 68 and 83 sec in two rats to reach 0°C following immersion in liquid nitrogen, while Stavinoha et al. (35) reported 31 sec for immersion in liquid nitrogen and 77 sec for dichlorodifluoromethane at liquid nitrogen temperature. Paradoxically, the latter method yielded higher values for mean brain acetylcholine. These were about the same as others have obtained by decapitation and rapid removal of the brain followed by freezing in liquid nitrogen (36,37). In a recent study (37) brain acetylcholine levels were found to be higher in rats which were sacrificed by decapitation than in those killed by immersion in liquid nitrogen. The former procedure revealed a clear-cut diurnal cycle while the latter did not, suggesting that data obtained by decapitation followed by rapid removal and freezing of the brain may be physiologically more relevant. Some workers have found that freezing *in situ* leads to lower levels (38). If a freezing procedure is followed, it is imperative that thawing should not be allowed to occur before homogenization in a medium which inactivates the synthetic and destructive enzymes. Freezing and thawing invariably causes a serious loss (39,40), presumably

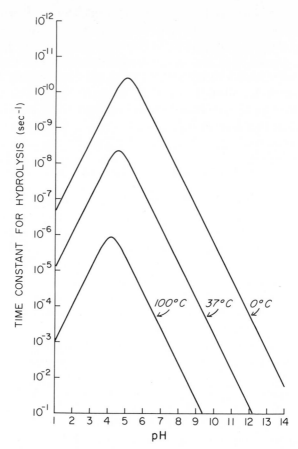

Fig. 1.   Stability of acetylcholine as a function of pH, calculated from the data of Butterworth et al. (41).   Maximum stability is attained at 100°C at pH 4.18, in agreement with Hofmann (157).   The corresponding optima at 37 and 0°C are pH 4.60 and 5.05.   The curves shown are for the sum of H^+ and OH^- catalyzed reactions; other nucleophiles in the solution may increase the rate significantly (e.g., phosphate, thiols, etc.)

by releasing the acetylcholine so that it becomes susceptible to attack by cholinesterase.

The hydrolysis of acetylcholine is catalyzed by both acid and base, and the stability is maximum at a pH of 4–5 (Fig. 1) (41,42).   Cholinesterases and choline acetyltransferase are inactive at this pH, and it is therefore usual to employ an acidic extraction medium which will yield a homogenate with a pH in this region.   Since the final pH of the homogenate will generally depend on the relative weight of tissue and extraction medium, it is important to check the final pH under the conditions used. The use of a large excess of acid (e.g., 10% trichloracetic acid) may reduce the pH to a point at which significant hydrolytic losses may occur. The use of a buffer at pH 4 as an extracting medium eliminates this problem, and the stability of acetylcholine at this pH is so great that the homogenate may be boiled without significant loss to release bound acetylcholine and remove protein (43–46). This procedure is particularly useful when the crude extract is to be bioassayed, since no organic solvents or toxic ions are introduced which interfere with the assay.

Trichloracetic acid may be conveniently removed from extracts prior to bioassay by repeatedly extracting with ether after first centrifuging to remove precipitated protein. Holmstedt (47,48) has used 0.4M perchloric acid as an extractant prior to bioassay; most of the perchlorate ion can then be removed by subsequent neutralization with potassium carbonate, since potassium perchlorate is sparingly soluble. However, the residual perchlorate ($\sim$ 2 mM) may be sufficient to cause significant potentiation of acetylcholine on the frog rectus abdominis preparation (49).

Most investigators have employed a combination of an organic solvent, usually ethanol or acetone, and acid, e.g., acetic or formic, to extract choline esters from tissue. The organic solvent helps to release bound acetylcholine, and facilitates subsequent concentration of the extract by evaporation or extraction. Release of the bound acetylcholine is not immediate, and it is necessary to store the homogenate close to 0°C for 20–60 min to achieve quantitative extraction. The most commonly employed extracting media are summarized in Table I.

III. CONCENTRATION OF THE EXTRACT

One of the advantages of using an organic extraction medium is the ease with which it may be concentrated by evaporation. If the organic solvent is acetone, it may be conveniently removed together with lipids by extraction with ether without loss of choline esters (25). Evaporative concentration of aqueous solutions is in general of little value because of the large quantity of salts which remain. Ion exchange columns may be used (50,51) but are also of limited use if the electrolyte concentration is high. Acetylcholine and other cationic compounds may be extracted into an organic phase as an ion pair with hydrophobic anions (28,52,53). A colorimetric estimation procedure has been described which is based on ion-pair extraction of acetylcholine into ethylene dichloride as a complex with bromphenol blue (52,54). Fonnum has recently reported

188   DONALD J. JENDEN AND L. B. CAMPBELL

TABLE I
Media Commonly Used for Extracting Tissues for
Acetylcholine and Related Compounds

Composition	Volume (ml/g)	Comments	Ref.
10% aqueous trichloracetic acid	2	Favored by many as giving highest yields. Excess acid can be removed by ether extraction. pH of homogenate generally <1; significant hydrolysis may occur.	30, 39, 73, 81, 109, 153, 154
0.1–0.4 4N aqueous perchloric acid	1.5–3	Perchlorate can be precipitated at (°C as potassium salt. Perchlorate causes sensitization of frog rectus abdominis.	48, 49, 93, 155
Citric acid/sodium phosphate buffer at pH 4	5–20	Acetylcholine released and protein precipitated by heating (100° for 2–10 min). Suitable for small tissue samples which may be dropped into buffer at 100°C. Claimed to extract no sensitizing material from brain.	43, 45, 46
95% ethanol with 0.2% acetic acid or formic acid	2.5–10	Final pH of homogenate usually ~4. Allows homogenization of frozen pulverized tissue below 0°C. Extract may readily be concentrated by evaporation.	30, 34, 98, 156
15% 1N formic acid, 85% acetone	2.5–20	Very high recoveries of acetylcholine claimed (152). Allows homogenization of frozen pulverized tissue below 0°C. Acetone can be removed by ether extraction.	25, 100, 152

that acetylcholine can be extracted into ethyl butyl ketone in the presence of sodium tetraphenylboride (53), a device which is suggested for scintillation counting of radioactive acetylcholine and may be of general value in its isolation from aqueous solution. The extraction is unfavorably influenced by the presence of salts in the aqueous solution. The proportion of acetylcholine extracted is reported to be independent of pH in the range 3–7, and of the relative volumes of aqueous and organic phases in the range 2:1 to 20:1. Considerable concentration can therefore be achieved while simultaneously removing salts. Acetylcholine can be recovered in the aqueous phase free of tetraphenylboride by shaking with 0.2N HCl.

The most general methods for concentrating acetylcholine from aqueous solution are based on precipitation of sparingly soluble salts. Quantitative precipitation can only be achieved from dilute solution if a coprecipitant is used; this is most commonly tetramethylammonium, tetraethylammonium, or choline in a concentration of about 1 mM. Quantitative precipitation of isotopically labeled choline esters from very dilute solution can conveniently be achieved by the addition of the unlabeled carrier compound (55).

The most commonly used precipitant is Reinecke salt, which was first used for acetylcholine by Kapfhammer and Bischoff (56). It has been extensively employed for isolation of choline esters from natural sources (13,14,57) and to concentrate and partially purify an extract prior to paper chromatography or bioassay. Bentley and Shaw (58) used reineckate precipitation in 60% ethanol in the presence of excess choline to precipitate acetylcholine while leaving in solution alkaloids such as atropine, hyoscine, morphine, and physostigmine. Histamine and epinephrine also remain in solution. This is a valuable means of separating choline esters from alkaloids which may interfere with bioassay. Precipitation from an alkaline solution may be used to separate choline from betaines, which are left in solution (59,60). Excess ammonium reineckate may be removed from the precipitate by washing with n-propanol (61). Although it is apparently unnecessary to remove the reineckate ion prior to bioassay (43,58,62), this is frequently desirable, and can conveniently be accomplished by treatment with an ion exchange resin in chloride form either on a column (9,31) or by batch treatment (25).

It has long been known that choline and acetylcholine form insoluble periodides and enneaiodides (63). Periodide has distinct advantages over reineckate for some purposes as a precipitant of acetylcholine and related compounds, but unaccountably has been relatively little used. Shaw (64) and Beattie (65) have described colorimetric estimation procedures based on the intense color of these complexes and their insolubility in aqueous solution. Periodide has recently been used as a precipitant in the presence of tetramethylammonium (66,67) or choline (68). It yields a much lower blank (69) in the radiometric procedure for choline acetyltransferase estimation described by McCaman and Hunt (70). Moreover, periodides have the general advantage that excess iodine can readily be removed from the precipitate by sublimation, leaving iodide salts. This is a highly desirable anion for pyrolysis gas chromatography (71,72) and for direct mass spectrometry of quaternary ammonium compounds (73). It is also preferable to reineckate for bioassay. The precipitation is normally carried out in an acidic medium; betaines and weak bases are left in solution at elevated pH (74), which may be useful in separating choline esters from alkaloids prior to bioassay, as in Bentley and Shaw's (58) procedure for reineckates.

Sodium tetraphenylboride (Kalignost) has been used both as a precipitant (18,75,76) of choline esters and as a means of extraction into organic solvents (53). When applied to tissue extracts this reagent has the disadvantage of yielding very bulky precipitates, probably because of their high potassium content. Potassium tetraphenylboride is very insoluble in water, a fact which is used in its analytical determination (77). The potassium salt is not extracted into ethyl butyl ketone under the conditions described by Fonnum (53).

Other sparingly soluble salts of choline esters employed to concentrate them from dilute aqueous solution or to characterize the esters by crystallization and mixed melting point include the hexylate (dipicrylamine salt) (78), phosphotungstate (79), aurichloride (2,10,78), and platinichloride (2,80). Whittaker (31) has summarized some physical properties of these and other derivatives which have been used in isolation and characterization.

IV. SEPARATION TECHNIQUES

As mentioned above, several procedures for acetylcholine estimation, both chemical and biological, omit any attempt at separation of different active components of the tissue except insofar as the extraction and concentration procedures accomplish this objective. This omission may be justifiable in the interest of simplicity, speed, and economy when a highly specific biological detector is used, but inevitably results in some degree of uncertainty as to the compound(s) being measured. Paper or thin-layer chromatography or paper electrophoresis have frequently been used before bioassay to enhance the specificity of the determination, and methods have recently been described for gas chromatography of specific derivatives of choline and choline esters. Ion exchange column chromatography may be of value (57,81), and is of particular importance for the separation of acetylcholine from acetylcoenzyme A in radiometric methods for choline acetyltransferase assay. Acetylcoenzyme A is anionic and is not retained on a cation exchange resin (9,82).

1. Paper Chromatography

Methods for paper chromatography of choline esters and related compounds have been extensively reviewed by Whittaker (31). The same solvent systems may be used for thin-layer chromatography on cellulose powder, which is faster and gives somewhat better resolution. Both techniques have serious limitations. The R_f values are very sensitive to temperature, the amount of loading, and particularly salts and other materials which are present in tissue extracts. This has led to major errors in the identification of pharmacologically active components eluted from paper chromatograms (22,23). The associated ions have a major influence on the running of these compounds, and in particular trichloracetate and tetraphenylboride alter the R_f in most solvents. For this reason periodide precipitation may be particularly useful if paper chromatography is to be performed subsequently. Acidic solvents may give multiple spots because of partial replacement of the associated ion by the solvent ion. Basic solvents give poor results because of partial hydrolysis. The solvent systems which are now most commonly used are n-butanol:ethanol:acetic acid:water (8:2:1:3) (83) and water-saturated butanol (84). Table II summarizes R_f values obtained with these solvents. Localization of the spots can be achieved by several reagents which are tabulated by Whittaker (31). The most useful of these is probably iodine vapor (84,85), since this leaves the esters unchanged and they can subsequently be eluted for quantitation by bioassay. Iodoplatinate is not mentioned by Whittaker, but gives particularly good visualization, the spots appearing blue on a red background (50,86). It is prepared by dissolving 1 g of platinic chloride in 10 ml water, mixing the solution with 250 ml of 4% potassium iodide, and making up the volume to 500 ml with water (87). A microbiological method has been described (88) which is capable of detecting as little as 30 ng of choline on a paper chromatogram.

2. Paper Electrophoresis

Many of the disadvantages of paper chromatography can be avoided by the use of paper electrophoresis. It is fast (1–4 hr) and reproducible, gives very high resolution, and is apparently unaffected by associated ions and other components of tissue extracts which alter the R_f unpredictably in paper chromatography (89). Paper electrophoresis was first used for choline esters by Chefurka and Smallman (90), who resolved choline from acetylcholine on Whatman 3MM paper soaked in lithium sulfate of ionic strength 0.1. Two important factors in the selection of a suitable buffer are volatility of the buffer and the stability of the esters. For these reasons most investigators have used solutions of acetic acid or buffers composed of acetic acid and a volatile amine (4,9,26,27,91,92). Henschler (91) employed a triethylamine/acetic acid

TABLE II
R_f Values of Choline and Choline Esters (22,23,57,76,84)

Substance	Butanol/Ethanol acetic acid/water (8:2:1:3) Ascending	Water saturated n-butanol Ascending	Descending
Acetylcholine chloride	0.35	0.14	0.09
Acetylcholine perchlorate		0.30	0.22
Acetylcholine tetraphenylborate	0.37		
Acetyl-β-methylcholine chloride	0.46	0.19	0.15
Acetyl-β-methylcholine perchlorate			0.31
Acetyl-β-methylcholine tetraphenylborate	0.44		
Propionylcholine chloride		0.22	0.17
Propionylcholine perchlorate		0.42	0.32
Propionylcholine tetraphenylborate	0.51		
Butyrylcholine chloride		0.28	0.22
Choline chloride	0.27	0.09	0.06
Choline perchlorate		0.24	0.21
Choline tetraphenylborate	0.29		
Benzoylcholine chloride		0.28	0.23
Lactylcholine chloride		0.18	
Lactylcholine perchlorate		0.34	
Acrylylcholine chloride	0.78		
Dimethylacrylylcholine chloride	0.70	0.35	

buffer at pH 4.6, μ = 0.03 on Schliecher and Schüll 2043b paper at a voltage gradient of 20 V/cm at 15°C, and obtained complete separation of choline from acetylcholine, propionylcholine, butyrylcholine, isovalerylcholine, acetylthiocholine, benzoylcholine, and carbaminoylcholine; formylcholine ran with choline, and isobutyrylcholine with butyrylcholine. Ryall et al. (27) employed a more concentrated triethylamine/acetic acid buffer at pH 3.9 (1.2 and 2.5% v/v), which did not resolve acetylcholine from propionylcholine and butyrylcholine but gave good separation from acetylcarnitine. The most satisfactory buffer so far described is $1.5M$ acetic acid, $0.75M$ formic acid (pH 2.0), in which

Potter and Murphy (89) obtained the excellent separations summarized in Table III, using 18 V/cm on Whatman 1 for 1 hr. This separation was subsequently incorporated into the radiometric procedure for acetylcholine estimation described by Feigenson and Saelens (93). Friesen et al. (4), using ammonium acetate or sodium phosphate buffers between pH 3 and 6, found that immediate application of a high voltage gradient

TABLE III
Electrophoretic Mobilities of Choline Esters and
Some Commonly Used Drugs (89)[a]

Substance	Mobility (cm/hr)
Nicotine	12.0
Hexamethonium	11.8
Tetramethylammonium	10.6
Succinylcholine	10.3
N-Methylethanolamine	9.8
Choline	9.5
β-methylcholine	9.3
Thiocholine	8.8
Decamethonium	8.7
Acetylcholine	8.5
Acetyl-β-methylcholine	8.3
Carbamylcholine	8.2
Tetraethylammonium	8.2
Propionylcholine	8.0
Acetylthiocholine	7.8
N,N-Dimethylethanolamine	7.6
Triethylcholine	7.5
Carnitine	7.2
Butyrylcholine	7.2
Hemicholinium-3	7.0
Pilocarpine	6.9
Benzoylcholine	6.7
Acetylcarnitine	6.5
Neostigmine	5.9
d-Tubocurarine	5.8
Physostigmine	5.1
Cocaine	5.1
Atropine	4.9
Scopolamine	4.8
Betaine	4.0
Acetylcoenzyme A	0.5

[a] Electrophoresis was performed at room temperature at 18 V/cm for 1 hr on Whatman 1 paper in 1.5 M acetic acid-0.75 M formic acid buffer at pH 2.

caused poor resolution and tailing which could be avoided if a lower voltage (3–5 V/cm) was used for the first hour, followed by 17–27 V/cm for 1.5–5 hr. This technique may be particularly useful when concentrated tissue extracts are being analyzed because of the greater likelihood of ionic interactions influencing the migration.

3. Gas Chromatography

Gas chromatography is capable of higher resolution than any other chromatographic system, and has the additional advantage of a wide selection of detectors, some of which are extremely sensitive. It cannot be applied directly to choline and its esters because the compounds to be chromatographed must have a significant vapor pressure, and it is therefore necessary to prepare volatile derivatives in quantitative yield which are specific for the esters from which they are prepared. The first derivatization procedure to be described was reduction of the ester group with potassium borohydride, yielding ethanol in the case of acetylcholine (94). This reaction was first used with lithium chloride as a catalyst by Cooper (95) in an estimation procedure for acetylcholine in which the ethanol was oxidized by alcohol dehydrogenase, the reaction being followed fluorimetrically by the coupled reduction of DPN.

$$\text{acetylcholine} \xrightarrow[\text{LiCl}]{\text{KBH}_4} \text{ethanol} \xrightarrow[\text{alcohol dehydrogenase}]{\text{DPN} \quad \text{DPNH}} \text{acetaldehyde}$$

The procedure was not suitable for estimations on brain because of interfering fluorescent material in brain. Stavinoha and co-workers (26,94) pointed out that ethanol could readily be quantitated by gas chromatography, and successfully applied the procedure to brain using calcium chloride in place of lithium chloride. Acetylcarnitine yielded only small amounts of ethanol, presumably because of the deactivating effect of the carboxylate group on the acyl moiety (96,97). It can be removed by prior paper electrophoresis. A distinct advantage of this procedure over Cooper's fluorimetric method (95) is that propionylcholine and butyrylcholine can be simultaneously estimated as propanol and butanol. The lower limit of accuracy using the original method is about 0.5 μg (3 nmoles) acetylcholine; this could undoubtedly be lowered by improving the gas chromatographic conditions and scaling down the volumes so that a greater proportion of the final volume could be injected into the gas chromatograph. Porapak Q (28) and Porapak QS (98) afford greater efficiency in the analysis of ethanol than the 20% Carbowax 6000 on HDMS-treated Chromasorb W employed by Stavinoha (26). The values obtained for levels of acetylcholine in rat brain were consistently higher than by bioassay (guinea pig ileum), and a similar discrepancy was observed by Howes et al. (99) using the same techniques in mice. They suggested that the bioassay results might have been affected by inhibitory materials in the brain extracts. The difference could also be due to the increased sensitization produced by alkaline boiling of the brain extracts (100) in tissue controls conducted according to Feldberg (101).

Cranmer (98) employed hydrolysis by sodium hydroxide to yield a volatile derivative (acetic acid) for gas chromatographic determination of acetylcholine, and claimed recoveries of 98–100%, with a lower limit of accurate detectability of 0.5 μg (3 nmoles) acetylcholine per rat brain. Hydrolysis by acetylcholinesterase yielded identical results.

Both these methods use derivatives of choline esters which are not inherently specific in the sense that the compounds to be analyzed are the only possible sources. The specificity of the method is therefore dependent on the initial processing of the tissue and the reaction conditions used, which presumably are chosen to remove components such as other acetate esters which would yield the same derivative as acetylcholine. In the case of Cranmer's method (98), free acetate ion in the tissue extracts would also be expected to interfere. Clearly it is preferable to utilize a derivative which is uniquely characteristic of the compound to be estimated. Two procedures have been described which are based on the removal of a methyl group from the quaternary nitrogen atom of choline and its esters, yielding volatile tertiary amines. The first of these employs a reaction described by Shamma et al. (102), in which sodium benzenethiolate (thiophenoxide) in aprotic solvents and at low temperatures causes a nucleophilic displacement of the tertiary amine, yielding methyl phenyl sulfide as a biproduct:

$$R_2 - \overset{R_1}{\underset{R_3}{\overset{|}{\underset{|}{N^{\oplus}}}}} - CH_3 + {}^{\ominus}S\text{\textbenzene} \rightarrow R_2 - \overset{R_1}{\underset{R_3}{\overset{|}{\underset{|}{N}}}} + CH_3S\text{\textbenzene}$$

Although earlier work indicated that thiolates attack ester groups and would therefore be unsuitable (103,104), it was found that quantitative yields of dimethylaminoethanol or the corresponding ester could be obtained by refluxing with 50 mM sodium benzenethiolate in anhydrous butanone under a nitrogen atmosphere (105–107). It was reported subsequently (25) that the tedious procedure originally described to remove impurities from the benzenethiolate could be greatly simplified if 25 mM benzenethiol was also included in the reaction mixture. A solvent extraction procedure is used after reaction to concentrate the amines for gas chromatography and to remove benzenethiol and methyl phenyl sulfide. The procedure has been applied successfully to brain extracts, and the identity of the acetylcholine derivative was confirmed rigorously by combined gas chromatography and mass spectrometry (Section V-2-c) (5). Dimethylaminoethyl acetate is not a component of normal rat brain, since omission of the demethylation step resulted in disappearance of the corresponding peak. Good resolution of the amines was obtained on a 6-ft, $\frac{1}{4}$-in. column of Polypak I (Par I) coated with 1% phenyldiethanolamine succinate. The detection limit using a flame ionization detector is about 10 ng (50 pmoles).

Szilagyi, Schmidt, and Green (71,108) have described a simpler procedure in which choline and its esters may be analyzed by pyrolysis gas chromatography. The procedure appears generally applicable to onium compounds (72). A solution of the compounds as iodides or chlorides is evaporated on the pyrolyzer ribbon (Barber-Coleman Model 5180) and subsequently pyrolyzed at 500°C for 15 sec (72) or 730°C for 30 sec (71). Pyrolysis products are swept into a column of 1% phenyldiethanolamine succinate on Par I in the stream of nitrogen used as a carrier. The procedure lends itself to concentration of relatively large volumes of solution in the pyrolyzer, and the sensitivity is about 5 ng (25 pmoles) (66). It has been adapted to the estimation of acetylcholine in tissue perfusates by preliminary precipitation with periodide (66), using butyrylcholine as an internal standard. However, so far there has been no report of its use on tissue extracts in which some preliminary separation may be required to prevent interference by other compounds during pyrolysis.

V. DETECTION AND MEASUREMENT

In the last few years several chemical methods have been described which are greatly superior to earlier procedures for choline ester measurement in both sensitivity and specificity. Nevertheless, the most sensitive bioassay methods still exceed the capability of any purely chemical procedure with the exception of combined gas chromatography and mass spectrometry, which few investigators have at their disposal. There remains a place for bioassay in situations where the utmost sensitivity is required, and perhaps in certain kinds of routine work in which the identity of the active component (e.g., of a tissue perfusate) is established and a large number of estimations of relatively low accuracy must be made. A simplified type of bioassay may then be run concurrently with the experiment generating the samples, requiring only a few minutes per sample.

1. Biological Methods

Although many biological test objects are known which have extreme sensitivity to acetylcholine, the specificity is low compared to the best chemical procedures now available, and it is impossible to determine more than one substance simultaneously using a single bioassay method. The results are also notoriously susceptible to artifacts arising from unidentified components of tissue extracts. Drugs which may have been used in the experiment are also likely to appear in tissue extracts, and may enhance or reduce the sensitivity of the test tissue. A number of supplemental techniques may be used to circumvent these problems, at some cost in time and size of the sample required:

1. Preliminary separation of the compounds to be determined by paper electrophoresis, followed by elution and bioassay. Although this allows the biological and/or chemical determination of more than one component of the extract and minimizes the risk of interference by drugs, quantitative recovery is virtually impossible to achieve, particularly with very small quantities.

2. Parallel bioassay (27,109), i.e., simultaneous estimation of the "equivalent acetylcholine activity" on several test objects. If these give similar results, it provides strong circumstantial evidence for the identity of the active component as acetylcholine. If the results on different test organs give different values of "acetylcholine activity," and the identity of two active components is known, it is frequently possible to infer the concentrations of these components.

3. Modification of the test object by added drugs may be used to enhance both its sensitivity and its specificity, e.g., the treatment of guinea pig ileum with mipafox to enhance sensitivity (110,111) and with mepyramine and tryptamine to prevent responses to histamine and 5-hydroxytryptamine, respectively (112). However, it is not possible to differentiate between different choline esters in this way. Evidence as to the identity of the component producing the biological response can be obtained by showing that it is prevented by an appropriate blocking agent, e.g., tubocurarine for frog rectus abdominis and atropine for guinea pig ileum assays in the case of acetylcholine. Again, this will not differentiate satisfactorily between different choline esters.

4. Pretreatment of the extract. If the substance to be assayed is acetylcholine, its activity should disappear when boiled in $0.1N$ NaOH for a few minutes but remain unchanged when boiled at pH 4; it should rapidly disappear on treatment with acetylcholinesterase. These simple checks may be rapidly conducted and have been commonly employed to provide corroborative evidence of identity. Pretreatment of extracts with chymotrypsin has also been used to remove bradykinin and substance P prior to assay on the guinea pig ileum (34,112) and thereby increase the specificity of the assay. Whenever a tissue extract is to be assayed for acetylcholine it is essential to compare its activity with that of the reference compound made up in the same extract after the acetylcholine has been destroyed by basic hydrolysis (101,113) or treatment with acetylcholinesterase, in order to compensate for tissue components which may modify the response of the test organ. However, physostigmine may be destroyed by heating in alkaline solution (114), and the sensitization of the frog rectus abdominis by brain extracts is increased by heating in alkali (100). For these reasons treatment with cholinesterase followed by inactivation of the enzyme seems preferable to alkaline hydrolysis for tissue controls.

Details of the many methods available for biological assay of choline esters will not be reviewed here, since the accounts of MacIntosh and Perry (29) and Whittaker (31) are sufficiently detailed and for the most part remain up to date. Some of the more significant recent trends and developments will, however, be summarized.

The frog rectus abdominis, guinea pig ileum, and dorsal muscle of the leech are most commonly used. The frog rectus abdominis is simple to set up, reliable, and reproducible. It is sensitized with physostigmine ($10^{-5}M$) or neostigmine ($10^{-6}M$), and can be further sensitized by a factor of 5–10 by acetone (0.5–1.0%) (115,118). The threshold concentration of acetylcholine when treated in this way is about $10^{-8}M$. The smallest quantity which can be assayed can be further reduced by using the superfusion technique (119) in place of the conventional muscle bath.

The increasing popularity of the guinea pig ileum as a test object is due in part to the finding that morphine suppresses the tonic contraction and spontaneous motility resulting from treatment with anticholinesterases, apparently by reducing acetylcholine output from nerve terminals (110,120–122). This allows the use of anticholinesterase treatment to enhance the sensitivity of the assay. Mipafox [bis-(isopropylamino)-fluorophosphine oxide] is preferable to DFP, neostigmine, or tetraethylpyrophosphate as an anticholinesterase (111). The ileum is incubated in Krebs solution with mipafox (10 μg/ml) and morphine sulfate (50 μg/ml) for 1 hr, after which the mipafox is washed out. Morphine is kept in the solution throughout the assay. Maximum sensitivity is in the range 10^{-11}–$10^{-9}M$(111). The specificity of the assay may be improved by including an antihistamine (43,112,123,124) and a serotonin antagonist (112). Mepyramine maleate and tryptamine hydrochloride are suitable (112). Spontaneous motility of the ileum can be minimized by storing the ileum in the refrigerator overnight (62) and by conducting the assay at 30°C.

The dorsal muscle of the leech, while very sensitive after treatment with eserine ($30\,\mu M$), has until recently been unpopular because of its slow recovery between responses and unstable baseline. Both the response and recovery are faster when isometric rather than isotonic recording is used, and a stable baseline is then readily achieved (125). Morphine ($200\,\mu M$) facilitates relaxation leaving the sensitivity to acetylcholine unchanged (126). Threshold responses can be obtained at about $5 \times 10^{-9}M$ acetylcholine, and the preparation is more specific than the guinea pig ileum, being unaffected by histamine and 5-hydroxy-tryptamine. Adaptation of Gaddum's microbath (127) to very small strips of the longitudinal body wall muscle of the leech allows the detection of less than 1 pmole of acetylcholine, and the accurate assay (5%) of 10 pmoles (128).

Since the original description of the heart of the clam *Venus mercenaria* for the bioassay of acetylcholine (129), it has been of considerable interest because of its great sensitivity. Several other species of clam may also be used (130–132). Although generally at least as sensitive to 5-hydroxy-tryptamine as to acetylcholine, methysergide ($5\,\mu M$) (34,133) may be used to block the stimulant effect of 5-hydroxytryptamine, hence making it more specific. Sensitivity to most other neurohumors is negligible. A detailed survey of the species of clam suitable for this assay, the relative potencies of different choline esters, and recommended procedure for the assay has recently been published (132). The sensitivity is variable but the threshold quantity for an inhibitory effect is generally 1–10 pmoles. The action of choline esters is unaffected by atropine but is blocked by mytolon [2,5 bis-(3'diethylaminopropylamine)-benzoquinone bis-benzyl chloride]. The use of other invertebrate muscles for bioassay purposes has also been reviewed (134).

2. Chemical Methods

Most chemical methods for the detection of choline esters depend upon reactions of one of the two functional groups in the molecule, and require preliminary separation to achieve reasonable specificity. Early methods utilizing colored complexes with the quaternary ammonium groups (52,64,65) are not sufficiently specific to be useful analytically. A recent method based on colorimetric measurement of complex formation with gallocyanine also gives positive results with a variety of compounds (135).

The reaction of hydroxylamine with esters in alkaline solution was utilized by Hestrin (136) for a colorimetric procedure based on the absorption of ferric hydroxamate complex at 540 nm. Acetylcholine, propionylcholine, and butyrylcholine but not carbaminoylcholine give a positive color reaction, and thiocholine esters may also be estimated in this way (137). The reaction has been adapted to develop spots on paper chromatograms (84). By using a preliminary step in which quaternary ammonium compounds are precipitated (75,76,79), the specificity of the reaction for choline esters can be improved and a modification for small quantities has been described (138), but its sensitivity is not sufficient for most biochemical purposes (10 nmoles). Maslova (139,140) has used polarography instead of colorimetry to estimate the amount of ferric ion complexed with hydroxamate, and claims very high sensitivity (0.5 pmoles). The procedure was applied to estimate acetylcholine concentrations in the superior cervical ganglion of rabbits, yielding lower values than bioassay (141), but no details of specificity or recovery are given.

A. FLUORIMETRY

Fellman (67) used a related reaction between the ester group and hydrazine to develop a very sensitive fluorimetric procedure for the estimation of acetylcholine. Both its cationic nature and the activating effect of the neighboring ammonium group on the acyl moiety (142) are ingeniously exploited to achieve greater selectivity. After precipitation with periodide, acetylcholine is absorbed onto a cation exchange column, where it is reacted with hydrazine for a controlled time. The acetylhydrazide is eluted from the column and coupled with salicylaldehyde to yield the intensely fluorescing salicylhydrazone.

Excess salicylaldehyde is removed by reduction with potassium borohydride. The limit of detectability is 200 pmoles. Acetylcholine, propionylcholine, and butyrylcholine all yield the same fluorescence on a molar basis, but acetylcarnitine, acetylcoenzyme A, phosphorylcholine, and benzoylcholine do not respond significantly. The method was applied to estimate rat brain acetylcholine levels, and the values obtained were in reasonable agreement with other techniques. Recovery of added acetylcholine from brain homogenates varied from 75 to 100% and an internal standard was recommended, but no details of its use were given. Recovery from simple salt solutions was 100%.

B. ISOTOPIC METHODS

If the choline ester to be determined is isotopically labeled, the problem reduces to one of using an appropriate separation procedure followed by counting of the fractionated material. Paper electrophoresis is probably the method of choice for separation. The compounds to be analyzed should be added as carriers before the separation to minimize adsorption and elution losses, and to permit location of the band. This is particularly important if paper chromatography is used for the separation because the R_f is likely to be seriously altered by components of tissue extracts (22,23). Further confirmation of the identity of the compounds can be obtained by isotope dilution analysis (6). Carrier is added to the eluted material and it is repeatedly crystallized. The tetrachloroaurate salts may readily be recrystallized from water for this purpose, and a constant specific activity on repeated recrystallization constitutes strong presumptive evidence of identity of the labeled compound with the added carrier. Schuberth et al. (9) have recently described a procedure for measuring acetylcholine turnover rates in mouse brain *in vivo*, in which labeled choline and acetylcholine are measured in the same extract by paper electrophoresis followed by combustion of the strips and liquid scintillation counting. Total choline levels were measured by a radiochemical method similar to that described by Smith and Saelens (143). The aqueous brain extract was adjusted to pH 7.3, concentrated and incubated at 37°C for 1 hr with choline acetyltransferase and acetyl-^{3}H-coenzyme A. Carrier acetylcholine was added to stop the reaction and the solution was treated with saturated ammonium reineckate. The *n*-propanol-washed precipitate was dissolved in 50% acetone/water and passed through a Dowex 2-X8 ion exchange column in the chloride form (cf. 82). The eluate was then concentrated and measured by liquid scintillation. A linear relation was found between choline and radioactive acetylcholine formed (200–1200 pmoles) with an overall recovery of 63.7%.

A similar principle was employed by Feigenson and Saelens (93) to develop a radiometric assay for acetylcholine. Acetylcholine was separated from the aqueous brain extract by paper electrophoresis, eluted, and hydrolyzed with ammonium hydroxide. After incubation of the sample for 1 hr at 37°C with choline acetyltransferase and acetyl-^{14}C-coenzyme A, it was again subjected to paper electrophoresis under the same conditions.   The acetylcholine was then eluted and counted by liquid scintillation.   This method has a very high inherent specificity because of the electrophoretic separations and the specificity of choline acetyltransferase.   The separation of choline from acetylcholine in the electrophoresis step was 99.6%.   The final count was linear with the acetylcholine in the sample, with a limit of detection of about 50 pmoles and an absolute recovery of 77.6%.   It was successfully applied to extracts of mouse brain, with results in agreement with other methods. It can presumably be adapted to the estimation of other choline esters by eluting the appropriate strip after the first electrophoretic run.   One of the drawbacks of the method as described is the necessity of preparing choline acetyltransferase freshly every day.   The preparation of this enzyme from human placenta described by Schuberth (144) is stable for weeks when stored at $-20°C$, and its use would overcome this problem.

C. MASS SPECTROMETRY

Mass spectrometry of choline esters has recently received some attention because of the very high sensitivity and specificity of the mass spectrometer as a detector, and the development of methods for coupling it to the effluent from a gas chromatograph.   Johnston et al. (73) studied the mass spectra of choline esters and related compounds to provide a basis for their estimation and characterization in trace amounts.   When heated to about 200°C under high vacuum, the halide salts dissociate into the methyl halide and corresponding tertiary amine.   Mass spectra of the amines are characteristic of each compound, and some of the data published by Johnston et al. (73) are reproduced in Table IV.   This should be considered a basis for approximate comparison only, since details of the mass spectrum vary with the instrument used, temperature, inlet system, electron energy, and other variables; definitive identification of a compound can be made only by comparison with the authentic reference compound under identical conditions.   The base peak for all the compounds listed has a mass/charge ratio of 58, and is attributed to the $(CH_3)_2N^+{=}CH_2$ ion.   Molecular ions have a low abundance in every case.   A sample of about 2 nmoles of acetylcholine bromide was necessary to obtain a clearly interpretable spectrum (Perkin Elmer RMU-6D single focusing system), but it was possible to detect as little as 100 pmoles from the peaks at m/e 50 and 52, due to CH_3Cl^+.

Mass spectra are of little interpretive value unless they are obtained with pure substances, when they can provide rigorous evidence of identity. The mass spectrometer is therefore ideally complemented by the gas chromatograph, which can separate complex mixtures into pure com-

TABLE IV

Mass Spectra of Choline, Choline Derivatives, and Betaines[a]

m/e	Choline	Acetyl-choline	Propionyl-choline	Butyryl-choline	Acetyl-β-methylcholine	Acetylthio-choline	DL-Carnitine	Acetyl-DL-carnitine
42							28.0	57.7
55							24.0	24.1
58	100	100	100	100	100	100	100	100
59							42.0	34.5
70					1.3			
71	1.3	8.3	14.5	20.0	1.6	5.3		
72					1.4	3.6		
74								
84		2.7					2.0	0.7
87		2.1					15.2	7.7
88			1.5	1.6	1.9	1.2		
89	5.3[b]							
101			2.0					
102					0.4	0.8	1.0	0.2
103					0.9	1.1		
104						0.8		
115				1.8				
130								
131					0.1			
145		1.4[b]						
147								
159			1.0[b]	0.6[b]	0.5[b]	1.2[b]		

[a] Spectra measured with a Hitachi Perkin-Elmer RMU-6D single focusing mass spectrometer. Ionizing electron energy of 70 eV. Sample oven rapidly heated to 190°C. Reservoir, orifice leak, and ionization chamber maintained at 180°C. Peak intensity expressed as percentage of base peak (m/e 58).
[b] Molecule ion (73).

203

ponents, but provides no information as to their identities. Methods for interfacing these instruments have recently been devised (145), and such an integrated system can be used to identify every component of a mixture under ideal conditions. For these reasons it appears probable that mass spectrometry will be of maximum value for the analysis of choline esters in biological materials only when appropriately coupled with a suitable separative technique, of which gas chromatography is undoubtedly the most powerful at this time. It is also necessary that the volatile derivative employed for the gas chromatography should uniquely reflect the choline ester from which it was formed.

Hammar et al. (5) have employed an integrated gas chromatograph/ mass spectrometer system to achieve the first rigorous identification of acetylcholine in fresh mammalian brain, using the benzenethiolate N-demethylation procedure described by Jenden et al. (25,106). Mass spectra obtained from authentic dimethylaminoethyl acetate and from a derivatized rat brain extract at the same retention time were identical. By focusing the mass spectrometer on a single mass/charge ratio instead of scanning the spectrum in the conventional manner, it is possible to achieve a much greater sensitivity (146), and in the study referred to, a chromatographic analysis was made with the mass spectrometer focussed on m/e 58. This is the base peak which is common to all choline esters in Table IV. The sensitivity was such that 5 pmoles could readily be detected, and it was possible to conclude that propionylcholine and butyrylcholine, if present at all, did not exist in fresh rat brain in this quantity.

Mass spectrometry offers unexcelled opportunities as an extremely specific and sensitive method of detection when used in conjunction with a gas chromatograph. Although at the moment variable losses in the molecular separator and ion source limit its value as a quantitative instrument, these also can probably be compensated for by the use of authentic reference compounds labeled with stable isotopes as internal standards. When employed for the analysis of choline esters in tissues, these techniques should exceed the sensitivity of the most refined bioassays while retaining a high order of specificity.

VI. RECOMMENDED PROCEDURES

1. Fluorimetric Method (67)

Reagents. *1. Potassium periodide solution.* Shake 0.5 g potassium iodide and 1.0 g iodine with 10 ml water at room temperature for at least 1 hr. Remove excess iodine by centrifuging.

2. Tetramethylammonium bromide solution (10 mM). Dissolve 15.4 mg of tetramethylammonium bromide in 10 ml water. Keep in a refrigerator for no more than a week.

3. Hydrazine solution (2M). Make up 0.64 ml redistilled hydrazine to 10 ml with water.

4. Salicylaldehyde solution (4%). Dissolve 0.2 ml salicylaldehyde in 5 ml dimethylformamide (spectroquality), and store in a dark bottle.

5. Potassium borohydride solution (0.2M). Dissolve 108 mg potassium borohydride in 10 ml dimethylformamide. Centrifuge and store the supernatant for no more than a week.

6. Biorex 70 ion exchange resin (50–100 mesh). Wash repeatedly with distilled water. Recycle three times between pH 10 (NaOH) and pH 1 (HCl), rinsing 3–4 times with water after each step. Adjust to pH 8 with 0.5N NaOH, wash in a large column with 3M NaCl (4 liters of solution on 1 lb resin) and finally rinse with distilled water until the effluent is free of Cl⁻ (silver nitrate test). Prepare columns 7 × 0.7 cm.

7. Hydrochloric acid, 0.07N.

8. Sodium hydroxide, 2N.

Procedure. To a 10 ml sample of aqueous solution, add 0.1 ml Reagent *2* and 0.4 ml Reagent *1* in that order, mixing after each addition. Allow to stand at 0°C for 20 min, centrifuge at 0°C for 15 min at 1000g, and remove the supernatant. Dissolve the precipitate in 5 ml ether, add 5 ml water, and shake. Discard the ether layer and extract three times with 5 ml ether to remove the remaining iodine; warm to 100°C for 10 min to drive off residual ether and pour the sample on to the resin column. Rinse with 20 ml water, being careful not to disturb the top of the column. When just dry, add 0.1 ml Reagent *3* to the column and wait precisely 2 min. Add 2.5 ml Reagent *7* to the column, discard the first 0.5 ml, and collect 2 ml of eluate. Add 0.1 ml Reagent *4* to the eluate and place in a water bath at 37°C for 30 min. Add 0.1 ml Reagent *5* and wait 8 min. Add 0.1 ml Reagent *8*, mix, and read in an Aminco-Bowman spectro-photofluorometer at 370 nm/475 nm. The result should be compared to a blank and acetylcholine standards in the appropriate concentration range, which must be run with each group of determinations.

The sensitivity of this method is about 200 pmoles, and is limited by the reagent and tissue blank. Since recovery from brain extracts is not quantitative, it is most useful for estimations of acetylcholine in tissue perfusates. The principle and limitations of the method are discussed above (Section V-2-A).

2. Gas Chromatography (25,106)

Reagents. *1. Hexyltrimethylammonium bromide* (10 mM). Dissolve 224 mg hexyltrimethylammonium bromide in 100 ml water. Store in the refrigerator.

2. Extracting medium. To 15 ml 1N formic acid in water, add 50 μl Reagent *1* and make up to 100 ml with acetone.

3. Ammonium reineckate solution. Shake 200 mg Reinecke salt in 10 ml ice cold water for 45 min. Filter. Make up this solution immediately before use.

4. Tetraethylammonium chloride solution (10 mM). Dissolve 166 mg tetraethylammonium chloride in 100 ml water. Store in the refrigerator.

5. Biorex 9 suspension. Stir 50 g Biorex 9 (chloride form) in 250 ml absolute methanol for 20 min. Allow to settle and decant supernatant. Repeat three times. Filter and transfer resin to an open dish in a desiccator over magnesium perchlorate. Dry overnight under vacuum, and store in a desiccator. Weigh out 300 mg and suspend in 10 ml methanol redistilled from sodium. Store the suspension for no more than 1 month, tightly stoppered.

6. Butanone, redistilled. To 500 ml reagent grade butanone, add 50 g molecular sieve, Linde Type 3A, $\frac{1}{8}$ in. and 10 g anhydrous potassium carbonate. Allow to stand for 30 min, shaking occasionally. Distill slowly using a 40-cm glass helices column, discarding the first 50 ml, until 300 ml is collected.

7. Sodium benzenethiolate. Dissolve 33 g (0.3 mole) redistilled thiophenol in 150 ml absolute methanol, and cool in an ice bath. Dissolve 4.6 g (0.2 mole) sodium metal under a stream of nitrogen, adding the metal in small pieces. Remove from the ice bath, add 300 ml toluene and distill slowly, replacing toluene as necessary to maintain the volume at about 400 ml until a total of 600 ml has been added. Sodium benzenethiolate will crystallize out in glistening plates as the methanol is distilled. Filter the suspension under dry nitrogen while still hot, and wash with boiling toluene. Transfer to an open dish and dry over magnesium perchlorate in a desiccator under vacuum. Store in a desiccator. This reagent slowly deteriorates over a period of 1–2 years after preparation.

8. Demethylation reagent. This is made up immediately before use. Add 5 ml Reagent *6* to 7.5 mg glacial acetic acid and mix. Add this solution to a tube containing 50 mg Reagent *7*, displace the air from the tube with a stream of dry nitrogen, stopper tightly, and shake to dissolve. There is some residual turbidity due to the sodium acetate formed, but this can be ignored.

9. Citric acid solution, 0.5M. Dissolve 9.6 g citric acid in 100 ml water. Store in the refrigerator for no more than 1 month.

10. Chloroform. Shake 200 ml spectroquality chloroform with an equal volume of $2N$ NH_4OH until both phases are clear. Remove the aqueous phase in a separatory funnel and wash the chloroform twice with an equal volume of $2N$ H_2SO_4 and three times with distilled water. Add 1% absolute ethanol to the chloroform and store in a dark bottle.

11. Ammonium citrate (2M)/ammonium hydroxide (7.5M) buffer. Dissolve 45.2 g ammonium citrate $[(NH_4)_2HC_6H_5O_7]$ in 20 ml water. Add slowly 50 ml $15N$ ammonium hydroxide, stirring, and make up to 100 ml with water.

Procedure. The procedure to be described has been used primarily for rat brains. For smaller structures the volumes can be appropriately scaled down throughout. When scaled down ten times it is convenient to run the demethylation reaction and subsequent extraction steps in disposable tubes, made from 12 cm lengths of Corning #234040 borosilicate glass tubing (2.4 mm i.d.) sealed at one end. The other end is sealed during the reaction and cut off to allow extraction. The entire chloroform phase from the last step can then be injected into the gas chromatograph. Scaling in the following description is appropriate for 1–2 g of tissue containing 1–50 nmoles of acetylcholine.

Excise the tissue as rapidly as possible, drop it into liquid nitrogen, and pulverize it in a stainless steel mortar at the same temperature. Transfer the powder to a cold tared Potter-Elvejhem homogenizer tube, weigh it, and add precisely 4 ml Reagent *2* at 0°C. Homogenize and keep at 0°C for 30 min. Transfer to a centrifuge tube and centrifuge at $36,000g$ for 30 min at 0°C. Decant the supernatant solution into a graduated 12 ml conical centrifuge tube and extract with an approximately equal volume of ether. Evaporate the remaining ether and acetone with a stream of dry nitrogen, leaving a volume of about 1 ml of aqueous solution.

Add 0.1 ml Reagent *4* and a volume of Reagent *3* equal to the total volume of solution. Mix thoroughly after each addition, and leave at 0°C for 45 min for the reineckates to precipitate. Centrifuge (International Clinical Centrifuge Model CL, used for this and subsequent steps), discard the supernatant, and dry the precipitate in a desiccator under vacuum. Add 0.5 ml Reagent *5*, triturate for a few minutes, and centrifuge. Transfer the supernatant to a clean 12 ml conical centrifuge tube and evaporate the methanol with a stream of dry nitrogen.

Add 0.5 ml Reagent *8*, displace air from the tube with nitrogen, cap tightly using a Teflon cap liner, and place in a water bath at 80°C for 45 min, shaking every 5–10 min.

Cool, add 0.1 ml Reagent 9 and 2 ml pentane (spectroquality).   Shake, centrifuge, and discard upper phase.   Wash the aqueous (lower) phase twice with 1 ml pentane to remove traces of benzenethiol and methyl phenyl sulfide.   Evaporate the remaining traces of pentane with a stream of nitrogen.   Add 50 μl Reagent 10, 0.1 ml Reagent 11, shake, and centrifuge.   Withdraw 5–10 μl from the lower (chloroform) phase with a Hamilton syringe (#701N) and inject into the gas chromatograph.

Gas Chromatography.   A gas chromatograph equipped with flame ioniza-tion detector is required, and the column must be silanized to avoid tailing of peaks and losses on column.   Satisfactory resolution is obtained with a 6 ft × $\frac{1}{4}$ in. column packed with 1% phenyldiethanolamine succinate (PDEAS) on Polypak 1 (= Par 1).   The analysis is run iso-thermally at a temperature of 150–175°C, depending on the individual column, with a nitrogen flow rate of 45 ml/min.   The flow characteristics of Polypak can be improved by mixing it with GasChrom Q in a 3:1 ratio.

Higher efficiency was obtained with the PDEAS replaced by a new liquid phase which was described as bis-(succinimidoethyl)-dodecylamine (5,25).   Subsequent work has indicated that this is a linear polymer of dodecyldiethylene triamine (Eastman #10464) and succinic acid.   It is best prepared by refluxing equimolar amounts of dodecyldiethyl-enetriamine and dimethyl succinate in toluene, followed by removal of the methanol formed in the reaction by distillation of the solvent.   The same material gives good resolution when coated (2%) with OV 101 (5%), on GasChrom Q at a column temperature of 100°C.

The quantity of acetylcholine, propionylcholine, or butyrylcholine in the samples is deduced from the ratio of the height of the corresponding peak to the height of the peak due to the internal standard, hexyltri-methylammonium (Fig. 2).   Although fluctuations in oven temperature and aging of the column may cause changes in the peak heights, the peak height ratios are constant for a given column when the relative molar quantities are the same.   Standards are run at weekly intervals, in which 0.1 or 1 ml samples of aqueous solutions containing hexyltrimethyl-ammonium and acetylcholine in 0.1M ammonium acetate buffer at pH 4.0 are analyzed from the reineckate precipitation step.   The hexyltrimethyl-ammonium concentration is fixed at 25 μM, and the acetylcholine is 25 and 50 μM.   The peak height ratio is linearly related to the mole ratio, and the calibration factor rarely changes by more than 1–2% from week to week.

The principle on which this procedure is based has been discussed above (Section IV-3).   It is somewhat complex but gives reliable results in the

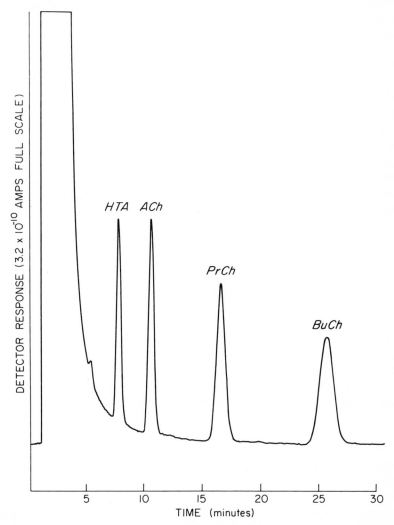

Fig. 2. Gas chromatogram of N-demethyl derivates of hexyltrimethylammonium (HTA), acetylcholine (ACh), propionylcholine (PrCh), and butyrylcholine (BuCh). Amounts were all 2.5 nM except HTA, which was 1.25 nM. F&M 5750A gas chromatograph with flame ionization detector. Silanized glass column, 6 ft $\times \frac{1}{4}$ in. o.d., packed with Polypak I and GasChrom Q (3:1) coated with 1% dodecyldiethylenetriamine/succinamide polymer (see text). Column temperature, 160°C; injection port and detector temperatures, 200°C; N_2 flow 45 ml/min; oxygen and hydrogen 185 and 30 ml/min, respectively.

range of 1–50 nM/g of tissue, with a detection limit of about 50 pmoles when scaled down to allow injection of the whole sample, and a standard deviation of the same amount in replicate measurements on larger samples. The very high resolution afforded by gas chromatography makes it quite specific, and both the sensitivity and specificity can be enhanced by using specialized detectors, of which the mass spectrometer offers most promise (see Section V-2-c).

3. Biological Assay: Guinea Pig Ileum

Reagents. *1. Morphine sulfate solution* (15 mM). Dissolve 250 mg morphine sulfate in water and make up to 50 ml.

2. Pyrilamine maleate solution (1 mM). Dissolve 40 mg pyrilamine maleate in water and make up to 100 ml.

3. Krebs solution with morphine sulfate (15 μM) *and pyrilamine maleate* (1 μM). Stock Krebs solution has the following composition:

NaCl	7.17 g/l
KCl	0.356 g/l
$Na_2HPO_4 \cdot 2H_2O$	2.32 g/l
$NaH_2PO_4 \cdot H_2O$	0.449 g/l
$CaCl_2$ (anhydrous)	0.284 g/l
$MgSO_4 \cdot 7H_2O$	0.294 g/l

In order to avoid precipitation, it is best to dissolve all the salts except magnesium sulfate and calcium chloride in about 800 ml water, then dissolve the last two compounds separately and mix the aqueous solutions before making up to a liter. This solution may be kept 1–2 weeks without refrigeration. Immediately before use, add 2 g glucose, 1 ml Reagent *1*, and 1 ml Reagent *2* to each liter of stock Krebs solution.

4. Mipafox solution. Dissolve 1.0 mg mipafox [bis-(isopropylamino-fluorophosphine oxide] in Krebs solution and make up to 100 ml. Use immediately.

5. Acetylcholine standard solutions. Acetylcholine perchlorate is a convenient salt to use as a primary standard because unlike halides it is not hygroscopic (147). Make up a 10 mM solution of acetylcholine in water by dissolving 181 mg of acetylcholine chloride or 245 mg acetylcholine perchlorate in water and making up to 100 ml. Store in the refrigerator for no more than 1 week. Make subdilutions of this solution immediately before use in 0.154M sodium chloride (9 g/l). High dilutions are best prepared in silanized glassware to avoid errors due to adsorption on glass surfaces. Concentrations of the acetylcholine standards should be a geometric series (1.0, 1.6, 2.5, 4.0, 6.3, 10, 16, and

25 nM) since the response of the tissue in the working range is linear with log (dose). Depending on the sensitivity of the individual preparation, this series may have to be extended downward, but it is unlikely that higher concentrations will be required.

Procedure. Healthy adult guinea pigs of either sex weighing 600–800 g should be used. Stun and exsanguinate the guinea pig and remove 10–15 cm of terminal ileum. Transfer to a dish containing Krebs solution and remove the contents and attached mesentery. Take a section 2–3 cm long and incubate in 100 ml Reagent 4 for 1 hr at 37°C, with oxygen bubbling through the solution. The remainder of the segment can be stored in the refrigerator immersed in Krebs solution for use the following day. At the end of the 1 hr incubation wash the strip of ileum with Krebs solution, and arrange in a tissue bath at 30°C containing 5 ml oxygenated Krebs solution for isotonic recording with a transducer and potentiometric recorder or a kymograph. Leave undisturbed for 30 min before proceeding.

Test the responses to the standard solutions by adding 0.5 ml, leaving for 15 sec, and replacing with fresh Krebs solution. Replace the solution again after 1 min, and leave for 1 min more before adding the next standard. Determine the standard concentrations which consistently produce between 10 and 70% of the maximum contraction. Dilute the sample to be assayed if necessary so that the addition of 0.5 ml to the bath produces a response in the same range. Then choose two standard concentrations which produce a smaller (S_1) and larger (S_2) response than the unknown and obtain a series of responses in the order $S_1 U S_2$ $S_2 U S_1$. The concentration of the unknown is then estimated by linear interpolation on log dose, i.e.,

$$\log_{10} X_u = \log_{10} X_1 + 0.2 \frac{Y_u - Y_1}{Y_2 - Y_1}$$

where X is the concentration, Y is the mean of the two responses to it, and subscripts u, 1, and 2 refer to U_1, S_1, and S_2, respectively. More complex assay designs can be used, details of which are provided in standard texts on bioassay (148–150).

The procedure described is sufficiently accurate for most purposes and minimizes the quantity of sample needed for the assay. It is generally possible to assay 5 pmoles with an accuracy of 10–20%, and it is not unusual to obtain ten times this sensitivity. The preparation can be used repeatedly for a series of assays, each of which can be completed in about 20 min once its sensitivity range is established. However, speed and sensitivity are achieved at the cost of serious limitations in terms of specificity. Positive responses are obtained to other choline esters, although acetylcholine is the most active by a factor of 50–100, to 5-hydroxytryptamine and to various polypeptides which may be present in tissue extracts. Drugs which are present in the extract may interfere with the assay. The specificity may be enhanced by various manoeuvers (see Sections III and V-1). If the sample to be assayed is a tissue extract rather than a perfusate, sensitizing material may be present in it. This effect should be allowed for by making up the final acetylcholine standard solutions S_1 and S_2 in the extract itself after this has been heated (100°C) for 1 min at pH 10 (NaOH) and neutralized (113).

For these reasons the guinea pig ileum assay is recommended only when the sensitivity required exceeds that available with chemical techniques, or when the speed and sensitivity justify the loss of specificity and accuracy in a particular application.

If mipafox is not available, it may be prepared as follows (151): Add isopropylamine (42.9 g) in chloroform (50 ml) slowly to phosphorus oxychloride (28.6 g) in chloroform (100 ml), at −10°C, with stirring. Allow the temperature to rise to 0°C, and add a solution of potassium fluoride (29 g) and potassium carbonate (0.5 g) in water (29 ml), with stirring. Raise the temperature to 40°C for 30 min, remove the chloroform layer, and evaporate in vacuum until the product begins to crystallize. Add 300 ml light petroleum (b.p. 40–60°C) and reflux to dissolve. On cooling, crystals of bis-(isopropylamino)-fluorophosphine oxide, m.p. 65°C, are obtained (19 g).

Acknowledgment

The valuable assistance of Miss Julie A. Kuenzel in compiling a bibliography is gratefully acknowledged.

References

1. H. H. Dale and H. W. Dudley, *J. Physiol.* (London), *68*, 97 (1929).
2. A. J. Ewins, *Biochem. J.*, *8*, 44 (1914).
3. A. J. D. Friesen, J. W. Kemp, and D. M. Woodbury, *Science*, *145*, 157 (1964).
4. A. J. D. Friesen, J. W. Kemp, and D. M. Woodbury, *J. Pharmacol. Exp. Ther.*, *148*, 312 (1965).
5. C.-G. Hammar, I. Hanin, B. Holmstedt, R. J. Kitz, D. J. Jenden, and B. Karlén, *Nature*, *220*, 915 (1968).
6. L. W. Chakrin and F. E. Shideman, *Int. J. Neuropharmacol.*, *7*, 337 (1968).
7. C. O. Hebb, G. M. Ling, E. G. McGeer, P. L. McGeer, and D. Perkins, *Nature*, *204*, 1309 (1964).
8. E. T. Browning, M. A. Abdel-Rahman, and M. P. Schulman, *Pharmacologist*, *8*, 214 (1966).
9. J. Schuberth, B. Sparf, and A. Sundwall, *J. Neurochem.*, *16*, 695 (1969).
10. E. Stedman and E. Stedman, *Biochem. J.*, *31*, 817 (1937).
11. D. Henschler, *Hoppe-Seyler's Z. Physiol. Chem.*, *305*, 97 (1956).

12. E. A. Hosein, P. Proulx, and R. Ara, *Biochem. J.*, *83*, 341 (1962).
13. J. Banister, V. P. Whittaker, and S. Wijesundera, *J. Physiol.* (London), *121*, 55 (1953).
14. V. P. Whittaker, *Biochem. Pharmacol.*, *1*, 342 (1959).
15. V. P. Whittaker, *Biochem. J.*, *71*, 32 (1959).
16. V. Erspamer and O. Benati, *Biochem. Z.*, *324*, 66 (1953).
17. K. Kuriaki, T. Yakushiji, T. Noro, T. Shimizu, and S. Saji, *Nature*, *181*, 1336 (1958).
18. H. Kewitz, *Arch. Exp. Pathol. Pharmakol.*, *237*, 308 (1959).
19. E. A. Hosein and R. Ara, *J. Pharmacol. Exp. Ther.*, *135*, 230 (1962).
20. E. A. Hosein and A. Orzeck, *Nature*, *210*, 731 (1966).
21. E. A. Hosein and P. Proulx, *Nature*, *187*, 321 (1960).
22. G. Pepeu, K. F. Schmidt, and N. J. Giarman, *Biochem. Pharmacol.*, *12*, 385 (1963).
23. J. Crossland and P. H. Redfern, *Life Sci.* (Oxford), *10*, 711 (1963).
24. C. Hebb and D. Morris, *Nature*, *214*, 284 (1967).
25. I. Hanin and D. J. Jenden, *Biochem. Pharmacol.*, *18*, 837 (1969).
26. W. B. Stavinoha and L. C. Ryan, *J. Pharmacol. Exp. Ther.*, *150*, 231 (1965).
27. R. W. Ryall, N. Stone and J. C. Watkins, *J. Neurochem.*, *11*, 621 (1964).
28. G. A. R. Johnston, H. J. Lloyd and N. Stone, *J. Neurochem.*, *15*, 361 (1968).
29. F. C. MacIntosh and W. L. M. Perry, *Methods Med. Res.*, *3*, 78 (1950).
30. J. Crossland, *Methods Med. Res.*, *9*, 125 (1961).
31. V. P. Whittaker, in *Handbuch der Experimentellen Pharmakologie*, Vol. 15, D. Eichler and A. Farah, Eds., *Cholinesterases and Anticholinesterase Agents*, G. B. Koelle, Sub-Ed., Springer-Verlag, Berlin, 1963, p. 1.
32. C. O. Hebb and K. Krnjević, in *Neurochemistry*, 2nd ed., K. A. C. Elliott, Ed., Thomas, Springfield, 1962, p. 452.
33. I. Hanin, *Advan. Biochem. Psychopharmacol.*, *1*, 111 (1969).
34. R. Takahashi and M. H. Aprison, *J. Neurochem.*, *11*, 887 (1964).
35. W. B. Stavinoha, B. R. Endecott, and L. C. Ryan, *Pharmacologist*, *9*, 252 (1967).
36. L. B. Campbell, I. Hanin, and D. J. Jenden, *Biochem. Pharmacol.*, 19,2053 (1970).
37. I. Hanin, R. Massarelli, and E. Costa, *Physiologist*, *12*, 246 (1969).
38. K. A. C. Elliott and N. Henderson, *Amer. J. Physiol.*, *165*, 365 (1951).
39. J. Crossland, *J. Physiol.* (London), *114*, 318 (1951).
40. J. Crossland, H. M. Pappius, and K. A. C. Elliott, *Amer. J. Physiol.*, *183*, 27 (1955).
41. J. Butterworth, D. D. Eley, and G. S. Stone, *Biochem. J.*, *53*, 30 (1953).
42. L. E. Tammelin, *Svensk Kem. Tidskr.*, *70*, 157 (1958).
43. C. Bianchi, L. Beani, and A. Bolleti, *Experientia*, *22*, 596 (1966).
44. L. L. Simpson, F. de Balbian-Verster, and J. T. Tapp., *Exp. Neurol.*, *19*, 199 (1967).
45. L. Beani and C. Bianchi, *J. Pharm. Pharmacol.*, *15*, 281 (1963).
46. V. P. Whittaker, *Biochem. J.*, *72*, 694 (1959).
47. B. Holmstedt, G. Lundgren, and A. Sundwall, *Life Sci.* (Oxford), *10*, 731 (1963).
48. B. Holmstedt and G. Lundgren, in *Mechanisms of Release of Biogenic Amines*, U. S. von Euler, Ed., Pergamon, Oxford, 1966, p. 439.
49. I. Hanin and D. J. Jenden, *Experientia*, *22*, 537 (1966).
50. R. M. Marchbanks, *Biochem. J.*, *110*, 533 (1968).
51. C. W. Sheppard, W. E. Cohn, and P. J. Mathias, *Arch. Biochem. Biophys.*, *47*, 475 (1953).

214   DONALD J. JENDEN AND L. B. CAMPBELL

52. R. Mitchell and B. B. Clark, *Proc. Soc. Exp. Biol. Med.*, *81*, 105 (1952).
53. F. Fonnum, *Biochem. Pharmacol.*, *17*, 2503 (1968).
54. R. Mitchell, A. P. Truant, and B. B. Clark, *Proc. Soc. Exp. Biol. Med.*, *81*, 5 (1952).
55. J. K. Saelens and W. R. Stoll, *J. Pharmacol. Exp. Ther.*, *147*, 336 (1965).
56. J. Kapfhammer and C. Bischoff, *Hoppe-Seyler's Z. Physiol. Chem.*, *191*, 179 (1930).
57. M. J. Keyl, I. A. Michaelson and V. P. Whittaker, *J. Physiol.* (London), *139*, 434 (1957).
58. G. A. Bentley and F. H. Shaw, *J. Pharmacol. Exp. Ther.*, *106*, 193 (1952).
59. E. Strack and H. Schwaneberg, *Hoppe-Seyler's Z. Physiol. Chem.*, *245*, 11 (1936).
60. H. M. Bregoff, E. Roberts and C. C. Delwiche, *J. Biol. Chem.*, *205*, 565 (1953).
61. D. Glick, *J. Biochem.*, *156*, 643 (1944).
62. M. H. Aprison and P. Nathan, *Arch. Biochem. Biophys.*, *66*, 388 (1957).
63. F. J. Booth, *Biochem. J.*, *29*, 2064 (1935).
64. F. H. Shaw, *Biochem. J.*, *32*, 1002 (1938).
65. F. J. R. Beattie, *Biochem. J.*, *30*, 1554 (1936).
66. D. E. Schmidt, P. I. A. Szilagyi, D. L. Alkon, and J. P. Green, *Science, 165*, 1370 (1969).
67. J. H. Fellman, *J. Neurochem.*, *16*, 135 (1969).
68. R. E. McCaman, M. W. McCaman, and M. L. Stafford, *J. Biol. Chem.*, *241*, 930 (1966).
69. A. M. Goldberg, A. A. Kaita, and R. E. McCaman, *J. Neurochem.*, *16*, 823 (1969).
70. R. E. McCaman and J. M. Hunt, *J. Neurochem.*, *12*, 253 (1965).
71. P. I. A. Szilagyi, D. E. Schmidt, and J. P. Green, *Anal. Chem.*, *40*, 2009 (1968).
72. D. E. Schmidt, P. I. A. Szilagyi, and J. P. Green, *J. Chromatogr. Sci.*, *7*, 248 (1969).
73. G. A. R. Johnston, A. C. K. Triffett, and J. A. Wunderlich, *Anal. Chem.*, *40*, 1837 (1968).
74. J. S. Wall, D. D. Christianson, R. J. Dimler, and F. R. Senti, *Anal. Chem.*, *32*, 870 (1960).
75. P. Marquardt and G. Vogg, *Hoppe-Seyler's Z. Physiol. Chem.*, *291*, 143 (1952).
76. K.-B. Augustinsson and M. Grahn, *Acta Physiol. Scand.*, *32*, 174 (1954).
77. P. Raff and W. Brotz, *Z. Anal. Chem.*, *133*, 16 (1952).
78. D. Ackermann and H. Maner, *Hoppe-Seyler's Z. Physiol. Chem.*, *279*, 114 (1943).
79. W. E. Stone, *Arch. Biochem. Biophys.*, *59*, 193 (1955).
80. H. W. Dudley, *J. Chem. Soc.*, *1931*, 763.
81. J. E. Gardiner and V. P. Whittaker, *Biochem. J.*, *58*, 24 (1954).
82. B. K. Schrier and L. Shuster, *J. Neurochem.*, *14*, 977 (1967).
83. K.-B. Augustinsson and M. Grahn, *Acta Chem. Scand.*, *7*, 906 (1953).
84. V. P. Whittaker and S. Wijesundera, *Biochem. J.*, *51*, 348 (1952).
85. G. Brante, *Nature, 163*, 651 (1949).
86. M. Wallach, A. Goldberg, and F. E. Shideman, *Int. J. Neuropharmacol.*, *6*, 317 (1967).
87. G. J. Mannering, A. C. Dixon, N. V. Carroll, and O. B. Cope, *J. Lab. Clin. Med.*, *44*, 292 (1954).
88. L. M. Lewin and N. Marcus, *Anal. Biochem.*, *10*, 96 (1965).
89. L. T. Potter and W. Murphy, *Biochem. Pharmacol.*, *16*, 1386 (1967).
90. W. Chefurka and B. N. Smallman, *Can. J. Biochem. Physiol.*, *34*, 731 (1956).

91. D. Henschler, *Hoppe-Seyler's Z. Physiol. Chem.*, *305*, 34 (1956).
92. N. Frontali, *J. Insect Physiol.*, *1*, 319 (1958).
93. M. E. Feigenson and J. K. Saelens, *Biochem. Pharmacol.*, *18*, 1479 (1969).
94. W. B. Stavinoha, L. C. Ryan and E. L. Treat, *Life Sci.* (Oxford), *3*, 689 (1964).
95. J. R. Cooper, *Biochem. Pharmacol.*, *13*, 795 (1964).
96. W. D. Thomitzek and E. Strack, *Acta Biol. Med. Germ.*, *13*, 447 (1964).
97. J. H. Fellman and T. S. Fujita, *Biochem. Biophys. Acta*, *97*, 590 (1965).
98. M. F. Cranmer, *Life Sci.* (Oxford), *7*, 995 (1968).
99. J. F. Howes, L. S. Harris, W. L. Dewey, and C. A. Voyda, *J. Pharmacol. Exp. Ther.*, *169*, 23 (1969).
100. I. H. Stockley, *J. Pharm. Pharmacol.*, *21*, 302 (1969).
101. W. Feldberg, *J. Physiol.* (London), *103*, 367 (1945).
102. M. Shamma, N. C. Deno, and J. F. Remar, *Tetrahedron Lett.*, *13*, 1375 (1966).
103. J. C. Sheehan and G. D. Davies, *J. Org. Chem.*, *29*, 2006 (1964).
104. W. R. Vaughan and J. Banman, *J. Org. Chem.*, *27*, 739 (1962).
105. D. J. Jenden, S. I. Lamb, and I. Hanin, *Fed. Proc.*, *26*, 296 (1967).
106. D. J. Jenden, I. Hanin and S. I. Lamb, *Anal. Chem.*, *40*, 125 (1968).
107. I. Hanin, Doctoral Dissertation, University of California (University Microfilms No. 68-11, 871), 1968.
108. P. I. A. Szilagyi, D. Schmidt, and J. P. Green, *Fed. Proc.*, *27*, 471 (1968).
109. H. C. Chang and J. H. Gaddum, *J. Physiol.* (London), *79*, 255 (1933).
110. L. C. Blaber and A. W. Cuthbert, *J. Pharm. Pharmacol.*, *13*, 445 (1961).
111. A. T. Birmingham, *J. Pharm. Pharmacol.*, *13*, 510 (1961).
112. M. Toru, J. N. Hingtgen, and M. H. Aprison, *Life Sci.* (Oxford), *5*, 181 (1966).
113. W. Feldberg and C. Hebb, *J. Physiol.* (London), *106*, 8 (1947).
114. S. Ellis, F. L. Planchte, and O. H. Straus, *J. Pharmacol. Exp. Ther.*, *79*, 295 (1944).
115. H. C. Chang, T.-M. Lin, and T.-Y. Lin, *Chin. J. Physiol.*, *17*, 145 (1949).
116. J. E. Davis, *Amer. J. Physiol.*, *162*, 616 (1950).
117. J. Morley and M. Schachter, *J. Physiol.* (London), *157*, 1P (1961).
118. H. Kalant and W. Grose, *J. Pharmacol. Exp. Ther.*, *158*, 386 (1967).
119. A. Ahmed and N. R. W. Taylor, *J. Pharm. Pharmacol.*, *9*, 536 (1957).
120. W. D. M. Paton, *Brit. J. Pharmacol.*, *11*, 119 (1957).
121. W. Schaumann, *Brit. J. Pharmacol.*, *12*, 115 (1957).
122. E. S. Johnson, *J. Pharm. Pharmacol.*, *15*, 69 (1963).
123. C. C. Chang, H. C. Cheng, and T. F. Chen, *Japan J. Physiol.*, *17*, 505 (1967).
124. C. R. Diniz and J. M. Torres, *Toxicon*, *5*, 277 (1968).
125. T. Forrester, *J. Physiol.* (London), *187*, 12P (1966).
126. M. F. Murnaghan, *Nature*, *182*, 317 (1958).
127. J. H. Gaddum and R. P. Stephenson, *Brit. J. Pharmacol.*, *13*, 493 (1958).
128. J. C. Szerb, *J. Physiol.* (London), *158*, 8P (1961).
129. J. H. Welsh, *J. Neurophysiol.*, *6*, 329 (1943).
130. B. Hughes, *Brit. J. Pharmacol.*, *10*, 36 (1955).
131. R. J. Ladd and G. D. Thorburn, *Aust. J. Exp. Biol. Med. Sci.*, *33*, 207 (1955).
132. E. Florey, *Comp. Biochem. Physiol.*, *20*, 365 (1967).
133. R. E. Loveland, *Comp. Biochem. Physiol.*, *9*, 95 (1963).
134. J. H. Welsh, *Nature*, *173*, 955 (1954).
135. J. R. Steinhart, *Biochim. Biophys. Acta*, *158*, 171 (1968).
136. S. Hestrin, *J. Biol. Chem.*, *180*, 249 (1949).
137. E. Heilbrounn, *Acta Chem. Scand.*, *10*, 337 (1956).

138. R. P. MacDonald, C. Gerber, and A. Nielsen, *Amer. J. Clin. Pathol.*, *25*, 1367 (1955).
139. A. F. Maslova, *Vop. Med. Khim.*, *10*, 311 (1964).
140. A. F. Maslova, *Fed. Proc.*, *24*, 548 (1965).
141. W. Feldberg, *J. Physiol.* (London), *101*, 432 (1943).
142. J. H. Fellman and T. S. Fujita, *Nature*, *211*, 848 (1966).
143. J. C. Smith and J. K. Saelens, *Fed. Proc.*, *26*, 296 (1967).
144. J. Schuberth, *Biochim. Biophys. Acta*, *122*, 470 (1966).
145. R. Ryhage, *Anal. Chem.*, *36*, 759 (1964).
146. C.-G. Hammar, B. Holmstedt, and R. Ryhage, *Anal. Biochem.*, *25*, 532 (1968).
147. F. K. Bell and C. J. Carr, *J. Amer. Pharm. Ass.*, *36*, 272 (1947).
148. C. I. Bliss, *The Statistics of Bioassay*, Academic Press, New York, 1952.
149. D. J. Finney, *Statistical Method in Biological Assay*, 2nd ed., Haffner, New York, 1964.
150. J. H. Burn, *Biological Standardization*, 2nd ed., Oxford Univ. Press, London, 1950.
151. D. F. Heath, *J. Chem. Soc.*, *1956*, 3796.
152. M. Toru and M. H. Aprison, *J. Neurochem.*, *13*, 1533 (1966).
153. O. Galehr and F. Plattner, *Pflügers Arch.*, *218*, 506 (1928).
154. S. E. Lewis and B. N. Smallman, *J. Physiol.* (London), *134*, 241 (1956).
155. D. Morris, G. Bull and C. O. Hebb, *Nature*, *207*, 1295 (1965).
156. W. E. Stone, *Arch. Biochem. Biophys.*, *59*, 181 (1955).
157. von E. Hofmann, *Helv. Chim. Acta*, *13*, 138 (1930).

Determination of Activity of Cholinesterases

KLAS-BERTIL AUGUSTINSSON,

Biochemical Institute, University of Stockholm, Sweden

I. INTRODUCTION

Cholinesterases (ChE) constitute a group of esterases which hydrolyze choline esters at a higher rate than other esters, provided the hydrolysis rates are compared at optimum and controlled conditions regarding substrate concentration, pH, ionic strength, temperature, etc. All esterases of this specificity, with a few exceptions among lower vertebrates and invertebrates, are inhibited by 10 μM eserine and are also much more sensitive to quaternary ammonium salts than are other types of esterases.

This affinity of ChE for cationic substrates and inhibitors is the most characteristic feature of these enzymes and suggests that the active center of ChE, in contrast to other esterases, includes a negatively charged group ("anionic site") or alternatively a van der Waals center in addition to the ester-binding group ("esteratic site"). The existence of a second site in the active surface in ChE seems to be the structural characteristic of these enzymes, and determines in many cases the action specificity by binding and orienting the substrate. The mechanism of action of the second "nonesteratic" site in the formation of enzyme-substrate or enzyme-inhibitor complexes differs among various types of ChE.

In *acetylcholinesterases* (AChE, EC 3.1.1.7), which have acetylcholine as their natural substrate, this site is anionic in character and is one of the features which distinguishes these enzymes from other types of *cholinesterases* (EC 3.1.1.8). In these esterases the dominant type of force involved in the reactions of substrates and inhibitors with the second site are van der Waals forces (1). The structure of the esteratic site, on the other hand, does not seem to be much different from that in other esterases, especially those which are sensitive to organophosphorus compounds (2). Consequently, cholinesterases including the acetyl-cholinesterases can hydrolyze *a priori* an ester, irrespective of the presence of a positively charged or alkylated group in the substrate molecule. The esteratic site primarily determines the substrate specificity and the catalytic function, particularly regarding the acid radical of the ester. Referring to the substrates hydrolyzed at the highest rate, one can distinguish between butyrylcholinesterases (BuChE), propionylcholines-terases (PrChE), benzoylcholinesterases (EC 3.1.1.9), etc. Such terms carry no implications as to their physiological substrates, which are still unknown for these enzymes.

Vertebrate blood sera generally contain multiple forms of cholinesterase, and evidence for the electrophoretic heterogeneity of the enzyme in human serum has been demonstrated by several workers (3,4). In most instances, the observations with human serum indicated the presence of three or four zones of cholinesterase activity. These isoenzymes have been the subject of intensive studies during the last years from genetic and pharmacogenetic points of view. They differ in regard to sensitivity to dibucaine or NaF and seem to be genetically determined. Further studies have shown additional esterase components, altogether six or seven are present in certain sera, demonstrated by both one- and two-dimensional electrophoresis. These aspects of the polymorphism of serum ChE is discussed in two excellent books by Goedde et al. (5) and by Latner and Skillen (4). The methods used for selective determination of the variants of the serum ChE are mentioned below.

TABLE I

Principles of Various Assay Methods for Cholinesterases
Using a Choline or Thiocholine Ester[a]

1. Reaction with hydroxylamine—$FeCl_3$ (Hestrin) (Section VI-1)
2. Schönemann (Section VI-2)
3. Spectrophotometric; benzoylcholine as substrate (Kalow and Lindsay) (Section VI-3)
4. Radioisotopic (^{14}C-carboxy-ACh) (Section VII-1)

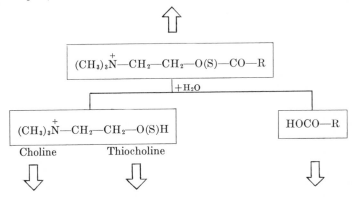

$$(CH_3)_3\overset{+}{N}—CH_2—CH_2—O(S)—CO—R$$

$+H_2O$

$$(CH_3)_3\overset{+}{N}—CH_2—CH_2—O(S)H \qquad HOCO—R$$

Choline   Thiocholine

1. Microbiological	1. Reaction with dithiobis-(2-nitrobenzoate) (Ellman) (Section V-2-B)	1. Gasometric (Warburg) (Section II-1)
		2. Change in pH (Michel and Acholest) (Section III-1)
2. *Neurospora* (Section V-1)	2. Polarographic (Section V-2-A)	3. Titrimetric (automatic) (Section IV)
	3. Iodometric titration (Section V)	4. Radioisotopic (^{14}C- or ^{3}H-acetyl) (Section VII-2)
	4. Nitroprusside-Na (Section V)	

[a] The use of noncholine esters is discussed in Section VIII.

It would seem clear from the above that a choline (or thiocholine) ester is the choice of substrate in quantitative determination of either type of ChE; the principles of the methods used are summarized in Table I. A number of noncholine esters, however, have been introduced, particularly to make possible the application of certain simple technical principles in ChE determination for screening tests assaying esterase inhibitors. Substrates frequently used in this connection are summarized in Table II. The main problem is to determine which esterases are involved in the catalytic reaction and whether an ester is actually split by a ChE alone. Actually, a number of esterase types may be concerned

TABLE II

Substrates Recommended and Frequently Used in Cholinesterase Determinations[a]

Ester substrate	Active esterases	Choice of method
Choline esters	Eserine resistant esterases do not split these esters, with rare exceptions	Great variety of choice of method: manometric, colorimetric, or pH methods
Acetylcholine useful salts: iodide, bromide, chloride (highly hygroscopic), perchlorate; radioactive: ^{14}C-acetyl and ^{3}H-acetyl	All types of cholinesterases, particularly AChE of the nervous system, and other conducting structures, erythrocytes and the blood plasma of certain lower vertebrates	Various
Propionylcholine useful salts: iodide, bromide, chloride (highly hygroscopic)	Various types of cholinesterases, particularly AChE and cholinesterases present in rat, several avian species, and frog	Various
n-Butyrylcholine useful salts: iodide, bromide, chloride (highly hygroscopic)	BuChE present in the blood plasma of mammals (e.g., human, dog, horse, cat); is generally not split by AChE, which is inhibited instead	Various
Succinyldicholine useful salt: iodide	Normal, but not the "atypical" type (or dibucaine-resistant esterase) of BuChE present in human blood plasma	Various
Benzoylcholine useful salt: chloride	A cholinesterase present in human and other mammalian blood plasmas and tissues. BuChE probably also active	Spectrophotometry at 240 nm where the ester absorbs but not its reaction products
Acetyl-β-methylcholine useful salt: iodide	AChE, generally not other types of cholinesterase	Great variety of choice of method
Thiocholine esters	Eserine-resistant esterases do not generally split these esters	
Acetylthiocholine useful salt: iodide	Generally the same specificity pattern as for those cholinesterases which split acetylcholine, but is usually split at a somewhat higher rate than acetylcholine. Certain eserine-resistant esterases can hydrolyze this ester at low rates	The thiocholine produced determined by the Ellman method, iodometrically, polarographically, or by other techniques. Frequently used for histochemical application

220

TABLE II (*Continued*)

Ester substrate	Active esterases	Choice of method
Butyrylthiocholine useful salt: iodide	Generally the same specificity pattern as for butyrylcholine	The thiocholine produced by various techniques. See acetylthiocholine
Phenyl esters Phenyl acetate	Arylesterases of mammalian tissues, but also hydrolyzed by both carboxylesterases, acetylesterases, cholinesterases, and AChE and a number of enzymes of unknown specificity	Various methods available, particularly those based on the determination of the liberated phenol
Nitrophenyl esters o- and p-Nitrophenyl acetate	A number of esterases belonging to various groups of hydrolases both resistant and sensitive to organophosphorus compounds; also proteinases (e.g., chymotrypsin, trypsin), acylase, and carbonic anhydrase	The production of the yellow-colored nitrophenylate ion is followed spectrophotometrically at 400 nm
Naphthyl esters α- or β-Naphthyl acetate α- or β-Naphthyl butyrate	Various types of esterases, particularly the carboxylesterases, serum cholinesterase (βAc > αBu, βBu > αAc), acetylcholinesterase (βAc only), acylase (βAc higher than the others), lipase (βAc highest rate), and phosphatase (βAc > αAc)	Naphthol produced is determined spectrophotometrically after coupling with a suitable diazonium salt, or fluorimetrically. Histochemical and tracing techniques for nonspecific esterases
Indoxyl esters Indoxyl acetate	Serum cholinesterases at a higher rate than acetylcholinesterases; carboxylesterases, and arylesterases. Also acylase, lipase, and acid phosphatase, but not chymotrypsin	The production of indigo blue is followed spectrophotometrically or that of indigo white fluorimetrically
N-Methylindoxyl butyrate	Human serum BuChE	Fluorimetric (λ_ex = 430 nm, λ_em = 501 nm)
Indophenyl esters Indophenyl acetate	Preferably AChE	The production of the blue-purple product is followed spectrophotometrically at 625 nm

ᵃ Enzymes listed: acetylcholinesterase (AChE; EC 3.1.1.7), cholinesterase (e.g., BuChE; EC 3.1.1.8), benzoylcholinesterase (EC 3.1.1.9), carboxylesterase (EC 3.1.1.1.), arylesterase (EC 3.1.1.2), lipase (EC 3.1.1.3), acetylesterase (EC 3.1.1.6), alkaline (EC 3.1.3.1) and acid phosphatase (EC 3.1.3.2), chymotrypsin (EC 3.4.4.5. and 6), acylase, carbonic anhydrase (EC 4.2.1.1).

when a noncholine ester is used as substrate.   They are briefly mentioned below (Section VIII).

ChE belong to a wider group of esterases which are readily inhibited by organophosphorus compounds.   Such esterases have been called B-esterases and include also the *carboxylesterases* (EC 3.1.1.1.; previously called aliesterases).   An essentially similar reaction mechanism is probably valid for these enzymes (cholinesterases and carboxylesterases); the esterase reacts with the ester to produce an intermediary acyl-enzyme complex, which can react with a variety of acyl acceptors, including water. The inhibition of these esterases by certain phosphoryl, carbamyl, and sulfonyl derivatives can be explained by an analogous mechanism (2,6,7).   The carboxylesterases differ from ChE in being unable to hydrolyze choline esters and in being resistant to 10 μM eserine.   Otherwise there seems to be a close relation between the two types of esterases, which are suggested to have a common phylogenetic origin (8).   A third group of hydrolases sensitive to organophosphorus compounds is constituted by the lipases (EC 3.1.1.3).

Esterases being resistant to organophosphorus compounds were designated A-esterases.   Like the carboxylesterases, they hydrolyze carboxylic esters other than choline esters; since aromatic esters are split at particularly high rates, these enzymes are now called *arylesterases* (EC 3.1.1.2) (9,10).   Some of them are probably responsible also for the hydrolysis of certain organophosphate esters (e.g., DFP, paraoxon) (11). The arylesterases, being typical SH-enzymes, are inhibited by chelating agents and heavy metal ions (e.g., Hg^{2+}).   A second type of esterase resistant to organophosphorus compounds and present in animal tissues has been designated as C-esterase (12).   This esterase differs from arylesterase in being resistant to chelating agents and Hg^{2+}.   As with arylesterase, it acts preferentially on acetic esters and is better called *acetylesterase* (EC 3.1.1.6).

Because of the widespread interest in ChE from various points of view, it is not surprising that numerous methods have been described for their determination (cf. Table I).   These methods were critically reviewed in 1957 in Vol. 5 of the *Methods of Biochemical Analysis* (13).   The present account of methods of assaying these enzymes is therefore restricted to the improvements of earlier methods and to the development of new principles in estimating ChE activity with reference to the usefulness of the methods under different conditions, among others in assaying certain esterase inhibitors with enzymatic methods, in determining the genetic variation of isoenzyme patterns, and in quantitative ultramicrochemical studies of enzyme distribution.   In the meantime, the subject has been critically reviewed by Witter in 1963 (14), who discussed the problems mainly common to the methods for assaying blood cholinesterases.   In addition, some brief reviews have recently been published (7,15–21), most of them being restricted to certain aspects and details of the methods discussed.

II. GASOMETRIC METHODS

The enzymatic hydrolysis of the ester proceeds in a bicarbonate-carbonic acid buffer in a closed vessel attached to a manometer.   The carbonic acid concentration is fixed by the pressure of CO_2 in an atmosphere of N_2 and CO_2.   The acid liberated reacts with the bicarbonate ion to form carbonic acid, which dissociates, and CO_2 is given off in equivalent amounts to the acid produced and estimated manometrically, usually at constant volume.

1. Warburg Technique

One of the most frequently used methods of assaying ChE activity is the Warburg manometric technique, which has been described in a large number of modifications (13).   A detailed discussion of the theory and operating practice is available in the book by Umbreit et al. (22), and details of the procedure for the determination of ChE in plasma, erythrocytes, and crude or purified enzyme preparations are given in the previous review (13).   The general method has not been improved in any important details and existing procedures have been employed, except in a few cases (23–25).

In most methods described, the medium contains not only bicarbonate (25 mM) but also Ca^{2+} or Mg^{2+} (1.5–50 mM) and Na^+ (100–150 mM) and/or K^+ (2 mM).   The presence of these extra ions in the reaction mixture is still a matter of controversy (26).   Simplified flasks and manometers (27) do not seem to be more convenient to use than the standard apparatus (13).   The total volume of the reaction mixture is generally 2–3 ml.   The enzyme is either placed in the main compartment of the flask or in the side bulb.

Among the many studies in which the Warburg technique has been used, the following may be mentioned: normal variation of ChE levels in plasma and erythrocytes in rats (26,28) and humans (29,30), in stored blood (12,25), in blood samples dried on filter paper (23); temperature dependence of serum ChE using various substrates (31); comparison of the kinetics of the hydrolysis of ACh and other choline esters by liver (32), human serum (33), and rat plasma and erythrocytes (34); AChE in erythrocytes of patients with paroxysmal nocturnal hemoglobinuria (35) and of rabbits in relation to erythropoiesis (36); effect *in vitro* and *in vivo* of organophosphates (37,38) and carbamates (24,39).   The results obtained with this technique agreed fairly satisfactorily when compared with those obtained with the Hestrin method (30,38), the Michel method (30), the titrimetric method (24,40), and the Acholest test (40–43). In studies of histochemical localization of ChE in kidney (44) and chick embryos (45), the Warburg technique was used as a complement for quantitative assay.

It is sometimes difficult to evaluate the ChE inhibiting effect of certain inhibitors (e.g., carbamates) because of the rapid reversibility of the ChE inhibitor complex in the presence of certain substrates. These difficulties are largely responsible for the inability to evaluate *in vitro* assays of *in vivo* inhibition, especially in nervous tissue, and correlate these values to the toxicologic and pharmacologic properties of a reversible esterase inhibitor. Baron et al. (24) therefore used an abbreviated Warburg technique to measure the maximal level of inhibition produced by carbamates before dissociation of the enzyme-inhibitor complex; a sequence of manometer readings was made at 30-sec intervals for 5 min after substrate introduction and the reaction velocity determined by the method of least squares programmed in a computer. Studies with this modified manometric technique showed that the intensity of the cholinergic effects following carbamate-induced intoxication correlated with the degree of brain AChE inhibition, and the toxic effects produced were due primarily to this inhibition.

The modification mentioned in the previous review (13), including the automatic recording devices and a simplified apparatus (27), have not been studied further.

The Warburg manometric technique is still one of the most accurate and reliable methods available of assaying ChE activity; the accuracy can be within $\pm 2\%$. The method is useful in a variety of experimental condition including those with inhibitors and activators, and is of great versatility since rate curves are obtained. One disadvantage of the technique is that the pH (7.2–7.7) of the medium cannot be varied. Moreover, the rate of hydrolysis cannot be studied at substrate concentrations below 0.4 mM because of insufficient gas liberated. Still another limitation is the difficulty of assaying reversible esterase inhibitors, as mentioned above.

2. Microgasometric (Diver) Techniques*

The Cartesian diver method, extensively investigated by Linderstrøm-Lang, Holter, and Zeuthen at the Biological Institute of the Carlsberg Foundation in Copenhagen, was applied for the first time to the assay

* I thank Drs. Ezio Giacobini and Miro Brzin for valuable comments to this section.

of ChE by Linderstrøm-Lang and Glick (46). The principle of this ultramicrogasometric technique was discussed in the middle of the seventeenth century when it was discovered in 1648 by R. Magiotti (47), a pupil of Galileo Galilei in Pisa. It is similar to that utilized in the Warburg technique, i.e., any change in the amount of gas contained in a vessel (the diver) requires a corresponding change in pressure to maintain the gas volume constant in order to keep the vessel floating in a medium at a fixed level. In other words, the pressure changes may be taken as a measurement of the changes in the amount of gas in the diver.

The diver technique (Table III) is a rather involved and time-consuming procedure, requiring repeated close manual adjustments by a trained person. This greatly limits the number of samples examined in one single experiment. A technical difficulty involved is mixing solutions inside the diver. For this purpose the laser beam technique has been tested. New types of diver techniques, such as the magnetic diver (see below), do not overcome these limitations which, however, have been minimized recently by the use of automatic recording devices. In addition, the instrumentation of this technique is rather costly.

The diver technique, however, is the most sensitive method available for the measurements of gaseous exchange in biological material. In addition, it is particularly useful for metabolic studies with living cells, for which few methods are known.

Ampulla Diver Technique. The diver floatation technique has been improved considerably during recent years, particularly by Zeuthen and his co-workers. Zajicek and Zeuthen (49,56) used a special "ampulla" diver for the quantitative determination of ChE in individual cells with a sensitivity of approximately $3 \times 10^{-4}\,\mu l\ CO_2$ per hour per single cell (megacaryocytes, sympathetic ganglion of rat) (50) in a reaction mixture of 0.2–0.3 μl (accuracy $\pm 5\%$). This ampulla technique was used with minor modifications by Giacobini in a series of investigations, which are summarized in two monographs (68,69), on the distribution and localization of ChE in single nerve cells isolated from sympathetic (50,51) and spinal ganglions (53), in single anterior horn cells (52), and in the cytoplasm and nucleoplasm of isolated nerve cells (54,68). In the latter study the diver technique was improved to measure $1 \times 10^{-6}\,\mu l\ CO_2\ hr^{-1}$ in pico-liter volumes of cellular material (54).

The ampulla diver technique was used by Brzin and Zeuthen (57) in following the hydrolysis of acetylthiocholin by single end plates from mouse gastrocnemius muscle in parallel runs measuring the precipitation of thiocholine, formed in equivalent amounts to acetic acid, by copper in a medium containing Cu-glycinate. The latter reaction is the basis of a frequently used histochemical method for ChE, introduced by Koelle

TABLE III
Studies on Cholinesterases Using the Cartesian Diver Technique or its Various Improvements

Author	Year	Type of technique used	Studies performed	Ref.
Linderstrøm-Lang and Glick	1938	Cartesian diver	Serum and gastric mucosa. Sensitivity $10^{-3}\mu l\ hr^{-1}$	46
Boell and Shen	1944	Cartesian diver	Embryonic development of the nervous system of *Amblystoma*	48
Zajicek and Zeuthen	1956	Ampulla diver	Single somatic cells (megakaryoblasts, megakaryocytes). Sensitivity $3{-}10^{-4}$ $\mu l\ hr^{-1}$	49
Giacobini and Zajicek	1956	Ampulla diver	Individual nerve cells. Short	50
Giacobini	1957	Ampulla diver	Single neurones and their constituent parts (cell body, neurite) from sympathetic ganglia (frog, rat). Details of method	51
Giacobina and Holmstedt	1958	Ampulla diver	Regions of spinal cord (anterior horn cells). Selective detn of ACh and BuChE	52
Giacobini	1959	Ampulla diver	Individual spinal ganglion cells. Sensitivity $10^{-4}{-}10^{-6}$ $\mu l\ hr^{-1}$	53
Giacobini	1959	Ampulla diver	Cytoplasm and nucleoplasm of isolated nerve cells. Sensitivity 10^{-6} $\mu l\ hr^{-1}$	54
Zeuthen	1961	Cartesian diver balance	The Cartesian diver balance described in detail. Not ChE	55
Zajicek and Zeuthen	1961	Ampulla diver	Details of the technique for ChE detn in individual cells	56
Brzin and Zeuthen	1961	Ampulla diver and the Cartesian diver balance	Single end plates of muscle splits acetylthiocholine and under proper conditions, deposits Cu-thiocholine at rates which can be followed continuously with the diver balance (gravimetric) and compared in parallel experiments with the ampulla-diver method (gasometric)	57
Brzin et al.	1964	Magnetic diver	First description. Not ChE. Sensitivity 10^{-6} $\mu l\ hr^{-1}$	58
Brzin and Zeuthen	1964	Ampulla diver supported by the magnetic balance	Use of the ampulla diver at constant pressure measuring the changes in reduced weight using the magnetic diver balance supporting the ampulla. Not ChE	59
Brzin et al.	1965	Magnetic diver balance	Detn of ChE activity per unit surface area of conducting membrane	60
Brzin et al.	1966	Magnetic diver balance	AChE detn in the frog sympathetic and dorsal root ganglia	61
Larsson and Løvtrup	1966	Automated magnetic diver balance	First automated devise. Not ChE. Sensitivity 10^{-3} $\mu l\ hr^{-1}$	62
Hamberger et al.	1967	Automated microdiver diver balance	Automatic recording unit for the microdiver technique, using a servomotor by impulses from an optical "niveau-monitor"	63
Brzin and Dettbarn	1967	Magnetic diver balance	AChE in different regions of myelinated nerve fibers of frog	64
Brzin et al.	1967	Magnetic diver balance	Detn of AChE in individual neurons of cervical sympathetic ganglia followed by the electron microscopic-cytochemical technique applied to the same unfixed isolated neuron	65
Tennyson et al.	1967	Magnetic diver balance	AChE in sympathetic and dorsal root ganglia (frog) and development of ChE in the nervous system of embryonic rabbit and human; followed by electron microscopic-cytochemical localization of AChE	66
Mitchard et al.	1969	Automated magnetic diver balance	An improvement of the method of Larsson and Løvtrup (62). Sensitivity 10^{-5} $\mu l\ hr^{-1}$	67

and Friedenwald (70).   The Cu-thiocholine precipitated was measured by the Cartesian *diver balance*, the principle of which is similar to that of the original diver technique, except that both volume and pressure are changed.   It was first used by Zeuthen (55) to determine the submerged or reduced weight of large living cells.   In the study performed by Brzin and Zeuthen (57), it was found that only $\frac{1}{3}$–$\frac{1}{2}$ of the thiocholine formed during the reaction is precipitated (the rest is suggested to be its soluble disulfide).   The diver balance (gravimetric) method for ChE used for this special purpose is therefore not quantitative.

The ampulla diver method requires repeated and manual manometrical adjustments.   Due to the demands of continuous recording and the advantage of autoregulation of the equilibrium pressure, an automatic recording unit has been developed (63).   An optical "niveau-monitor" ascertains the position of the diver, containing the sample, and transfers its impulses through an amplifier to a regulation device, a servo system, which controls the pressure in the floating vessel and is connected to a recorder, registering the pressure as a function of time.   This automated diver technique, first devised for respiration studies, has not yet been applied to the determinations of ChE.

Magnetic Diver Balance.   Because the diver balance is limited in its use by the leakage of gas through the open tail, Brzin et al. (58) tried to make the balance more stable by replacing the compressible air bubble with a tiny permanent magnet enclosed in the diver and balancing the diver floating in a medium of lower density than itself by controlling a magnetic field around the diver.   In the method described, the density of the floating medium is constant and changes in the buoyancy (floating power) are compensated by a magnetic force (for details, see Ref. 62).   The sensitivity of an ampulla-diver was found to be considerably better (10^{-7} μl CO_2 hr^{-1}) when the diver was supported by such a floating magnetic diver balance (59).   This technical device has been further developed and simplified, and it was used by Brzin and his co-workers for the determination of ChE activity per unit surface and of a conducting membrane (60) in the frog sympathetic and dorsal root ganglia (61,65).   In these studies it was also possible to apply the electron-microscopic-cytochemical technique to the same isolated ganglion or neuron as was used in the gasometric measurement.

The first automated magnetic diver balance was reported by Larsson and Løvtrup (62), who adopted the principle of the electromagnetic balance mentioned above.   The current produced by the photocell is continuously recorded and directly related to the buoyancy changes in the diver.   The sensitivity of this self-recording magnetic diver is about 10^{-3} μl hr^{-1}.   A modified apparatus, based upon the same principle as that used by Larsson and Løvtrup, was recently described by Mitchard et al. (67). By reintroducing a modified ampulla diver, the sensitivity of the fully automated magnetic diver balance, registering both temperature and gas-volume variations, could be increased to the order of 10^{-5} μl hr^{-1}, during which time the recorder was stable and could be measured to an accuracy of 1 mm. This technique has not yet been adapted to ChE determinations.

Some fundamental differences between the Cartesian and the magnetic diver seem worthwhile mentioning. In the ampulla, operating as a Cartesian diver with bubbles <0.1 μl, the capillary pressure of the gas bubble gives rise to a substantial difference in the amount of gas dissolved in the fluid present in the ampulla and in the floatation medium surrounding the tip of the tail. The concentration gradient formed produces a rapid diffusion of gas out of the diver. Moreover, the pulsating pressure changes facilitate this diffusion. These phenomena make the Cartesian ampulla diver practically useless for the determination of low ChE activities ($<5 \times 10^{-6}$ μM of ACh or 1×10^{-4} μl CO_2 per sample). Measurements of smaller amounts of CO_2 by this technique seem to present practical difficulties.

When the ampulla is operating as a magnetic diver, it becomes a constant pressure instrument and the saturation of fluids with gas does not change during the measurement. In addition, the column of fluid in the tail of the ampulla does not exchange during the readings so that it is more effective in limiting gas diffusion.

In contrast to the Cartesian diver, where the gas charge of the diver determines the sensitivity of the instrument, the buoyancy of the magnetically operating ampulla is controlled by the magnetic force which is independent of the size of gas bubble in the ampulla. By selecting a suitably large bubble, the gas loss due to the overpressure in the ampulla can be considerably decreased without affecting the sensitivity. Consequently the magnetic diver is much more sensitive than the Cartesian diver and is more suitable when small gas volume changes are expected, particularly when CO_2 is evolved (Brzin, personal communication).

3. Other Gasometric Techniques

Other gasometric procedures seem to offer no advantages over the Warburg method. Simplified manometers and flasks (27) are no more convenient to use than the standard apparatus. An early adaptation to ChE of the classical gasometric method using the Barcroft differential apparatus or the gas analyzer of Van Slyke and Neill (13) has not been tested further in recent years.

The principle of the manometric technique has been used recently in a radiometric modification (71). Sodium bicarbonate-^{14}C is used to react with acetic acid liberated during the hydrolysis of ACh. The resulting $^{14}CO_2$ is trapped and quantitatively measured by liquid scintillation counting. This procedure requires only a simple reaction vessel, easily constructed in the laboratory, and an incubator-shaker apparatus. The disadvantage of this technique, used so far only to rather gross determinations, is that it does not give a distinct picture of the course of the reaction itself.

III. CHANGE IN pH

The change in pH due to the production of acid is either measured electrometrically with a pH meter or determined by the change in color of an indicator.

1. Electrometric Measurement According to Michel

The enzyme is allowed to act on acetylcholine in a standard buffer solution for a fixed time interval (usually 1–2 hr). The pH of the mixture is measured using the glass electrode at the beginning and the end of this time interval. The rate of change of pH (usually Δ pH/hr) is a measure of enzymatic activity. The original method, developed by Michel in 1949, has been frequently used since then both following the original details proposed and also with slight modifications (13). It is also the basis for several indicator methods described below.

The buffer solution, used by almost all authors, is a phosphate buffer of pH 8.0, which is more dilute for the determination of plasma ChE than for the erythrocyte enzyme. This buffer is designed so that the decrease in activity of the esterase with pH over the range 8.1–6.0 is almost compensated for by a decrease in the buffer capacity. Correction factors tabulated in several publications have been worked out for non-enzymatic hydrolysis of the substrate and for slight deviations in the ratio of change in buffer capacity (contributed also by the tissue used) to change in esterase activity as the pH is lowered. To avoid precipitation of phosphate, Mg^{2+} is not added, and to avoid complex formation with the Mg^{2+} and/or Ca^{2+} present in the blood, citrate or oxalate should be substituted by heparin as an anticoagulant.

Michel's method, 20 years after its first description, is still one of the most popular methods for routine determination of blood ChE. In most cases the details proposed by Michel have been followed with minor modifications regarding the volume of the reaction mixture, the composition of the buffer solution, enzyme dilutions, and microsample methods for collecting and handling the blood samples. Some of the more recent modifications are summarized in Table IV.

Blood ChE values of normal persons (72–74) and of laboratory and domestic animals (75–79) have been reported from studies with this technique. Practical considerations involved in a blood ChE activity testing program, in connection with the toxic effects of organophosphate insecticides, have been discussed (80) in connection with the reliability of the test and the validity of the results obtained with Michel's method. The critical evaluation of these measurements as a valuable laboratory tool in preventive medicine has been discussed (81). The values obtained in most cases are comparable with those obtained by other techniques for ChE assay (30,82), a special study being performed by regressing the mean values obtained with the more modern pH-stat method on the mean Michel values (83). A high degree of correlation was found between Michel's method and an automated (Hestrin) procedure when whole blood and plasma samples were used and good correlation was also obtained with erythrocytes (84).

The absolute accuracy and the sensitivity of Michel's method is not as great as those of the titrimetric and Warburg methods. This is due primarily to the fact that one measures the pH, which is a logarithmic function of the acid concentration, rather than determining the acid production itself. In contradistinction to other more accurate methods, the activity is not measured at several time intervals, i.e., the change in pH with respect to time is not usually recorded for each assay. An improvement of the technique in this respect is the description by Tammelin and co-workers (13) of a recording pH meter, which simultaneously registers six enzyme reactions and is calibrated to relate the pH of the solution to the rate of addition of acetic acid. This apparatus has not been used outside the Swedish group but the increased accuracy of the electrometric procedure obtained by its use is a good reason for further applications. A rate-recording modification of Michel's original procedure was recently described (73). It is without the advantages of Tammelin's calibration technique, but shortens the reaction time to less than 5 min and allows for the immediate detection of any deviation from linearity in the reaction trace. The use of continuous flow electrode vessels has also been described (85). A method has been proposed for recording the kinetics of enzyme reaction based upon the measurements of high-frequency conductivity of a solution containing ChE and ACh (86).

2. Change in Color of an Indicator; Acholest

Principle. The change in pH resulting from the production of acid in a solution of a choline ester incubated with ChE can be estimated by the

TABLE IV
Modifications of Michel Electrometric Method Suggested Since 1957

Author	Year	Material	Modifications suggested	Studies performed and/or advantages compared with the original procedure	Ref.
Greenberg and Calvert	1957	Plasma	20 μl plasma is transferred directly to the substrate with a 20-ml Sahli hemoglobinometer pipet instead of diluting it and removing an aliquot	Decrease in time to make a large number of detn (20 samples in duplicate at one time)	87
Frawley and Fuyat	1957	Dog plasma and erythrocytes	Similar to those described earlier for use with rat blood (Frawley et al., 1952), except a higher concn of washed red cells	Effect of low dietary levels of parathion and systox on blood ChE in dogs	88
Williams et al.	1957	Dog plasma and erythrocytes; rat whole blood	Same as above. Slight modifications of substrate concn suggested for rat whole blood	Details of method reported	75
Stubbs and Fales	1960	Plasma and erythrocytes	Capillary blood samples and unwashed red cells (20 μl)	The same precision as for the Michel micro method (venous blood); values obtained significantly higher with the micromethod. High degree of correlation between the two methods	89
Burnett	1960	Serum	Same reagents as the original ones; larger total reaction vol (40 ml buffer, 4 ml ACh, and 0.4 ml serum)	Value of serum ChE as an indicator of liver function	90
Patchett and Batchelder	1960	Plasma	1.0 ml of buffer + 1.0 ml plasma (or plasma-inhibitor mixture) + 0.2 ml of ACh. Incubation for 120 min	Detn of organophosphate residues	91
Melichar	1961	Serum	Introduction of continuous flow electrode vessels. Similar to that described earlier by Tammelin (1953)	45 samples tested in 2 hr	85
Burman	1962	Erythrocytes	Michel's original buffer in a greater dilution. Reaction mixture similar to that used by Tammelin	The fall in pH in 20 tubes recorded in 1 hr. Less than 1 ml of blood needed	92
Johnston and Huff	1965	Frozen plasma	Micro modification	Stability of ChE in frozen plasma	93
Lee	1966	Plasma and erythrocytes	0.16 ml plasma or unwashed red cell pack. The rate of pH change is recorded as a linear function during 5 min	Facilitation of detection of errors. Michel correction factors were not necessary to be used	73
Witter et al.	1966	Plasma and erythrocytes	Initial pH reading omitted	Twice as many samples analyzed within a given length of time	94
Meinecke and Oettel	1966, 1967	Plasma and erythrocytes	Identical with the original Michel method, except the separation of erythrocytes from plasma after the suspension of 0.1 ml blood in saline	Cf. discussions on this technique by Meinecke and Oettel (95) and Pilz (96)	74, 79
Mestres et al.	1967	Serum	Slight changes in concn. of ACh and serum	Detn of organophosphates (<0.05 μg) in fruit juice	97
Callahan and Kruckenberg	1967	Erythrocytes of various animals	Eightfold increase in erythrocyte volume for samples from certain animals		78

231

change in color of an indicator rather than with a pH meter. This change in color will approximate to a linear function of the decrease in pH if the pK of the buffer solution (e.g., phosphate-barbital) is no more than 0.1 unit from that of the indicator used. It would be advisable to standardize with the acid and plot μM of acid against the photometer readings.

Most of the procedures employing an acid-base indicator have been devised for serum (or plasma); only a few are available for whole blood or erythrocytes. An obvious complication using this technique is that the indicator (e.g., phenol red) can combine with protein, particularly serum albumin. These errors are usually minimized when dilute buffers and serum solutions are used. They can also be overcome if the variations in initial absorbance (at 535 nm), caused by binding of the indicator to serum proteins, are allowed for by a calibration based on the ratio of the final to the initial absorbance (98). In this and similar modifications (99–101) using phenol red (13) a more concentrated buffer and larger amounts of serum in the reaction mixture are permitted. This, in its turn, decreases the interference due to CO_2 or other contaminants, and makes the concentration of the indicator a relatively noncritical factor. In addition, it must be confirmed that the indicator, when present during the reaction, has no inhibiting effect on the enzyme.

In addition to phenol red, a number of other indicators have been tested (13). One of the first published methods for serum ChE describes the use of m-nitrophenol, employed also in recent years (102). Cresol red (103,104) and bromocresol purple (105), and even lithmus paper (106), are other indicators tested in this connection, but they seem to have no advantage over phenol red or *bromothymol blue*, which is the indicator of choice in screening tests and in the so-called "Acholest" technique (see below). The latter indicator has been used for the quantitative assay of erythrocyte AChE as well as plasma ChE (107,108), because in the region of its absorption (620 nm) the interference from hemoglobin is small. When whole blood and plasma are tested in parallel runs, the erythrocyte activity can be obtained by difference (107). The indicator concentration does not become a critical variable when a calibration curve and the ratio of the final to the initial absorbance is used in the calculation (98). It seems, however, that a procedure using isolated erythrocytes would be preferable to the calculation of this activity by difference.

A number of techniques have been described with bromothymol blue as an indicator. Except those mentioned in the previous review (13), the direct reading procedure has been employed only rarely in recent years (109–115). The technique is available commercially as a test kit

(116) and provides reasonable correlation with measurements of plasma ChE activity by the Michel method (82).

Automated Procedure. An automated continuous-flow system (Auto-Analyzer) for wet-chemical analysis was applied by Winter (117,118) for the first time to the analysis of serum ChE. It is based upon the colorimetric estimation of the change in pH of a buffered system (similar to that described by Michel in 1949) indicated by the color change of phenol red. Measured amounts of barbiturate-phosphate buffer, substrate (e.g., acetylcholine), and enzyme (e.g., serum) are mixed automatically and incubated for a standard length of time (at 37°C). Phenol red is then added, the reaction mixture dialyzed, and the absorption is measured at 535 nm and recorded. Up to 40 tests can be performed per hour with excellent reproducibility, and the entire procedure is automated. Because continuous dialysis is a feature of the analytical system used, it is possible to use this method for the determination on various animal tissues. It should be of great value when large numbers of routine assays of ChE have to be made. The serum ChE activity of a normal population has not yet been determined with this procedure. It was not tested with erythrocytes, but should be useful for that material as well. The procedure was recently applied to the assay of whole blood ChE *in vivo* with monkeys kept coupled to the AutoAnalyzer for up to 4 hr (119); by this technique it was possible to study the rate of decline in ChE levels following the oral administration of parathion.

A possible improvement of this technique should be a calibration curve with buffers of known pH and with serum of known ChE activity, expressed in μmoles of acid liberated per unit time.

This automation of ChE-activity determination using the Auto-Analyzer is the basis of a method for the determination of water-soluble organophosphorus insecticide residues in plant material (117,120,121). It has been employed in actual (122) and several simulated (120) residue problems. The method has been modified to include an automated elution-filtration technique for analysis of organophosphorus compounds on thin-layer chromatographic scrapings (123). A logical approach to the recently introduced multichannel system would be to split the sample stream automatically and analyze simultaneously for total organically-bound phosphorus and ChE-inhibition properties (124).

Acholest and Other Field Tests. The principle of the indicator methods discussed above has been used for the development of screening tests, which are easy to perform and require only simple equipment and few manipulations and reagents. In the earliest test, the color of a solution of hemolyzed blood (a drop from the finger tip), acetylcholine, and bromothymol blue is inspected visually after a short incubation under standard conditions. This procedure has been varied by matching the indicator color with the corresponding $\Delta pH/hr$ and per cent of normal ChE activity values, or by using the time for a fixed color change to occur as the measure of enzyme activity. References to these tests are given in the previous review (13,14). Further simplifications of this technique, including the use of specific substrates, have been published during the last years (110,111,125,126), but they do not seem to be any great improvement. Modifications have been described to make the test suitable for detecting residues of organophosphorus compounds in plant material (112,127,128). An agar-diffusion test, similar to the cup-plate assay method for antibiotics, has been developed also for use in detecting residues of these insecticides (112,129). The assay plates are prepared from a solution of agar containing blood and bromothymol blue. In each hole made in the plates, 0.1 ml of the sample solution is placed, and after 18 hr of diffusion the plate is sprayed with a solution of ACh. After about 30 min the greenish and transparent zones appear clearly against the orange background and can be measured easily. Calibration curves are made by plotting the zone diameters (in mm) against the log concentration of the ChE inhibiting compound.

These field tests are, however, not as accurate as the laboratory tests and the error is probably 10–25%.

A most popular and useful variant of the screening test, also based upon the indicator technique, is to use a filter paper impregnated with a choline ester and bromothymol blue and subsequently dried. Serum or plasma to be tested is added to this paper. The activity is obtained by comparing the color from initial blue to greenish yellow at various time intervals and at a fixed temperature with the color of standard papers prepared from indicator and buffers of known pH. This technique was described first by Herzfeld and Stumpf (130), and some years later a similar procedure was published by Sailer and Braunsteiner (43), to whom the first description was unknown. The test-paper, Acholest, is now commercially available (Österreichische Stickstoffwerke A. G., Linz/Donau, Austria), and can be easily obtained from various firms.

The Acholest method is suitable only for estimation of serum or plasma ChE (Table V), unlike the method employing liquid reagents with bromothymol blue as indicator which can measure whole blood ChE activity as well. The test paper method is easy to perform and, although it is not a method of great precision, it is suitable for routine clinical work (43,82,136,137,144,145) because it provides a fair quantitative indication of the degree of reduction in plasma ChE activity, for example, in cases of liver diseases (131,132,133,146) and of poisoning by organophosphorus compounds (41,142). The method seems to be useful (147) for screening the presence of ChE inhibition. The Acholest should not be used with stored blood, mainly because of the marked variation in plasma pH with the type of anticoagulant used and duration of blood storage (40). The method is useful in the field and has been tested under tropical and other unfavorable conditions (23,140). The results obtained correlate well with those obtained with the Warburg method (41,42,43,130), Michel method (82,132,134,136,138), the titration method (133,141), and with the Hestrin method (135). To obtain comparable results it is necessary to control the temperature or to correct the results for room temperature (134,139,140). It is also important to note the initial pH of the sample and the various factors which can influence it, e.g., the albumin concentration (135).

An improvement in the technique seems to be the use of a graduated continuous color band, from which the ChE activity (expressed in μmoles of substrate split per minute per milliliter plasma) can be read as the color change of the test paper after 6 min of incubation (141,148). This test paper (Merchotest) contains a pair of indicators (phenol red and naphtholphthaleine) of different pH ranges because the pH range of a single indicator is too narrow. A minor modification is to use small disks punched from original Acholest test papers (143). Another modification avoids the end-point reading by using a wet standard-color reference (136). Phenol red, instead of bromothymol blue, has also been tested as an indicator in the test-paper method (142).

A similar method related to the Acholest procedure is the use of an agar medium containing bromothymol blue and a choline ester (138). This technique was tested with whole blood and ACh or acetyl-β-methylcholine as substrate and found to be adequate as a fast screening technique for differentiating plasma and erythrocyte cholinesterases. Its advantages were claimed to include easier interpretation than the test-paper technique, speed, no requirement for special equipment, and no interference from the red cells with the reading.

The principle of these procedures have been utilized for the detection of ChE inhibitors on thin-layer (silica gel) chromatograms (149–151). The chromatogram is first sprayed with human plasma (undiluted) and then with a mixture of acetylcholine and bromothymol blue (or bromophenol blue).

IV. TITRIMETRIC METHODS

Principle. These procedures are based upon the determination of the acid liberated during the hydrolysis of the ester (in most instances a choline ester) by titration with standard alkali at constant pH using either an indicator or a potentiometer. Such methods were frequently used (without automation) before the introduction of more sensitive and convenient techniques. A compehensive summary of these early

TABLE V

Studies with the Acholest Test

Author	Year	Indicator	Substrate	Comparator (incubation time in min)	Enzyme	Studies performed	Ref.
Herzfeld and Stumpf	1955	BTB[a]	ACh-Br	Color scale, 10–60	Serum	Compared with Warburg	130
Sailer and Braunsteiner	1959	?	?	Yellow, normal range 5–19	Serum	Compared with Warburg. Liver diseases	43
Pietschmann	1960			5–19	Serum	Liver diseases	131
Churchill-Davidson and Griffiths	1961	BTB	ACh-Br	12–20	Serum	Compared with Michel	132
Jabsa et al.	1961	Sailer and Braunsteiner		6–17	Serum	Compared with Warburg	42
Lang and Intsesuloglu	1962	Sailer and Braunsteiner		6–17	Serum	Compared with titrimetric. 1000 patients, including those with liver cirrhosis	133
Bergman et al.	1962	Acholest		7–18	Stored blood	Not useful	40
Richterich	1962	Acholest			Serum	Compared with Michel. Relation to enzyme concn and temp	134
Vincent et al.	1963	Acholest		10–33	Serum	Effect of pH and albumin concn. Compared with Hestrin	135
Wang	1963	BTB	ACh-Br	3–8	Plasma	Compared with Michel. Clinical value.	136
Wang and Henschel	1967					Improved method	137

Author	Year	Indicator	Substrate	pH range	Blood fraction	Remarks	Ref.
Davidson and Adie	1965	BTB	ACh-Cl, MeCh[a]	6–20		Compared with Michel, 10 min incubation. Agar gel	138
Pleština	1966	Acholest			Plasma	Various temp	139
Holmes and Jankowsky	1966	Acholest		8–18	Serum	Compared with Michel and whole blood BTB test	82
Holmstedt and Oudart	1966	Acholest		7–15.5	Plasma	Tested in the field in Africa	140
Härtel et al.	1967	Phenol red + naphtholphthalein			Serum	Compared with automatic titration. Graduated continuous color band to read directly moles of acid found per ml	141
Fischl et al.	1968	Phenol red	ACh	Yellow = no org P; Pink violet = org P	Serum	Detection of org P poisoning in vivo	142
Fristedt and Övrum	1968	Acholest		6–18	Serum	Compared with Warburg	41
Braid and Nix	1968	Acholest, disks from		10–30	Serum	Only 21 μliter of serum needed in a triplicate run	143

[a] BTB, bromothymol blue; MeCh, acetyl-β-methylcholine.

237

procedures is included in the recent review (13). The introduction of an automatic recording titrator, however, made this procedure accurate, rapid, and convenient.

There are a number of shortcomings and problems inherent in the use of an indicator (13), and other methods are to be preferred even if the experiment is a preliminary one where great exactness is not required. Only a few studies using this technique have been reported recently (152–154). Several difficulties are overcome if a potentiometric method is used to measure the pH, a glass electrode being the best choice, although other types of electrodes have also been tested (155). This technique has been used for a comparative study of the activities of horse serum obtained with either ACh or tributyrin as substrate (156). It was also used for the assay of ChE in stored blood (40), in serum of normal individuals (157) compared with the Acholest test (141), and in the tissues of the snail *Helix* (158), and for the assay of esterase activity of various fractions during purification of acetylcholinesterase from the electric tissue of *Electrophorus* (159–161). It is useful as a general procedure for assaying various types of hydrolases (162).

In a slight modification of the general procedure the pH is allowed to fall for a standard length of time, stopped by the addition of an inhibitor, and then a titration is performed (163,164). The change in pH during the assay is an obvious disadvantage of this technique. In another modification, constant pH is maintained by removing H^+ electrolytically (165).

The possibility of carrying out the titration by following changes in conductivity during the enzyme reaction was mentioned in the previous review (13). This principle has been recently applied to the assay of serum ChE using a recording assembly (166), but it does not seem to have any significant advantages over the procedure described below.

Automatic Recording Titration. One of the best methods for ChE determination makes use of a recording potentiometric titration. Details of this procedure will therefore be presented, based upon the original works summarized in Table VI and essentially those reported by Jensen-Holm et al. (168), Jørgensen (169), and Nabb and Whitfield (177). Excellent descriptions of the apparatus and illustrations of the various components of the system have been published (180–182). A specially designed inhibition reaction vessel has been described by Main and Iverson (183); it permits incubation time as short as 1.2 sec at any convenient temperature.

Apparatus. *Titrator.* Radiometer TTT 1, with scale expander, or a combination of a pH meter (e.g., Radiometer PHM 28 or 26 with scale expander) and a titrator unit (Radiometer TTT 11).

TABLE VI
Some Characteristics of Titrimetric Methods Employing an Automatic Recording Titrator

Author	Year	Reaction mixture, total vol (ml)	pH	Medium	Enzyme	Substrate	Reaction time (min)[a]	NaOH (mM)	Studies performed	Ref
Wilson and Cabib	1956	40–350	7.00	0.1M NaCl, 1% gelatin, 10^{-4}M EDTA	Purified AChE[b]	ACh, aminoethylacetate	5	46.6	Detn of enthalpies and entropies of activation	167
Jensen-Holm et al.	1959	25	7.90–7.95	0.15M NaCl	Whole blood, tissue homogenate	ACh	3	100	Demonstration of spontaneous nonspecific liberation of acid in dead tissues	168
Jørgensen	1959	12.5	7.40	0.15M NaCl	Serum	ACh	2–3	20	50 normal individuals	169
Heilbronn	1958 1959	40	8.00	0.10M KCl	Electric organ, serum fractions	Thiocholine esters	1–10	100	Specificity and kinetics. A second syringe (containing the substrate soln) introduced to study the activity at low substrate concn	170, 171
Tominz and Gazzaniga	1961	12.5	7.4	1% NaCl	Rat brain (homogenate)	ACh		20	Spontaneous liberation of acid in tissue homogenates, influence of air oxygen thereon	172
Delaunois	1962	20.4	7.4[c]	H$_2$O	Serum	Choline esters	10	10	Apparatus and chart diagrams illustrated	173
Rubinstein and Dietz	1963	20.5	7.2	0.15M NaCl	Human serum	ACh and benzoylcholine	10	10	Comparison of the hydrolysis rates of the two substrates by normal and abnormal sera. Comparison with the Hestrin method	174
Jensen-Holm	1965	50	7.40	0.15M NaCl	Plasma, erythrocytes, tissues	ACh, 1mM and 10mM	10	100	Separate detn of AChE and other types of ChE with ACh as single substrate	175
Casterline and Williams	1967	3.2	7.4	0.02M NaCl	Erythrocytes	ACh	10	10.3	ChE inhibition in erythrocytes of rats fed low levels of a carbamate. Comparison with the Warburg technique	176
Nabb and Whitfield	1967	4.9	8.0	0.15M	Erythrocytes and plasma	ACh	3	3	Comparison of dogs, rats, rabbits, and humans. Effect of substrate concn, pH, and temp	177
Kitz and Ginsburg	1968	100	7.00	0.1M NaCl, 0.02M MgCl$_2$, 0.005% gelatin, 10^{-6}M EDTA	Purified AChE[b]	ACh		46.6	A matched pair of syringes used, delivering NaOH and ACh bromide of the same concn, respectively. Kinetics of inhibition	178
Ballantyne	1968	4.0	7.4	H$_2$O	Adipose tissue (homogenate)	ACh	10	10	Comparison with histochemical studies. Activities measured in the presence of selective inhibitors	179

[a] Titrant usually standardized at 20–100mM, carbonate free.
[b] AChE prepared from the electric tissue of *Electrophorus electricus*.
[c] Calomel and antimony electrodes.

Titrigraph. Radiometer type SBR 2 as recording instrument equipped with a 0.5-ml syringe and micrometer SBU 1 or ABU 11. Chart speed 20 mm/min. Pen speed 60%/min, corresponding to a maximum titration rate of 6% μmoles/min.

Reaction Vessel. Thermostated assembly, total capacity 30–40 ml, heated by a circulating water bath (30 $\pm$ 0.1°C), with a magnetic stirrer and a glass-calomel electrode system (e.g., type GK 2301 C).

Reagents. *Medium Solution.* 0.1M NaCl and 0.02M CaCl$_2$. When highly purified enzyme preparations are used, 0.005% gelatin and 10^{-4}M EDTA should be present.

Substrate Solution. Acetylcholine iodide, 0.250M in distilled water.

Sodium Hydroxide. 20 mM NaOH (carbonate free) as titrant. An appropriate amount of ca. 1M NaOH, accurately standardized by titration with 0.1M potassium hydrogenphthalate (end point of the titration at pH 7.90–7.95).

Enzyme Solution. Plasma is diluted with 0.1M NaCl to a suitable concentration for ChE assay; erythrocyte hemolysate or whole blood is similarly treated. A known amount of fresh tissue, placed in glass vessels cooled to 0°C, is homogenized with 0.1M NaCl at the low temperature and added to the reaction mixture.

Procedure. *Determination.* The glass electrode and the titrant are checked, and the former placed in the reaction vessel which contains 24.0 ml of the medium solution (pre-warmed to 30°C). The enzyme solution, 0.5 ml, is added and the pH adjusted to 7.60 with the titrant. If the pH of the mixture is initially below 7.0, a 100 mM NaOH solution is added manually until the pH is almost 7.6, and the instrument is allowed to titrate to exactly pH 7.6. Nitrogen is allowed to blow into the vessel near the liquid surface during this and the subsequent steps (to exclude any action of CO$_2$). Titration is started to test whether any nonspecific spontaneous acid liberation takes place. This is rare with fresh blood, plasma, or erythrocytes, but often occurs with dead tissues. After the titrigraph curve has been plotted for some minutes, 0.5 ml of the substrate solution is added with a constriction pipet or from a suitable syringe. The titrigraph is switched to "on" when the valve relay clicks at the first indication of a slight decrease of pH below 7.6. The instrument is allowed to titrate the reaction for 3 min or until at least 3 min of the recording is linear.

The spontaneous hydrolysis of the substrate is determined in the same way in the absence of enzyme.

Calculation. The initial reaction rate is read from the slope of the straight line recorded by the titrigraph. The ordinate, which has been calibrated in μmoles of NaOH, indicates μmoles of liberated acid, and the abscissa indicates time in minutes. For calculation, the acid produced both by nonenzymatic hydrolysis of substrate and by nonspecific liberation from tissues are subtracted. The enzyme activity is expressed in μmoles of substrate hydrolyzed per minute per milliliter (plasma) or milligrams (tissue or protein).

Comment. The automated pH-stat method for assaying ChE is one of the most convenient and precise methods available. Its great advantage is that the course of acid liberation can be continuously monitored. It is rapid and simple, and the conditions of the reaction are easily reproduced. Since little or no buffer need be present, the method is flexible and may be varied with regard to enzyme and substrate concentration, pH, temperature, and salt concentration for use with various biological material. Since rate curves can be obtained within 2 min, this method provides one of the best means to study the kinetics of the initial stages of the reaction of esterases with substrates inhibitors and activators, so well illustrated by the many excellent studies by Wilson, Kitz, and their co-workers (178,184). There seem to be no theoretical reasons that would prevent the use of this automated procedure for determining ChE in almost all situations (177), thus making the method important in this field. It has been tested for the determination of AChE in normal and denervated ganglia of the cat (185) and other tissues (186).

There are a few precautions that must be taken in order to obtain accurate results. To prevent errors due to the absorption of CO_2, nitrogen must be blown over the reaction mixture. The volume of alkali added during the course of determination should not be large enough to significantly increase the volume of the reaction mixture. The ionic strength of the mixture should be maintained at 0.1–0.2. Serious errors may result from a spontaneous nonspecific acid liberation in tissue homogenates (168), which can be corrected for by a preliminary titration before the substrate is added.

Studies of the esterase inhibition by carbamates gave a good correlation with those obtained with an abbreviated manometric technique (24). From a comparative study between the pH-stat and the Michel methods, Pearson and Walker (83) recently concluded that the former method offers better control of the assay conditions, higher sensitivity, and better reproducibility. The mean pH-stat values were regressed on the mean Michel values, and from the analysis of variance ($P < 0.005$ for both the human plasma and erythrocyte enzymes) it was concluded that the values obtained experimentally by the two methods could be related in satisfactory equations.

V. METHODS BASED UPON THE PRODUCTION OF THIOCHOLINE (OR CHOLINE)

1. Choline

The rate of hydrolysis of a choline ester should be measurable by the rate of the appearance of choline.   The determination of choline, however, is not as convenient as that of the acid produced, particularly when choline is present in mixtures with its esters.   Two principles have been employed recently.

One method is based upon the microbiological estimation of choline using a choline-less mutant of *Neurospora crassa;* choline is a much more powerful growth promotor of this organism than acetylcholine (187,188). This method seems to be highly impractical (several days for growth of the mycelium, high content of choline in certain organs).   In the other method, choline is oxidized (in the presence of NAD†) by an excess of liver choline dehydrogenase (EC 1.1.99.1) to betaine aldehyde and NADH.   This reaction is linked to the reduction of cytochrome c in the presence of excess NADH-dehydrogenase (EC 1.6.99.3) and can be followed spectrophotometrically (by increase in absorption at 550 nm) (189).   Neither of these methods seem to have any advantage over those based upon the determination of thiocholine when one of its esters is used as substrate.

2. Thiocholine

The use of thiocholine esters (mainly acetyl- and butyrylthiocholine) as substrates has given improved results as it is comparatively easy to determine the free SH-groups produced during the hydrolysis.   These thioesters have been applied with great success to studies of the cytological localization of both acetylcholinesterase and other types of cholinesterases (70,190), and have also been used to detect ChE active components after electrophoretic or chromatographic separation on paper or thin layer (191).   In addition, it has been established that acetylthiocholine is a satisfactory substrate for the quantitative measurement of ChE activity.

The rate of hydrolysis of a thiocholine ester can be measured by the rate of decrease in absorbance of the solution of acetylthiocholine iodide at 250 nm (applications of the original method to assay ChE inhibitors (192,193), or that of the corresponding chloride at 229 nm (194,195). This method has the disadvantage that the most convenient salt of acetylthiocholine (i.e., the iodide; the chloride is much less stable) is unsuitable because of the intense light absorbancy of iodide in the lower portion of the UV spectrum.

The determination of the thiocholine produced can be accomplished in a number of ways: (1) iodometrically, (2) by oxidation by sodium nitroprusside (196–199), and (3) by decoloration of the blue 2,6-dichlorophenol-indophenol measured at 600 nm (195). These methods have been reviewed in more detail (14). Two further principles based on the appearance of thiocholine have been developed recently: (4) the increase of the anodic polarographic wave of thiocholine, and (5) the production of a yellow color with 5,5′-dithiobis-(2-nitrobenzoate). These methods are promising in several ways, particularly the last one, and therefore need further discussion.

A. POLAROGRAPHIC METHOD

This new electrochemical method was described first by Kramer et al. (200). A constant current of 25 μA is applied across two platinum electrodes immersed together with a saturated calomel electrode in 25.0 ml of substrate to which 1.0 ml of enzyme is added. The reduction in potential of the Pt anode vs. the calomel electrode (due to the production of thiocholine by hydrolysis at pH 7.40 in 0.1M tris buffer), is automatically recorded against time. From the slopes of the resulting depolarization curves, $\Delta E/\Delta t$, the relative rates of hydrolysis may be calculated. This method has been applied to the assay of various organophosphorus compounds as cholinesterase-inhibitors (201), to the elucidation of the kinetics of hydrolysis of several thiocholine iodide esters by AChE and serum ChE (202), and to the measurement of ChE activity with immobilized (insolubilized) enzyme (203).

A similar method was described independently and simultaneously by Fiserová-Bergerová (204,205). She used a Heyrovsky polarograph with a dropping mercury electrode and a saturated calomel electrode and recorded the rate of increase in the anodic (oxidizing) current, i.e., the polarographic wave of the thiocholine produced by hydrolysis, as a function of time. The method has been applied to kinetic studies of plasma ChE and erythrocyte AChE with acetylthiocholine as substrate (205) and to the evaluation of normal activity values in human blood (206). The validity of this polarographic method was determined (354) by comparing the rates of thiocholine production with those obtained spectrophotometrically according to Ellman (see below). A similar method measures the increase in potential of a Ag-electrode covered with a thin layer of Ag-mercaptide (207).

The major advantages of the polarographic method include the simplicity of the apparatus, accuracy, noninterference in the enzymic process, and the ability to follow fast reactions. Further improvements of the method could be accomplished by using a rotating mercury electrode and a fast mixing technique, such as the stop-flow system developed by Chance. For constant, reproducible results, some iodide must be present in the reaction mixture. If the chloride or bromide of acetylthiocholine is used, iodide (as electroactive mediator) must be added to a concentration of about 0.1 mM.

B. SPECTROPHOTOMETRIC METHOD ACCORDING TO ELLMAN

Principle. Ellman (208) originally proposed in 1959 the use of 5,5′-dithiobis-(2-nitrobenzoic acid) (DTNB) as a reagent for sulfhydryl groups. He and his co-workers (209) later utilized this reaction in a method for the determination of ChE activity in whole blood, erythrocytes, and tissue homogenates using acetylthiocholine as substrate. The thiocholine formed during the hydrolysis rapidly reacts with DTNB and releases a colored 5-thio-2-nitrobenzoate anion (Fig. 1), having an absorption maximum at 405–420 nm (210,211), which can be readily measured spectrophotometrically. The attention paid to this technique was not very great until recently. The first modifications of the original procedure were reported by Garry and Routh (212), followed in recent years by a number of applications to various materials (Table VII). One of the advantages of the method is its adaptability for microdeterminations and for routine analyses suitable for automated procedures (210,215,223,224).

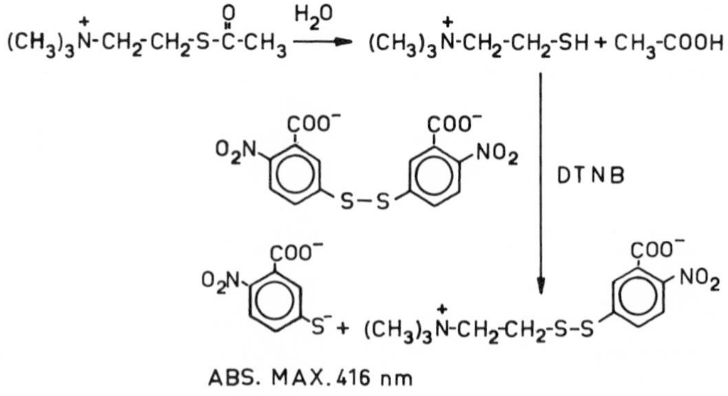

$$(CH_3)_3\overset{+}{N}\text{-}CH_2\text{-}CH_2\text{-}S\text{-}\overset{\overset{O}{\|}}{C}\text{-}CH_3 \xrightarrow{H_2O} (CH_3)_3\overset{+}{N}\text{-}CH_2\text{-}CH_2\text{-}SH + CH_3\text{-}COOH$$

ABS. MAX. 416 nm

Fig. 1.

The original method described by Ellman et al. (209) with minor modifications recommended later (211,212) will be presented below.

Apparatus. *Spectrophotometer*, having a temperature-controlled (25°C) sample compartment for cuvets (preferably with automatic cuvet positioner) connected to a log-linear recorder.
AutoAnalyzer (for details, see original papers referred to below).

Reagents. *Buffer Solutions.* Phosphate, 0.1M, pH 8.0 and 7.0 respectively. Also recommended: tris, 0.05M, pH 7.2.
DTNB-Buffer Solution, 5,5'-dithiobis-(2-nitrobenzoic acid), (mol wt = 396), 0.25 mM in phosphate or tris buffer. Ellman recommends a more stable stock solution (10 mM) prepared by dissolving 39.6 mg in 10 ml phosphate buffer (0.1 M) of pH 7.0 and adding 15 mg of $NaHCO_3$, in which the reagent is more stable than in a solution of pH 8.0.
Substrate. Acetylthiocholine iodide, 78 mM (21.67 mg/ml); stable for 10 days if kept refrigerated, but a fresh solution is best made each day.
Glutathione (for standardization), 0.813 mM. 25 mg in 100 ml distilled water.

Standardization Procedure. A series of glutathione standard solutions to give final concentrations in the range 0.01–0.1 mM and a solution of DTNB in buffer of pH 8.0 to give a final concentration of 0.25 mM in a final volume of 3.0 ml are prepared. The absorbances of the mixtures are determined at 412 nm.

Manual Procedure. 50.0 μl enzyme (e.g., blood serum) is mixed in the cuvet with 3.0 ml DTNB-buffer solution, and the reaction started by adding 20.0 μl of substrate. In the manual reading the increase in absorbance of the reaction mixture is read every minute (ΔA/min) for 3–5 min and the mean value calculated. With a recorder the increase in absorbance is recorded with a chart speed of 2 cm/min. When an automatic cuvet positioner is used, each cuvet is exposed to the light beam for 15 sec, and the procedure repeated twice. This procedure can be scaled down to microsize by using a microcell (total volume, 0.307 ml).
The ChE activity is expressed in international units (IU = μmoles of substrate hydrolyzed per minute per milliliter) and calculated in either of the following two ways:
(*1*) The difference in absorbance (ΔA) between the test sample and a suitable blank (without substrate or in the presence of an esterase inhibitor) is compared to a glutathione calibration curve to obtain the

TABLE VII

Modifications of the Ellman Spectrophotometric Method

Author	Year	Buffer soln Type	Buffer soln M	Buffer soln pH	Temp. (°C)	Total vol (ml)	Enzyme Source	Enzyme Concn[a]	Substrate (thiocholine iodide ester) Type	Substrate mM (final)	DTNB,[b] mM (final)	Measurement at nm	Studies performed	Ref.
Ellman et al.	1961	Phosphate	0.1	8.0	25	3.17[c]	Erythrocytes[b]	0.25 U	Acetyl[i]	0.5	0.33	412	Kinetics	209
Guth et al.	1964	Phosphate	0.1	8.0	37	0.055	Muscle fibers	Cross sections	Acetyl	5			See text	214
Garry and Routh	1965	Tris	0.05	7.4	37	4.52	Serum	20 μl	Acetyl	2	0.25	412	Effect of DTNB concn. Normal variation	212
Levine et al.	1966	Tris		7.4	37		Plasma Erythrocytes		Acetyl	5.3 / 2.7	0.25	420	Automated[d]	215
Buckley and Nowell	1966	Phosphate	0.2	7.0	24–25	0.2	Motor end plates	2.4–25	Acetyl	5	1.0[e]	412	After staining with Cu-thiocholine and dissecting	216
Weber	1966	Tris	0.1	7.4	25	2.22	Serum	20 μl	Acetyl	5	0.26[f]	405	Optimum conditions for pH and temp	217
Humiston and Wright	1967 (1965)	Tris	0.05	8.2	37	3	Plasma Erythrocytes		Acetyl	7.0 / 0.8	0.20 / 0.15	420	Automated.[d] Effect of pH, substrate concn, DTNB concn, specificity	210
Knedel and Böttger	1967	Phosphate	0.05	7.2	25	2.7[c]	Serum	10 μl	Butyryl	5	0.25[f]	405	Micromethod with 0.54 ml total volume and 2 liter serum	218
Bauman et al.	1967	Tris	0.1	7.4	30	3.0	Pads of serum ChE; homogenate[g]	0.5 ml	Butyryl	0.5	0.33	412	Storage stability of ChE pads	203
Sakai	1967	Phosphate	0.05	7.4	30	3.0	Insect organs	0.5–10 mg	Acetyl, butyryl	0.4	0.5	412	Type of ChE present	219
Bufardeci and Buonsanto	1967	Phosphate	0.05	7.2	25	3.2	Serum	5 μl	Acetyl	5	0.27	405	Short report	220
Karahasanoğlu and Özand	1966, 1967	Phosphate	0.1	8	23	3	Serum	20 μl	Butyryl	0.083	0.067	412	Screening test also described	221

Voss Voss and Geiss-bühler	1966 1967	Phosphate	$\frac{1}{15}$	8.0	37		Erythrocytes[h] Plasma		Acetyl, butyryl	1.0	0.25[e]	420	Automated.[d] Detn of org phosphates; kinetics	222, 223, 224
Voss Voss and Schuler	1968 1967	Phosphate	$\frac{1}{15}$	7	30	5	Plasma	20 μl	Acetyl	0.7	0.26[e]	Filter	Fast procedure	225
Voss	1968	Phosphate	$\frac{1}{15}$	8.0	37		Peacock plasma		Acetyl	0.4	0.25	420	Automated.[d] Detn of carbamates	226
Juul	1968	Phosphate	$\frac{1}{15}$	6.1	22	3	Plasma	13 μl	Acetyl, butyryl	3.2	0.25	412	Total activity in combination with disk-electrophoresis separation	227
Szász	1968	Tris	0.05	7.2	25	2.1	Serum	5 μl	Acetyl, butyryl	2.0	0.5[f]	405	Kinetics	211
Klingman et al.	1968	Phosphate	0.1	8.0	37	2.0	Sympathetic ganglia of rat	0.12–0.31 mg	Acetyl	0.5	0.1	412	Differential assay of two ChE types using selective inhibitors	228
Leegwater and van Gend	1968	Barbital-fosfat	0.03	7.4	37		Serum,[j] erythrocytes		Acetyl, butyryl		0.25	420	Automated with dialyzer. Screening of organophosphorus pesticides in plant extracts.	229
Wilhelm	1968	Phosphate	0.1	7.4	22–25	3.17	Plasma	20 μl	Acetyl	0.1–10	0.32[f]	412	108 healthy subjects. Exposure to carbamates	230

[a] Amount of enzyme present in the reaction (i.e., final) mixture.
[b] 5,5′-Dithiobis-(2-nitrobenzoic acid) (Aldrich Chemical Co. 2369 No. 29th, Milwaukee 10, Wisconsin).
[c] Micromethod described as a modification.
[d] Flow diagram illustrated. Glutathione generally used as standard. Coupled with an automated total P system for organophosphorus insecticides (cf. Ref. 124).
[e] DTNB and substrate in the same stock solution.
[f] Stock solution made in the buffer solution.
[g] Starch gel immobilized horse ChE pads, incubated with α-amylase; supernatant used in enzyme determination.
[h] Purified preparation from.
[i] The same method used with the carbon analog of acetylthiocholine (3,3-dimethylbutyl thioacetate) as substrate (213).
[j] Comparative studies with serum and erythrocyte stroma from man, horse, dog, and rat.

concentration of SH groups liberated. The ChE activity is then calculated as follows:

$$IU = \frac{\Delta A}{\min} \times \frac{\mu M \text{ glutathione}}{A \text{ unit}} \times 50$$

(2) Since the molar absorptivity of the yellow anion is known (ϵ = 1.36 $\times$ 10^4 mol^{-1} cm^{-1}), the rates in absolute units can be calculated as follows:

$$IU = \frac{\Delta A}{\min} \times \frac{10^3}{\epsilon} \times \frac{(\text{total vol})}{(\text{sample vol})}$$

Modified Manual Procedure for Tissue Preparations (214). Frozen cross sections (28 μ) from muscle are incubated for 2 hr in phosphate buffer and 5 μl of substrate solution. The reaction is stopped by the addition of 5 μl of 6% HClO$_4$ and the protein precipitate sedimented by centrifugation. Ten μliters of supernatant is added to 400 μl of a 1:150 dilution of Ellman's stock reagent (made with the phosphate buffer). After 2 min the absorptivity is determined at 412 nm.

Automated Procedure. For details of the standardization procedure of this technique based upon the Ellman reaction, the original papers (124,210,215,224) should be consulted. In all these studies the optimal reaction conditions were established and the steps involved in determining the activity from the recorded peak height are well illustrated. The automated method has been used to study the effects of changes in enzyme reaction parameters (e.g., time, temperature, substrate concentration, pH, and presence or absence of inhibitors) and for the evaluation of kinetic data (224). It has been recently applied to residue analysis of insecticidal carbamates using peacock plasma as the enzyme source (226), and of organophosphorus pesticides using plasmas and erythrocyte stroma from various animal species (229). In the latter system a dialyzer was inserted to eliminate interfering colors from plant extracts.

A further development of the automatic determination of ChE-inhibiting organophosphorus compounds is the coupling of the automated system for ChE determination based on the Ellman reaction with an automated total phosphorus system, then producing a dual detection system with simultaneous recording on a two-pen recorder of two distinct parameters from each sample (124).

Comment. This colorimetric method has several advantages, particularly in comparison with other spectrophotometric methods. In contrast with the other frequently used colorimetric method, i.e., the Hestrin method, the Ellman method is based upon the measurement of reaction product instead of the remaining intact substrate, thus making it more sensitive. The method is dependent on changes in the visible region of the spectrum which is convenient when checking unusual changes in the absorbance (e.g., turbidity, spills on the photocell windows). Homogenates of tissue do not require any special handling (e.g., precipitation of protein before readings). Other main advantages are simplicity, high precision, pH constancy, flexibility, short incubation periods, and continuous increase in color density as a function of incubation time. The author shares the opinion of Voss (224) that the principle described by Ellman and his co-workers (209) represents one of the most straightforward methods developed for ChE experiments.

VI. METHODS BASED UPON CHEMICAL DETERMINATION OF THE DISAPPEARANCE OF ACETYLCHOLINE

Since there is no longer any reason to use the biological methods to assay residual ACh or other choline esters after incubation with the enzyme preparation, chemical methods for assaying the rate of disappearance of the substrate (a choline ester as well as other types of esters) have been improved considerably. The best known technique, which is frequently used, is an adaptation of the hydroxamic method for the determination of choline esters. A few other techniques based upon similar principles have been described and will be mentioned briefly below. It will be noted that the rate of reaction obtained by all these methods is based upon a difference between initial and unreacted substrate, making it impossible to record rate curves directly.

1. Hestrin Method

This method was introduced by Hestrin in 1949 as a sensitive chemical method for the assay of ACh and has been applied to micro- and ultramicrodetermination of cholinesterases, the best known modifications being summarized in Table VIII. The ester which remains after incubation with the enzyme is converted quantitatively by alkaline hydroxylamine into a hydroxamic acid, which is measured spectrophotometrically at 500–540 nm by means of the red-purple complex formed with $FeCl_2$ in acid solution. Details of the procedure have been reported in the previous review (13). The method has been applied frequently to the determination of the cholinesterase activity of plasma, erythrocytes, and whole blood (14). It is not suitable for kinetic studies.

TABLE VIII

Some Characteristics of Modifications of the Hestrin Colorimetric Method

Author	Year	Total vol[a] (ml)	Buffer soln Type	Buffer soln pH	Serum (plasma)[b] (μl)	Reaction time[c] (min)	Color intensity measured at nm	Studies performed	Ref.[d]
Metcalf	1951	1.0	Phosphate	7.2	5	30	540	Micromethod for serum, erythrocytes, and brain. Optimum conditions studied. Daily and normal variation in plasma ChE. Applications to phosphate poisoning	231
de la Huerga et al.	1952	2.2	Barbital	8.0	200	60	540	Relation to substrate and serum concn. 132 normal individuals	232, 233, 234, 235
Fleisher and Pope	1954	2.0	Phosphate	7.2	5[e]	10	540	Relation to blood concn and activity as a function of time. Sex and individual variation. Application to phosphate poisoning	236, 237, 238, 239
Bonting and Featherstone	1956	0.04	Barbital	8.0	0.2	60	500	Ultramicro assay, also with 50 μg of nervous tissue. Effect of buffer system (pH 5.5–10.0). Various choline esters tested	240, 241
Vincent (and Segonzac)	1958	2.0	Barbital	7.4	50	30	520	Serum ChE. Used also for ArE with phenyl acetate as substrate	242, 243
Pilz	1958	10.0	Borate	7.3	20[f]	30	490	Careful chemical studies preceded the development of this procedure. Optimum conditions studied in detail	244, 245, 246
Pilz	1965	30.0	Borate	7.2	50	120 PhAc;[g] 24 hr SuCh	490	Seven various substrates tested with serum and erythrocytes of 250 normal individuals	247
Pilz	1969	27.0	Barbital	8.6	80	30 or 60[h]	490	Detailed descriptions of the latest modifications recommended for esterase detn in whole blood, erythrocytes, and serum with ACh as single substrate	248
Kuroda et al.	1959	3.0	Phosphate	7.4	12.5	60	530	Portable micromethod	249
Nesheim and Cook	1959	3.0	Phosphate	7.2	80	30	540	ChE inhibition method, originally developed by Cook (1954), for the analysis of organophosphate insecticides; further improved	250, 251
Willgerodt et al.	1968	3.15	Tris	7.4	25 or 50	20	490	50 normal individuals	252

[a] After mixing substrate with enzyme. Acetylcholine chloride or bromide generally used as substrate (except as otherwise stated) of a final molar concn of 1–4 mM.
[b] Amount of serum or plasma (except as otherwise stated) present in the reaction mixture.
[c] Time for the hydrolysis of substrate after which the reaction is stopped by adding alkaline hydroxylamine solution. After 1–3 min, HCl is added and then the color is developed with $FeCl_3$.
[d] References, not mentioned in the previous review (13), are given also to studies which make use of the procedure or its slight modifications.
[e] Erythrocyte AChE in whole blood.
[f] When whole blood is used, pH 8.6 and 60 min incubation are recommended (246). Original method described also for erythrocytes.
[g] PhAc, phenyl acetate; SuCl, succinyldicholine.
[h] 30 min for whole blood and erythrocytes, 60 min for serum.

Modifications. Among the more recent modifications of this method, that described by Pilz (244) contains a number of details on variables and has been adapted to various esterase systems. His procedure is based upon detailed chemical studies (253,254) of the usefulness of the hydroxamic acid-Fe(III) reaction in ester determinations; the influence of pH on the reaction between hydroxylamine and the ester and on the Fe(III)-hydroxamate containing solution; the effect of the Fe(III) concentration on the absorption of the colored solution; the stability of the color, etc. This study was later extented to noncholine esters (255). The enzymatic method differs from most other methods based on ester disappearance in using a comparatively large reaction volume (10 or 30 ml) and borate buffer. The results published by Pilz in a number of diagrams seem to be very exact (96) (experimentally found data plotted in a diagram usually follow a straight line in a most unexpected way), although he has not given any data on the precision or reproducibility from duplicate runs (24,95,96). The same method has been used for routine assay of ChE in the whole blood of the rat (245) and of normal human individuals (246). The esterase activity of serum and erythrocytes of 250 normal persons was tested with seven various substrates using a standardized procedure in order to make all the activity values obtained comparable (247). The procedure has also been used to determine simultaneously both types of ChE present in rat blood at two different pH values (256). The advantages of the method have been further emphasized by Pilz (96) in a separate paper. An adaptation of the method to the μliter scale, using tris buffer and a total reaction volume of 3.15 ml, was recently described and used for the routine determination of serum ChE (252). The sensitivity of this technique in comparison with Pilz' original procedure is not clear.

Another variant of the method, using a barbital buffer, is that by Vincent and Segonzac (242,243). It has been referred to only in a few cases (257,258), and has been tested for assaying arylesterase with phenyl acetate as substrate (159). A paper describing a portable micromethod for blood serum ChE should be consulted for details (249).

The Hestrin method has been modified also for continuous automated assay of blood ChE using specific substrates (butyrylcholine-I and acetyl-β-methylcholine-Cl) for plasma and erythrocyte enzymes (260, 261). This permits a three channel system to be used for the simultaneous monitoring of total blood ChE (with acetylcholine-Br as substrate) and of the two individual components. The ChE values obtained with this automated technique were found to correlate highly (correlation coefficient, 0.89 for plasma) or well (0.82 for whole blood) or less well

(0.76 for erythrocytes) with those obtained with a manual (Michel) technique (84).

When the results obtained with the Hestrin method in its various modifications have been compared with those obtained with other techniques, fairly satisfactory agreement has been found, e.g., with the automatic titration method (168,174,232,236), with the Warburg gasometric method (30,38,231), and the electrometric method by Michel (30,24,95). This is further discussed below.

Applications. The method of Hestrin in its various modifications has been recommended for the determination of organophosphate esterase inhibitors (insecticides present as residues in food products and other materials). In the original method, presented by Cook (262), the HCl reagent is replaced by trichloroacetic acid (dissolved in HCl) as a plasma protein precipitant to aid in clarification for photometric readings. In a later modification (263), the Fe reagent is added after filtration (to avoid fading of the color) and photometric readings are made at 500 nm (instead of 540 nm). It was applied to experiments showing that bromine treatment converts thionophosphonates (e.g., parathion, chlorthion) and dithiophosphates (e.g., malathion) to *in vitro* ChE inhibitors, thus making the enzymic method suitable for the analysis of these derivatives (264,265). When this method was compared with the manometric, the Michel, and the titrimetric methods, a maximum difference of $\pm 15\%$ was obtained (266). Data have also been presented on the effect of enzyme inhibitor reaction time on inhibition (250), on the close relationship of the effect of various amounts of certain organophosphate inhibitors on fly mortality and *in vivo* inhibition of fly ChE (251), and on the presence of these insecticides in fruit juice (267).

Hestrin's method was also used to measure the concentration of thiocholine esters in the range of 0.05–2 μmoles/ml (268) and for quantitative analysis of physostigmine (269). It has also been tested for the estimation of various types of ChE using selective substrates (99,260,270). An application has been described to the assay of local anesthetic esters (e.g., procaine), for which the method is less rapid than for choline and aliphatic esters (271). The enzymic hydrolysis of these anesthetics by liver and plasma was studied, and it was found that the liver of the guinea pig is particularly active in this respect. The disappearance of ACh, acetyl-β-methylcholine, and ethylchloroacetate in a perfusion solution passed through the ganglion cervicale of the cat has also been studied by Hestrin's method (272).

Comments. This method is very convenient, simple, and quick, and therefore suitable for routine investigations of ChE levels in a comparatively small amount of biological material, crude tissue homogenates, or purified preparations. It is not affected by spontaneous nonspecific acid or alkali liberation, which are obstacles for several other methods. The experimental conditions can be varied widely, including wide ranges of pH and buffer solutions, and a variety of substrate (0.04–5 μmoles/ml), enzyme, and salt concentrations. Various substrates (choline as well as ordinary esters) can be employed, allowing the study of esterase specificity. Under most conditions described a straight-line relationship between enzyme concentration and enzyme activity is obtained, and the rate of substrate disappearance is linear with time.

The accuracy of the Hestrin method is less than for the manometric, electrometric, titrimetric, Ellman, or the radiometric methods since the activity is determined by difference; generally at least 30% of the ester should be hydrolyzed in order to obtain the most accurate results. In contrast to several other methods, it does not, in a simple determination, give a distinct picture of the course of the enzymic reaction itself (i.e., rate curves are not obtained). A technical difficulty results from the formation of gas bubbles after the addition of $FeCl_3$, which is best prevented by filtration (together with precipitated protein) (168). The inhibiting effect of a large excess of phosphate or borate on the color reaction (already noted by Hestrin) has not been further investigated.

One difficulty in the method has been discussed by several authors. After the addition of alkaline hydroxylamine, the reaction mixture is stable for several hours at room temperature; however, fading does occur after the addition of $FeCl_3$, and consequently the absorbancy must be determined a few minutes thereafter (14). In addition, the stability of the Fe(III) reagent itself for ACh determination in blood has been questioned (74,95), but this has been postulated to be without justification (96). This reagent, which is present in excess, has a slight absorbance at the wavelength (515 nm) at which the Fe(III)-hydroxamate has its maximum absorbancy. Therefore the color is usually read at 540 nm (Pilz measures at 490 nm). Both fading of the color and absorbance of the Fe(III) reagent itself can be reduced when HNO_3 is substituted for HCl (168,273). It is surprising that this finding, obviously an improvement on the general method, has not been further investigated.

2. Schönemann Reaction

Another reaction which has been proposed for ACh determination and assaying ChE activity is the so-called Schönemann reaction. Those phosphoryl and carbonyl compounds which are good phosphorylating or acylating agents (i.e., organophosphates, acid halides, and acid anhydrides) react with perhydroxyl ion (peroxide anion) to form peracids which can oxidize benzidine indicators (e.g., o-dianisidine) to the colored form. Certain esters also give this reaction, provided the carbonyl carbon is particularly electrophilic (ethyl acetoacetate reacts but not ethyl acetate). The carbonyl carbon of ACh has a high electrophilicity and should consequently give the Schönemann reaction. This was actually demonstrated by Aksnes and Sandberg (274), and by Hanker et al. (275) who found that other choline esters (e.g., acetyl-β-methylcholine and succinylcholine) form colored products as well.

In principle, this method of assaying ChE activity is similar to that of the Hestrin method, i.e., measurement of the rate of disappearance of ACh. The sensitivities of the two methods seem to be similar, 0.1–0.4 μmoles of ACh/ml when o-dianisidine is used as a redox indicator. The method has been used only once in esterase studies, i.e., for assaying the activity of a preparation of bovine erythrocyte AChE (275). Its usefulness in ChE assay, however, should be tested further. Since organophosphorus compounds (strong inhibitors of esterases), when present in sufficient quantity, give a positive Schönemann reaction, these should not be present. In addition, iodide ions give this reaction; accordingly iodides of the choline esters cannot be used in this method.

3. Benzoylcholine and Other Choline Esters as Substrates

Certain aromatic esters have UV spectra which are shifted toward longer wavelengths and have high absorptivity compared to the spectra of the hydrolysis products. This principle is the basis for a method described by Kalow and Lindsay (276) for human serum ChE and using benzoylcholine as substrate. The disappearance of the absorption of this ester at 240 nm is followed, which gives rate curves with each assay, which may be carried out in 4 min with 0.02 ml of serum. A close correlation was found between the results obtained with this and the gasometric method if benzoylcholine was the substrate, but not if ACh (gasometric) and benzoylcholine (optic) were compared.

The general usefulness of this method is limited because the principle is based upon the use of a single substrate. Benzoylcholine is a specific substrate for human serum ChE, but the specificity of esterases from other sources towards this ester is not so well investigated. The simplicity of the method, however, has made it suitable for kinetic and other studies. It was the method used by Kalow and his colleagues in their classical work of the so-called "dibucaine-resistant" variant or atypical form of human serum ChE (277,278), and several other workers in the field of genetic ChE variation have used the same method (5,279,280,281, 282). In the excellent monograph by Goedde et al. (5) the various methods being used in these studies are described in detail. A screening test for typing of serum ChE phenotypes is described below (Section VIII). The usefulness of the Kalow-Lindsay method in clinical laboratories has also been illustrated (248,283,284), as has its use in the ChE assay of living cell cultures (HeLa Cells) (285).

A method using suberyldicholine as substrate has been published in a paper, available only in abstract (286). This ester is said to be split specifically by serum ChE. The principle of the technique used is not clear.

VII. RADIOMETRIC ASSAY

The use of a labeled substrate for ChE determination provides a promising principle for a highly sensitive micromethod. Such a method was first described by Winteringham and Disney (287), who used acetyl 1-^{14}C-choline and measured the unhydrolyzed substrate after the acetic acid-^{14}C formed had been carefully volatilized. A more useful variant of this technique is based upon a direct assay of acetic acid-^{14}C formed by the enzymic reaction. The two methods are described in more detail below.

The enzymic hydrolysis of succinyldicholine and the first product of its degradation, succinylmonocholine, in human sera of different cholinesterase phenotypes were investigated by Goedde et al. (288), using ^{14}C-methyl-labeled substrates. Radioactivity of substrates and split products were estimated by a liquid scintillation spectrometer after separation by high voltage electrophoresis on paper. The excretion of ^{14}C-methyl-labeled succinyldicholine and succinylmonocholine in the urine of the rat after intraperitoneal application had been studied earlier (289).

Labeled compounds in ChE determination have also been described for a modified manometric technique making use of Na-bicarbonate-^{14}C (see Section II-3), and for histochemical techniques with labeled esterase-inhibitors (e.g., ^{3}H-DFP; not described in this review).

1. Assay of Unhydrolyzed Acetyl-^{14}C-Choline

The original radiometric method of Winteringham and Disney (287, 290) is based upon incubation (blood or tissues, 50 μg) with acetate-labeled ACH (0.47 mCi/mmole) for the required time (0.5–3 min). Enzyme action is halted by the addition of acid and excess of inhibitor (eserine) to a small sample on a cavity microscope slide. The mixture is then allowed to dry and counted for ^{14}C-activity under a thin end-window Geiger-Müller tube in the conventional way. Under these conditions

^{14}C-acetate is completely volatilized, and the unhydrolyzed ^{14}C-labeled substrate remains. Control assays in which the enzyme is absent or inactivated (by 0.1 mM eserine) are made simultaneously. The difference in residual ^{14}C-activity measured is a direct measure of ChE activity. The method, which enables ChE activity to be determined at low substrate concentration and with minimal sample dilution, has been used under field conditions to measure human blood ChE (10 μl of haemolyzed and diluted whole blood) (291,292), and also to assay samples taken from persons exposed to carbamate insecticides (293).

This method has been shown particularly useful in studies on the effect of carbamates which behave as reversible esterase inhibitors. In these studies it is important to use a low substrate concentration and a minimal dilution of the inhibited enzyme. Disney (294) compared the electrometric method with the radiometric one and found considerably greater inhibition by the carbamates, but not by organophosphates, when the radiometric method was used. This was explained mainly by the high substrate concentration and considerable dilution of the enzyme involved in most of the conventional methods, including the electrometric. A comprehensive study of the kinetics of the inhibition of AChE by carbamates over the substrate concentration range 50 nM–25 mM was published shortly thereafter by Winteringham and Fowler (295–299). The results obtained in these studies using the radiometric method emphasize the great importance of taking substrate and dilution effects into account when using blood ChE inhibition as an index of animal or human exposure to insecticidal carbamates.

Another application of the method is the study of the metabolism and significance of ACh in the brain of the housefly (298) and the enzymic determination of organophosphate esterase inhibitors (300). In the latter study, a cholinesterase inhibitor (Bidrin) belonging to this group could be determined in the range 0.05–0.6 ng and a carbamate inhibitor (Arprocarb) in the range 5–16 ng (present in 1 g of plant material).

The main obstacle to this method, though sensitive and precise, is that it requires a major portion of the substrate to be hydrolyzed (as with the Hestrin method). Moreover, before the remaining substrate can be assayed, the acetic acid-^{14}C formed must be volatilized carefully.

2. Assay of Acetic Acid-^{14}C Formed

A direct assay of acetic acid-^{14}C formed by ChE activity (in rat blood) with acetyl-1-^{14}C-choline as a substrate was first described by Reed et al. (301). The unhydrolyzed substrate is removed by adsorption onto a cation exchange resin (Amberlite CG-120), and the radioactivity remaining in the supernatant fluid (i.e., acetate-1-^{14}C), assayed with a liquid scintillation spectrometer, is a measure of the hydrolysis reaction. The method requires less than milligram quantities of fresh tissue for each analysis (cf. 40 mg for the Warburg method) and is more sensitive than that described by Winteringham and Disney (290). It was applied to examine the distribution of AChE in liver flukes (*Fasciola hepatica*) and gave data in good agreement with those obtained with the Warburg method (302).

The method described by Reed and his co-workers was recently modified in some details for the assay of AChE activity in rat brain (303). It permits accurate estimation of the activity in 0.2 mg of tissue. A low final substrate concentration (4 μM ACh) was reported in this modification and the incubation mixture consisted of 0.1 ml of substrate, 0.1 ml of phosphate buffer, and 0.2 ml of brain supernatant.

A second technique for separating the ^{14}C-acetate liberated enzymically was described by Potter (304). He used a toluene-isoamyl alcohol solvent system to extract the acetate, an aliquot of the organic phase was removed, "backwashed" with a fresh volume of dilute acid, and a final portion of the organic phase was placed directly in the counting vial for radioactivity measurement by liquid scintillation spectrometry. This method was used to measure AChE in 0.1 ng–3 mg of brain tissue. Potter described the use of radioactive substrates other than ACh, i.e., ^{3}H-acetylcholine, ^{3}H-acetyl-β-methylcholine, and ^{14}C-benzoylcholine, as well as the effect of inhibitors. This technique represents a major improvement, mainly because of higher sensitivity, compared with the two other radiometric methods described above. A similar procedure was recently used to measure ChE activity in rat brain homogenate and subcellular fractions as well as in human whole blood (305).

A third procedure differs from Potter's method in the way in which the products are separated from the substrate. It was recently described by McCaman et al. (306), who measured acetate-1-^{14}C after the selective removal of the unhydrolyzed substrate (acetyl-1-^{14}C-choline) as an insoluble reinecke salt. Unlike the extraction procedure, the precipitation of unhydrolyzed substrate does result in a quantitative recovery of ^{14}C-acetate. The sensitivity of the method is adequate for the assay of 50 ng (protein) of brain tissue. The convenience was illustrated by kinetic studies of the inhibition of rat plasma ChE by acetylthiocholine. A minor modification of this procedure was recently used by Koslow and Giacobini (307). In their studies of the ChE activity of single nerve cells, they used a lower substrate concentration (0.6 instead of 1.0 mM) to avoid the substrate inhibition produced when very low concentrations of enzyme (1–5 ng of tissue) were present.

The radiometric determination of ChE activity by either of the two last-mentioned procedures offers an opportunity for ultramicrochemical studies of enzyme distribution, as well as for more detailed investigation of the kinetics of reversible esterase inhibitors, and the assay of enzymes in unaltered fluids (i.e., without any dilution and without any addition other than substrate).

VIII. METHODS BASED UPON THE USE OF NONCHOLINE ESTERS

It is generally accepted that no ChE so far investigated has an absolute specificity for choline esters. In fact, all ChE also split ordinary esters, the various enzymes having distinct specificity patterns. Thus, for example, AChE splits acetic acid esters more rapidly than propionic or butyric acid esters, whereas human plasma ChE catalyzes the hydrolysis of butyric acid esters at a higher rate than the esters of the lower homologous acids. Other types exist which split propionic acid esters at the highest rate. This rule is valid for choline as well as for noncholine esters. Consequently, any acetate should be a more favorable substrate for AChE than for other ChE, and a butyrate can be expected to be a useful substrate for human serum ChE.

There are, however, a great many hydrolases which can be responsible for the reaction in a crude enzyme preparation when using a noncholine ester. The main problem in the use of noncholine esters in ChE studies, therefore, is to determine whether the ester in question is actually split by a ChE alone and/or by another esterase as well. The generalization may be made that any ester can be used as a substrate for assaying ChE activity, assuming that the preparation studied contains only ChE and that the specificity of the esterase activity is known in detail. For example, a preparation containing AChE as the only esterase can be studied with any ester split by this enzyme.

As long as choline esters with more or less selective specificity for various ChE are available, such esters are preferable to the less specific noncholine esters. In special cases, however, certain noncholine esters may be of great value, for instance, in histochemical detection of these enzymes and in detection of ChE in chromatograms or electropherograms. Particularly useful substrates are those which on hydrolysis give reaction products which possess characteristic colors or are easily detected by spectrophotometric, fluorometric, or radiometric techniques.

In the following section the usefulness of noncholine esters in ChE studies will be discussed, the presentation being restricted to the development of new principles and to the improvement of earlier methods (13). The literature cited in this section is not intended to be complete.

1. Phenyl Esters

The advantage in using a phenyl ester as substrate results from the various simple techniques that are available to assay the phenol liberated. These esters, however, are also hydrolyzed by a number of esterases and peptide-hydrolases of unknown specificity. Phenyl *acetate* is a more suitable substrate for the assay of arylesterases (3,8,10,308) and should not be used for ChE. Phenyl *butyrate* has been recommended as a substrate for human plasma ChE (309), particularly in studies where other enzymes (phosphatases, lipases, sulfatases, glucuronidase) are determined by the same principle.

Phenyl *benzoate*, a classical substrate introduced by Gomori, has been used as a simple colorimetric method for assaying human serum ChE (310), and was recently applied to the analysis of sera of homozygotes, for "usual" and "atypical" ChE genes, and of heterozygotes (311). The values obtained were usually higher than those obtained with a manometric technique. Arylesterase does not seem to interfere in this method, which therefore seems to be acceptable as a routine method for human plasma ChE, since carboxylesterase is normally absent in this material. Phenol produced is determined with the Folin-Ciocalteau's reagent (310) or α-aminoantipyrin (309). Methods based upon acid production are obviously also useful. The phenyl benzoate method suffers from the disadvantage of low substrate solubility and a relatively high K_m value.

2. Nitrophenyl Esters

A variety of carboxylic esters of p-nitrophenol have been used to assay blood serum esterases, utilizing the development of color (p-nitrophenolate ion measured at pH 7.6 and 400 nm) from the hydrolysis of the colorless esters. Most types of esterases, however, split these esters, the specificity of the reaction being influenced by the nature of the aryl substituent. The arylesterases, for example, hydrolyze p-nitrophenyl acetate more rapidly than the butyrate, the reverse specificity being valid for human serum ChE. p-Nitrophenyl acetate has been used as substrate in quantitative assay of blood esterases (312,313), but it is recommended that the type of esterase actually measured with this substrate is investigated first with selective inhibitors and substrates. Other esterases, e.g., carboxylesterases and ChE in rat tissues, split p-nitrophenyl propionate at the highest rate. This was utilized recently (314) in a rapid and accurate spectrophotometric method for these enzymes.

The effect of the orientation of the nitro group of nitrophenyl esters on ChE and arylesterase activity was studied by Main (315,316). He demonstrated that the nitro group orientation, rather than the nature of the aryl substituent, influenced the activity of human serum ChE, but that the reverse was true for the arylesterases. Under standard conditions (pH 7.6, 0.05M phosphate buffer, 1 mM substrate) the ratios between the activities of ChE and arylesterase (ArE) for the various substrates were as follows:

		ChE/ArE
o-nitrophenyl	butyrate	42
	acetate	2
p-nitrophenyl	butyrate	2.4
	acetate	0.3

In addition, human serum ChE has a greater affinity and greater activity towards o-nitrophenyl butyrate than toward acetylcholine, and in crude serum this enzyme is responsible for at least 96% of the enzymic hydrolysis of the aromatic ester. In the presence of 5% (v/v) butan-1-ol, the ChE-catalyzed hydrolysis of o-nitrophenyl butyrate is activated 3-fold, whereas the arylesterase activity is completely inhibited. As a consequence of these results, Main developed a rapid, sensitive spectrophotometric method suitable for routine analysis of human serum ChE (316) and applied the same technique to a kinetic study of this enzyme (315). When a suitable photometer equipped with an automatic cuvet positioner and recorder is used with this technique, it has been possible to carry out at least 60 analyses/hr (317).

o-Nitrophenyl butyrate as substrate and succinyldicholine as inhibitor have been used for detecting "atypical" serum ChE (318,319). It was also demonstrated that differences exist in enzyme-substrate inter-action between o-nitrophenyl butyrate and choline esters for the usual and atypical forms of human serum ChE. This noncholine ester was not suitable for measuring ChE activity after reactivation of alkyl-phosphate-poisoned enzyme by 2-PAM, since 2-PAM alone (under the conditions of the reaction) splits the substrate noncatalytically liberating o-nitrophenol (320).

3. Naphthyl Esters

Some of the most frequently used substrates for esterase determinations are α-naphthyl acetate and certain other esters of α- or β-naphthol. These are particularly useful in histochemical identifications and in detecting esterase activity after resolution by electrophoresis or chroma-tography (gel, paper, etc.) (321,322). A recent study on the isoenzyme status of esterases in certain vertebrate tissues was based on the use of α-naphthyl acetate and some related α- and β-naphthyl derivatives (323,

324). In addition to studies on this type of substrate reported earlier (13), β-naphthyl acetate or butyrate has been used to assay total esterase (aryl-, carboxyl-, and cholinesterase) in erythrocytes and serum (325). The procedure using α-naphthyl acetate can be used for quantitative determination of esterase activities after starch gel electrophoresis. The active fractions may be identified after staining of the gel by elution with a selective solvent (e.g., n-amyl acetate–ethanol, 1:1) (326), gel sections can be broken down by alternate freeze-thawing, or enzymically, or the intensity of staining of the gel strip may be evaluated with a densitometer. Yet another procedure is based upon hydrolysis of the starch gel by NaOH, which releases the dye. The dye is then taken up in an immiscible solvent and the color determined spectrophotometrically. This method was recently applied for a quantitative study of the isoenzymes of ChE in equine serum (327).

It will be remembered that the use of a naphthyl ester as a substrate for esterase activity will not give a picture of all esterases present, because some may not split this ester type. Furthermore, α-naphthyl acetate, for example, is a nonspecific substrate for several forms of esterase, for which it has different affinities (328). However, in comparisons between species and tissues, or similar studies, the relative specific activities towards α-naphthyl acetate or another particular substrate can be a useful measure, provided it is stated what activities are measured and what types of esterase are not detected.

The principle used in the procedure with a naphthyl ester is to couple the naphthol liberated during the reaction with a suitable diazonium salt (e.g., 4-amino-diphenylamino-diazonium sulfate, diazo-o-dianisidine, Fast Blue RR) to produce a colored compound (247,329). Special naphthol esters have also been reported to be useful substrates, e.g., the yellow 2-azobenzene-1-naphthyl acetate which by hydrolysis gives red 2-azobenzol-1-naphthol (330).

Still another variant of the procedure with naphthyl esters is based on the observation that the molecular forms of α- and β-naphthol are very fluorescent, whereas the esters are nonfluorescent. A fluorimetric method was described for the direct and rapid estimation of both ChE and acid phosphatase, and in addition a continuous fluorimetric system for the assay of ChE inhibitors was developed (331,332). The rate of change in the fluorescense of the reaction mixture due to production of α-naphthol (λ_{ex} = 330 nm, λ_{em} = 460–470 nm) and β-naphthol (λ_{ex} = 320 nm, λ_{em} = 410 nm) is measured and correlated with enzyme activity. This procedure has been successfully used in kinetic studies with α-naphthyl butyrate and acetate and certain selective inhibitors of AChE and human serum ChE (333).

Agar Diffusion Test for Serum Cholinesterase Typing. The use of α-naphthyl acetate as a substrate for serum ChE was recently found to be particularly useful in a screening test for typing of serum ChE phenotypes. The differentiation of the "atypical" form of human serum ChE from the normal ("usual") form is carried out generally by using certain inhibitors, e.g., dibucaine (a local anaesthetic), fluoride, or succinyldicholine, which are much stronger inhibitors of the normal type than of the atypical type of the enzyme (5). Certain alcyl alcohols have recently been studied in this respect (282). Another differential inhibitor, R02-0683 (dimethylcarbamate of 2-hydroxy-5-phenyl-benzyl trimethylammonium bromide), is more selective than dibucaine, and was used by Harris and Robson (334) for a screening test, the "agar diffusion test." Unlike the standard method for determination of ChE types by UV spectrophotometry using benzoylcholine as substrate (see Section VI-3), this screening test requires only simple equipment and permits rapid testing of a large number of sera for distinguishing "normal" from "atypical" (and "intermediate") ChE types.

The agar-diffusion test, as originally described (334), uses α-naphthyl acetate as a substrate with 5-chloro-o-toluidine as diazo reagent. A control agar (tris buffer, pH 7.4) contains no inhibitor and a second tray contains the inhibitor (0.1 μM R02-0683). Sera to be tested are suitably diluted and inserted into wells (4 mm in diameter) in the trays. After incubating overnight (37°C) the agar is flooded with a mixture containing 100 ml of $0.2M$ phosphate buffer (pH 7.1), 2 ml of a 1% solution of α-naphthyl acetate in 50% aqueous acetone, and 20 mg of 5-chloro-o-toluidine. After 2 hr at room temperature the color is developed optimally for discrimination. In the control tray, sera with normal activity give a sharp-edged brown circular zone (about 1.3 cm in diameter). In the inhibitor tray two types of reactions can be observed: (*1*) negative-sera (phenotype "usual") which show marked inhibition resulting in faint, diffuse zones, and (*2*) positive sera (phenotype "atypical" and "intermediate") which yield sharp-edged, brown circular zones comparable in diameter to the corresponding zones in the control tray. The usefulness of this simple test has been confirmed by other workers (335,336,337), and the results found to agree fairly well with those obtained with the standard uv method. The application of the procedure to ChE screening of potential insecticides and residues has also been reported (338).

α-Naphthyl acetate was also used as substrate in a test-tube screening technique with NaF (339) or R02-0683 as inhibitor (340). The latter method detects both homozygotes and heterozygotes for the gene determining the atypical gene, which is not possible with the agar-diffusion test.

4. Indoxyl Acetate

Various authors have reported the colorimetric assay of ChE based on the formation of indigo blue from indoxyl acetate (13). In addition to ChE, acylase, lipase, and acid phosphatase also affect the hydrolysis of indoxyl acetate. The butyrate has also been tested as substrate and, as expected, is not hydrolyzed by AChE (from erythrocytes) (341). Indoxyl acetate, together with R02-0683 as a discriminating inhibitor, has been used for the determination of serum ChE phenotypes (342,343) in a similar way to that described above for α-naphthyl acetate. It is also useful for detecting esterase active components on chromatograms (344) and for detecting nanogram quantities of organophosphorus pesticides (345,346) and carbamates (347) on thin-layer chromatograms. In this connection a number of substituted indoxyl acetates (5-bromo-, 5-bromo-4-chloro-, etc.) have been tried as well. These substrates seem to give more satisfactory results than α-naphthyl acetate. Spray solutions of esterase (e.g., serum, liver homogenate) and substrate are used at a pH of about 8 and give colored products on hydrolysis which are stable and intense; white spots indicate the sites of an organophosphate (or other) esterase inhibitor. The same principle can be used in a test-tube technique (348).

Guilbault and Kramer (331) have proposed the use of indoxyl acetate as a fluorimetric substrate for ChE. Hydrolysis of this ester yields indoxyl, which is oxidized by air first to indigo white then to indigo blue. Both indoxyl and indigo white are fluorescent, indigo blue is highly colored but nonfluorescent, as is also indoxyl acetate. In the method described, the rate of increase in the fluorescence ($\lambda_{ex} = 395$ nm, $\lambda_{em} = 470$ nm) is measured at pH 6.5 and correlated with enzyme activity. The pH of the solution must be <7, otherwise the indigo white formed is rapidly oxidized to indigo blue (at pH > 7). The fluorescence is detectable in as low concentrations as 0.01 μM and extremely small quantities of esterase can be measured. Other fluorogenic substrates were resorufin acetate and butyrate which were also useful in esterase determination in the same way as indoxyl acetate (331,349). In this case the production of resorufin is measured ($\lambda_{ex} = 540$–570 nm, $\lambda_{em} = 580$ nm).

N-Methyl indoxyl esters is not as easily converted to the corresponding indigo as the nor compound. For this reason they were not used for histochemical purposes (341). N-Methyl indoxyl acetate, however, has a prominent structural resemblance to ACh and is hydrolyzed at a higher rate by ChE than is indoxyl acetate. In addition, the free fluorescent product following ester hydrolysis is relatively more persistent. Consequently, N-methyl indoxyl esters should appear to be ideal fluorogenic substrates. This was actually demonstrated recently by Guilbault and co-workers (350), who synthesized and tested the acetate, propionate, and butyrate of N-methyl indoxyl as substrates for AChE (erythrocytes) and ChE (horse serum). Comparison of these substrates with other fluorogenic esters (indoxyl esters, resorufin acetate, β-carbonaphthoxy-choline, β-naphthyl acetate, esters of umbelliferone, and 4-methyl umbelliferone) indicated that N-methyl indoxyl butyrate was the best substrate for human serum ChE. This technique seems to be a simple and promising routine method for this enzyme.

5. Indophenyl Acetate

Another simple method is also based on the direct measurement of one of the products of enzymic hydrolysis and does not require any intermediate reactions or additional reagents. This technique uses indophenyl acetate which is hydrolyzed (at pH 8.0) by AChE at a comparatively high rate (351) to produce a highly colored product (abs max at 625 nm). The indophenyl acetate procedure was later applied for a direct colorimetric analysis of ChE-inhibiting insecticides, using various enzyme sources (erythrocytes, serum, bee brain) (352). The same principle using bee brain AChE was also employed for the detection of such insecticides following separation on thin-layer chromatograms (34). For the histochemical location of esterases after starch-gel electrophoresis, 2,6-dichloro-phenolindophenyl acetate has been described (353).

References*

1. K.-B. Augustinsson, *Biochim. Biophys. Acta*, *128*, 351 (1966).
2. J. A. Cohen and R. A. Oosterbaan, *Hand. Exp. Pharmakol. Ergänz.*, *15*, 299 (1963).
3. K.-B. Augustinsson, *Ann. N.Y. Acad. Sci.*, *94*, 844 (1961).
4. A. L. Latner and A. W. Skillen, *Isoenzymes in Biology and Medicine*, Academic Press, London and New York, 1968.
5. H. W. Goedde, A. Doenicke, and K. Altland, *Pseudocholinesterasen. Pharmakogenetik, Biochemie, Klinik*, Springer, Berlin, 1967.
6. R. D. O'Brien, *Toxic Phosphorus Esters*, Academic Press, New York, 1960.
7. E. Heilbronn-Wickström, *Svensk Kem. Tidskr.*, *77*, 3 (1965).
8. K.-B. Augustinsson, in *Homologous Enzymes and Biochemical Evolution*, N. Van Thoai and J. Roche Eds., Advanced Study Group of NATO, Paris, 1968, pp. 299–311.
9. K.-B. Augustinsson, *Acta Chem. Scand.*, *13*, 571 (1959).
10. K.-B. Augustinsson, and G. Ekedahl, *Acta Chem. Scand.*, *16*, 240 (1962).
11. L. A. Mounter, *Hand. Exp. Pharmakol. Ergänz.*, *15*, 486 (1963).

* References in the text and tables to author's name and year (no reference number) will be found in the previous review of 1957 (Ref. 13 in the present list).

12. F. Bergman, R. Segal, and S. Rimon, *Biochem. J.*, *67*, 481 (1957).
13. K.-B. Augustinsson, *Methods Biochem. Anal.*, *5*, 1 (1957).
14. R. F. Witter, *Arch. Environ. Health*, *6*, 537 (1963).
15. J. M. Barnes, W. J. Hayes, and K. Kay, *Bull. World Health Org.*, *16*, 41 (1957).
16. E. Reiner, *Arhiv Hig. Rada Toksikol.*, *9*, 25 (1958).
17. H. H. Golz, *Arch. Ind. Health*, *18*, 138 (1958).
18. K.-B. Augustinsson, *Hand. Exp. Pharmakol. Ergänz.*, *15*, 89 (1963).
19. J. C. Gage, *Residue Rev.*, *18*, 159 (1967).
20. W. Pilz, *Z. Klin. Chem.*, *5*, 1 (1967).
21. V. Simeon, *Arhiv Hig. Rada*, *18*, 383 (1967).
22. W. W. Umbreit, R. H. Burris, and J. F. Stauffer, *Manometric Techniques*, 4th ed., Burgess, Minneapolis, 1964.
23. K.-B. Augustinsson and B. Holmstedt, *Scand. J. Clin. Lab. Invest.*, *17*, 573 (1965).
24. R. L. Baron, J. L. Casterline, Jr., and R. Orzel, *Toxicol. Appl. Pharmacol.*, *9*, 6 (1966).
25. M. Rombach, *Bull. Soc. Pharm. Nancy*, *72*, 25 (1967).
26. R. F. Witter, *Toxicol. Appl. Pharmacol.*, *4*, 313 (1962).
27. F. J. Zapp, *Z. Physiol. Chem.*, *307*, 36 (1957).
28. J. Orfila, *C. R. Soc. Biol.*, *150*, 2204 (1956).
29. G. Agioutantis, *Arch. Iatrikon Epistimon*, *16*, 19 (1960).
30. G. Hecht and E. Stillger, *Z. Klin. Chem. Klin. Biochem.*, *5*, 156 (1967).
31. L. A. Mounter, *Enzymologia*, *21*, 67 (1959).
32. Y. Hagiwara, *J. Showa Med. Ass.*, *19*, 98 (1960).
33. H. Breuer and M. Schönfelder, *Clin. Chim. Acta*, *6*, 515 (1961).
34. W. Winterlin, G. Walker, and H. Frank, *J. Agr. Food Chem.*, *16*, 808 (1968).
35. J. V. Auditore, R. C. Hartmann, J. M. Flexner, and O. J. Balchum, *Arch. Pathol.*, *69*, 534 (1960).
36. S. P. Bhatnager, *Arch. Int. Pharmacodyn. Ther.*, *175*, 422 (1968).
37. K. P. Du Bois, F. K. Kinoshita, and J. P. Frawley, *Toxicol. Appl. Pharmacol.*, *12*, 273 (1968).
38. W. Schaumann, *Arch. Exp. Pathol. Pharmakol.*, *239*, 81 (1960).
39. H. H. Moorefield and E. R. Tefft, *Contrib. Boyce Thompson Inst.*, *19*, 295 (1958).
40. H. Bergman, R. Kilches, S. Sailer, K. Steinbereithner, and E. Vonkilch, *Anaesthesist*, *11*, 279 (1962).
41. B. Fristedt and P. Övrum, *Acta Med. Scand.*, *184*, 493 (1968).
42. Z. Jabsa, M. Schönfelder, and H. Breuer, *Klin. Wochschr.*, *39*, 966 (1961).
43. S. Sailer and H. Braunsteiner, *Klin. Wochschr.*, *37*, 986 (1959).
44. G. L. Marx and Carter, M. K., *Amer. J. Physiol.*, *204*, 124 (1963).
45. P. Fransson and M. P. Molinari-Tosatti, *Boll. Soc. Ital. Biol. Sper.*, *43*, 1799 (1967).
46. K. Linderstrøm-Lang and D. Glick, *C. R. Trav. Lab. Carlsberg, Sér. Chim.*, *22*, 300 (1938).
47. R. Magiotti, in *Renitenza Certissima dell'Agua alla Compressione Dichiarata con Varii Scherzi in Occasione di Altri Problemi Curiosi*, Moneta, Rome, 1648.
48. E. J. Boell and S.-C. Shen, *J. Exp. Zool.*, *97*, 21 (1944).
49. J. Zajicek and E. Zeuthen, *Exp. Cell Res.*, *11*, 568 (1956).
50. E. Giacobini and J. Zajicek, *Nature*, *177*, 185 (1956).
51. E. Giacobini, *Acta Physiol. Scand.*, *42*, *Suppl. 145*, 49 (1957); *J. Neurochem.*, *1*, 234 (1957).

52. E. Giacobini and B. Holmstedt, *Acta Physiol. Scand.*, *42*, 12 (1958).
53. E. Giacobini, *Acta Physiol. Scand.*, *45*, 238 (1959).
54. E. Giacobini, *Acta Physiol. Scand.*, *45*, 311 (1959).
55. E. Zeuthen, *Gen. Cytochem. Methods*, *2*, 61 (1961).
56. J. Zajicek and E. Zeuthen, *Gen. Cytochem. Methods*, *2*, 131 (1961).
57. M. Brzin and E. Zeuthen, *C. R. Trav. Lab. Carlsberg*, *32*, 139 (1961).
58. M. Brzin, M. Kovic, and S. Oman, *C. R. Trav. Lab. Carlsberg*, *34*, 407 (1964).
59. M. Brzin and E. Zeuthen, *C. R. Trav. Lab. Carlsberg*, *34*, 427 (1964).
60. M. Brzin, W.-D. Dettbarn, P. Rosenberg, and D. Nachmansohn, *J. Cell. Biol.*, *26*, 353 (1965).
61. M. Brzin, V. M. Tennyson, and P. E. Duffy, *J. Cell. Biol.*, *31*, 215 (1966).
62. S. Larsson and S. Løvtrup, *J. Exp. Biol.*, *44*, 47 (1966).
63. A. Hamberger, L. Hamberger, and S. Larsson, *Exp. Cell Res.*, *47*, 229 (1967).
64. M. Brzin and W.-D. Dettbarn, *J. Cell. Biol.*, *32*, 577 (1967).
65. M. Brzin, V. M. Tennyson, and P. E. Duff, *Int. J. Neuropharmacol.*, *6*, 265 (1967).
66. V. M. Tennyson, M. Brzin, and P. Duffy, *Progr. Brain Res.*, *29*, 41 (1967).
67. M. Mitchard, E. Giacobini, and B. Carlsson, *Anal. Biochem.*, *37*, 112 (1970).
68. E. Giacobini, *Acta Physiol. Scand.*, *45*, *Suppl. 156* (1959).
69. E. Giacobini, in *Neurosciences Research*, Vol. 1, S. Ehrenpreis and O. S. Solnitzky, Eds., Academic Press, New York, 1968, p. 1.
70. G. B. Koelle, *Hand. Exp. Pharmakol. Ergänz.*, *15*, 186 (1963).
71. G. W. Ivie, L. R. Green, and H. W. Dorough, *Bull. Environ. Contam. Toxicol.*, *2*, 34 (1967).
72. J. A. Rider, J. L. Hodges, Jr., J. Swader, and A. D. Wiggins, *J. Lab. Clin. Med.*, *50*, 376 (1957).
73. L. W. Lee, *Amer. J. Med. Technol.*, *32*, 255 (1966).
74. K. H. Meinecke and H. Oettel, *Arch. Toxikol.*, *21*, 321 (1966).
75. M. W. Williams, J. P. Frawley, H. N. Fuyat, and J. R. Blake, *J. Ass. Offic. Agr. Chem.*, *40*, 1118 (1957).
76. F. Hermenze and W. J. Goodwin, *J. Econ. Entomol.*, *52*, 66 (1959).
77. M. Stevanović and R. Jović, *Vojnosanit. Pregled*, *16*, 782 (1959); from *Chem. Abstr.*, *55*, 1766 (1961).
78. J. F. Callahan and S. M. Kruckenberg, *Amer. J. Vet. Res.*, *28*, 1509 (1967).
79. K. H. Meinecke and H. Oettel, *Arch. Toxikol.*, *22*, 244 (1967).
80. J. H. Wolfsie, *Arch. Ind. Health*, *16*, 403 (1957).
81. R. S. Ganelin, *Arizona Med.*, *21*, 710 (1964).
82. J. H. Holmes and L. Jankowsky, *Arch. Environ. Health*, *13*, 564 (1966).
83. J. R. Pearson and G. F. Walker, *Arch. Environ. Health*, *16*, 809 (1968).
84. M. M. Grainger, W. A. Groff, and R. I. Ellin, *Arch. Environ. Health*, *16*, 821 (1968).
85. B. Melichar, *Pracovni Lekar.*, *7*, 355 (1961).
86. V. S. Andreyev, V. I. Rosengart, and V. A. Torubarov, *Ukr. Biokhim. Zh.*, *37*, 920 (1965).
87. S. Greenberg and C. Calvert, *Med. Technol. Bull.*, *8*, 59 (1957).
88. J. P. Frawley and H. N. Fuyat, *J. Agr. Food Chem.*, *5*, 346 (1957).
89. J. L. Stubbs and J. T. Fales, *Amer. J. Med. Technol.*, *26*, 25 (1960).
90. W. Burnett, *Gut*, *1*, 294 (1960).
91. G. G. Patchett and G. H. Batchelder, *J. Agr. Food Chem.*, *8*, 54 (1960).
92. D. Burman, *Amer. J. Clin. Pathol.*, *37*, 134 (1962).

93. D. G. Johnston and W. C. Huff, *Clin. Chem.*, *11*, 729 (1965).
94. R. F. Witter, L. M. Grubbs, and W. L. Farrior, *Clin. Chim. Acta*, *13*, 76 (1966).
95. K. H. Meinecke and H. Oettel, *Arch. Toxikol.*, *22*, 198 (1966).
96. W. Pilz, *Arch. Toxikol.*, *22*, 192 (1966).
97. R. Mesters, C. Jeanjean, and Mrs Tourte, *Ann. Fals. Expert. Chim.*, *60*, 33 (1967).
98. W. T. Caraway, *Amer. J. Clin. Pathol.*, *26*, 945 (1956); *Tech. Bull. Registry Med. Technol.*, *26*, 159 (1956).
99. K. Obara, K. Esashi, and H. Ueda, *Sogo Igaku*, *13*, 1302 (1956); from *Chem. Abstr.*, *54*, 24997 (1960).
100. H. Miyake, *Fukuoka Igaku Zasshi*, *49*, 1991 (1958); from *Chem. Abstr.*, *52*, 20360 (1958).
101. Ruzdić, *Farm. Glasnik*, *14*, 227 (1958); from *Chem. Abstr.*, *52*, 17353 (1958).
102. F. Rappaport, J. Fischl, and N. Pinto, *Clin. Chim. Acta*, *4*, 227 (1959).
103. A. A. Pokrovskiĭ, *Voenno-Med. Zh.*, *1953*, Nr. 9, 61; from *Chem. Abstr.*, *51*, 18072 (1957).
104. G. A. Panosyan, *Izv. Akad. Nauk Arm. SSR Biol. i Sel'skokhoz. Nauki*, *11*, No. 6, 21 (1958); from *Chem. Abstr.*, *52*, 17357 (1958).
105. M. Brzin and T. Priversek, *Vesn. Solven. Kem. Drustva*, *4*, 119 (1957).
106. M. Kitamura, M. Sato, M. Tsuda, H. Takahashi, and I. Takeda, *Japan J. Clin. Pathol.*, *9*, 340 (1961).
107. H. G. Biggs, S. Carey, and D. B. Morrison, *Amer. J. Clin. Pathol.*, *30*, 181 (1958): *Tech. Bull. Registry Med. Technol.*, *28*, 137 (1958).
108. M. B. Schiller, *J. Brasil. Psiquiat.*, *7*, 207 (1958).
109. W. E. Robbins, T. L. Hopkins, and A. R. Roth, *J. Econ. Entomol.*, *51*, 326 (1958).
110. A. A. Pokrovskiĭ, *Voenno-Med. Zh.*, *1960*, Nr. 1, 34; from *Chem. Abstr.*, *54*, 22804 (1960).
111. A. A. Pokrovskiĭ, *Bull. Exp. Biol. Med.* (Eng. ed.), *51*, 730 (1961).
112. E. Sándi, *Nahrung*, *6*, 57 (1962).
113. G. Dabija, I. Marinescu, and C. Popa, *Lucr. Inst. Cercet. Vet. Bioprep. Pasteur*, *3*, 331 (1964); from *Chem. Abstr.*, *68*, 68018 (1968).
114. H. W. Gerarde, E. B. Hutchison, K. A. Locher, and H. H. Golz, *J. Occup. Med.*, *7*, 303 (1965).
115. B. Bäriswyl, *Kleintierpraxis*, *10*, 93 (1965).
116. E. F. Edson, *World Crops*, *10*, 49 (1958).
117. G. D. Winter, *Ann. N. Y. Acad. Sxi.*, *87*, 629 (1960).
118. G. D. Winter, *Ann. N. Y. Acad. Sci.*, *87*, 875 (1960).
119. D. M. Serrone, A. A. Stein, E. A. Menegaux, M. A. Gallo, and F. Coulston, *Technicon. Symp.*, *1965*, 586 (1966).
120. F. A. Gunther and D. E. Ott, *Residue Rev.*, *14*, 12 (1966).
121. D. E. Ott and F. A. Gunther, *J. Ass. Offic. Anal. Chem.*, *49*, 662 (1966).
122. D. E. Ott and F. A. Gunther, *J. Econ. Entomol*, *59*, 227 (1966).
123. D. E. Ott and F. A. Gunther, *J. Ass. Offic. Anal. Chem.*, *49*, 669 (1966).
124. D. E. Ott, *J. Agr. Food Chem.*, *16*, 874 (1968).
125. E. F. Edson and M. L. Fenwick, *Brit. Med. J.*, *i*, 1218 (1955).
126. K. Hiraki, M. Kitayama, and S. Shibata, *Bull. Yamaguchi Med. School*, *3*, 19 (1955).
127. D. H. Cox and B. R. Baker, *J. Amer. Vet. Med. Ass.*, *132*, 385 (1958).
128. E. Sándi, *Munkavédelem*, *6*, 26 (1960); from *Chem. Abstr.*, *55*, 12520 (1961).
129. E. Sándi and J. Wight, *Chem. Ind.* (London), *1961*, 1161.

130. E. Herzfeld and C. Stumpf, *Wien. Klin. Wochschr.*, *67*, 874 (1955).
131. H. Pietschmann, *Wien. Z. Inn. Med.*, *41*, 409 (1960).
132. H. C. Churchill-Davidson and W. J. Griffiths, *Brit. Med. J.*, *1961*, *ii*, 994 (1961).
133. W. Lang and G. Intsesuloglu, *Klin. Wochschr.*, *40*, 312 (1962).
134. R. Richterich, Schweiz. *Med. Wochschr.*, *92*, 263 (1962).
135. D. Vincent, G. Segonzac, and M. C. Marques-Vincent, *Ann. Biol. Clin.*, *21*, 481 (1963).
136. R. I. H. Wang, *J. Amer. Med. Assoc.*, *183*, 792 (1963).
137. R. I. H. Wang and E. O. Henschel, *Anesthesia Analegesia*, *46*, 281 (1967).
138. C. K. Davidson and P. A. Adie, *Anal. Biochem.*, *12*, 70 (1965).
139. R. Pleśtina, *Arhiv Hig. Rada*, *17*, 291 (1966).
140. B. Holmstedt and J.-L. Oudart, *Bull. Soc. Pathol. Exotique*, *59*, 411 (1966).
141. A. Härtel, W. Gross, and H. Lang, *Z. Klin. Chem. Klin. Biochem.*, *5*, 26 (1967).
142. J. Fischl, N. Pinto, and C. Gordon, *Clin. Chem.*, *14*, 371 (1968).
143. P. E. Braid and M. S. Nix, *Arch. Environ. Health*, *17*, 986 (1968).
144. J. Lapis, A. Mandat, and H. Szymańska, *Polski Tygod. Lekar.*, *16*, 1848 (1961).
145. H. Weidemann, *Anaesthesist*, *14*, 352 (1965).
146. H. Jäger, *Landarzt*, *42*, 798 (1966).
147. A. P. M. van Oudheusden, *Pharm. Weekblad*, *97*, 606 (1962).
148. F. Hoppe, *Med. Lab.*, *19*, 240 (1966).
149. P. J. Bunyan, *Analyst*, *89*, 615 (1964).
150. D. G. Crosby, E. Leitis, and W. L. Winterlin, *J. Agr. Food Chem.*, *13*, 204 (1965).
151. J. J. Menn and J. B. McBain, *Nature*, *209*, 1351 (1966).
152. F. Masaaki, *Seikagaku*, *29*, 318 (1957); from *Chem. Abstr.*, *54*, 18725 (1960).
153. R. M. Fournier, *Mém. Poundres*, *40*, 403 (1958).
154. A. Heyndrickx and A. Vercruysse, *Medel. Rijksfac. Landbouw-Wetensch.*, *32*, 835 (1967).
155. M. Morita, *Sogo Igaku*, *11*, 707 (1954); from *Chem. Abstr.*, *51*, 16655 (1957).
156. J. Chouteau, P. Rancien, and A. Karamanian, *Bull. Soc. Chim. Biol.*, *38*, 1329 (1956).
157. G. Locati, F. Acerboni, L. Macchi, and M. Giglio, *Ann. Ostet. Ginecol.*, *88*, 729 (1966).
158. E. Korn, *Comp. Biochem. Physiol.*, *28*, 923 (1969).
159. A. B. Hargreaves, in *Bioelectrogenesis, Proceedings of the Symposium on Comparative Bioelectrogenesis*, C. Chagas and A. Paes de Carvalho, Eds., Elsevier, Amsterdam, 1961, pp. 397–405.
160. A. B. Hargreaves, *Anais II Reuniao Divisao Quimica Organica Bioquimica Associacao Brasileira De Quimica*, Rio de Janeiro, 1962, pp. 123–131.
161. A. B. Hargreaves and M. G. Da Silva, *O Hospital*, *55*, 99 (1959).
162. M. Schwartz and T. C. Myers, *Anal. Chem.*, *30*, 1150 (1958).
163. A. J. Goshev, *Vop. Med. Chim.*, *4*, 149 (1958); from *Chem. Abstr.*, *52*, 15632 (1958).
164. A. Jasiński, *Acta Physiol. Polon.*, *10*, 647 (1959).
165. D. W. Einsel, Jr., H. J. Trurnit, S. D. Silver, and E. C. Steiner, *Anal. Chem.*, *28*, 408 (1956).
166. J. Loiselet and G. Srouji, *Bull. Soc. Chim. Biol.*, *50*, 219 (1968).
167. I. B. Wilson and E. Cabib, *J. Amer. Chem. Soc.*, *78*, 202 (1956).
168. J. Jensen-Holm, H. H. Lausen, K. Milthers, and K. O. Møller, *Acta Pharmacol. Toxicol.*, *15*, 384 (1959).
169. K. Jørgensen, *Scand. J. Clin. Lab. Invest.*, *11*, 282 (1959).
170. E. Heilbronn, *Acta Chem. Scand.*, *12*, 1879 (1958).

171. E. Heilbronn, *Acta Chem. Scand.*, *13*, 1547 (1959).
172. L. Tominz and P. Gazzaniga, *Folia Med.*, *44*, 1010 (1961).
173. A. L. Delaunois, *Arch. Int. Pharmacodyn. Ther.*, *140*, 351 (1962).
174. H. M. Rubinstein and A. A. Dietz, *J. Lab. Clin. Med.*, *61*, 979 (1963).
175. J. Jensen-Holm, *Acta Pharmacol. Toxicol.*, *23*, 73 (1965).
176. J. L. Casterline, Jr., and C. H. Williams, *J. Lab. Clin. Med.*, *69*, 325 (1967).
177. D. P. Nabb and F. Whitfield, *Arch. Environ. Health*, *15*, 147 (1967).
178. R. J. Kitz and S. Ginsburg, *Biochem. Pharmacol.*, *17*, 525 (1968).
179. B. Ballantyne, *Arch. Int. Pharmacodyn. Ther.*, *173*, 343 (1968).
180. J. B. Neilands and M. D. Cannon, *Anal. Chem.*, *27*, 29 (1955).
181. L. Larsson and B. Hansen, *Svensk Kem. Tidskr.*, *68*, 521 (1956).
182. J. G. Beasley, A. C. York, S. T. Christian, and W. A. Frase, *Med. Biol. Eng.*, *6*, 181 (1968).
183. A. R. Main and F. Iverson, *Biochem. J.*, *100*, 525 (1966).
184. I. B. Wilson, in *Drugs Affecting the Peripheral Nervous System*, A. Burger, Ed., Dekker, New York, 1967, p. 381.
185. B. Holmstedt, G. Lundgren, and F. Sjöqvist, *Acta Physiol. Scand.*, *57*, 235 (1963).
186. M. Salvini, L. Tominz, and P. Gazzaniga, *Boll. Soc. Ital. Biol. Sper.*, *37*, 581 (1961).
187. D. Neubert and D. Maibauer, *Arch. Exp. Pathol. Pharmacol.*, *233*, 163 (1958).
188. G. Schatzberg-Porath, J. Zahavy, and S. Gitter, *Nature*, *198*, 686 (1963).
189. H. Bockendahl, *Z. Physiol. Chem. 336*, 172 (1964).
190. A. G. E. Pearse, *Histochemistry, Theoretical and Applied*, Churchill, London, 1960 (2nd ed.) and 1969 (3rd ed.).
191. J. Bernsohn, K. D. Barron, and A. Hess, *Proc. Soc. Exp. Biol.*, *108*, 71 (1961).
192. A. A. Rubin, J. Mershon, M. E. Grelis, and I. I. A. Tabachnick, *Fed. Proc.*, *16*, 333 (1957).
193. I. I. A. Tabachnick, J. Mershon, M. E. Grelis, and A. A. Rubin, *Arch. Int. Pharmacodyn. Ther.*, *114*, 351 (1958).
194. E. M. Gal, *Fed. Proc.*, *16*, 298 (1957).
195. E. M. Gal and E. Roth, *Clin. Chim. Acta*, *2*, 316 (1957).
196. D. E. McOsker and L. J. Daniel, *Arch. Biochem. Biophys.*, *79*, 1 (1959).
197. I. J. Greenblatt, L. Rosenfeld, and M. Kenin, *Clin. Chem.*, *7*, 410 (1961).
198. A. Rizzoli, *Boll. Soc. Ital. Biol. Sper.*, *41*, 1173 (1965).
199. F. S. LaBella and S. Shin, *J. Neurochem.*, *15*, 335 (1968).
200. D. N. Kramer, P. L. Cannon, Jr., and G. G. Guilbault, *Anal. Chem.*, *34*, 842 (1962).
201. G. G. Guilbault, D. N. Kramer, and P. L. Cannon, Jr., *Anal. Chem.*, *34*, 1437 (1962).
202. G. G. Guilbault, D. N. Kramer, and P. L. Cannon, Jr., *Anal. Biochem.*, *5*, 208 (1963).
203. E. K. Bauman, L. H. Goodson, and J. R. Thomson, *Anal. Biochem.*, *19*, 587 (1967).
204. V. Fiserová-Bergerová, *Coll. Czech. Chem. Commun.*, *27*, 693 (1962).
205. V. Fiserová-Bergerová, *Coll. Czech. Chem. Commun.*, *28*, 3311 (1963).
206. V. Fiserová-Bergerová, *Arch. Environ. Health*, *9*, 438 (1964).
207. C. C. Curtain, *Anal. Biochem.*, *8*, 184 (1964).
208. G. L. Ellman, *Arch. Biochem. Biophys.*, *82*, 70 (1959).
209. G. L. Ellman, K. D. Courtney, V. Andres, Jr., and R. M. Featherstone, *Biochem. Pharmacol.*, *7*, 88 (1961).

210. C. G. Humiston and G. J. Wright, *Toxicol. Appl. Pharmacol.*, *10*, 467 (1967); *Clin. Chem.*, *11*, 802 (1965).
211. G. Szasz, *Clin. Chim. Acta*, *19*, 191 (1968).
212. P. J. Garry and J. I. Routh, *Clin. Chem.*, *11*, 91 (1965).
213. P. Aarseth, J. A. B. Barstad, O. Rogne, and S. Øksne, *Histochemie*, *15*, 229 (1968).
214. L. Guth, R. W. Albers, and W. C. Brown, *Exp. Neurol.*, *10*, 236 (1964).
215. J. B. Levine, R. A. Scheidt, and V. A. Nelson, *Technicon Symp.*, *1965*, 582 (1966).
216. G. A. Buckley and P. T. Nowell, *J. Pharm. Pharmacol.*, *18*, 146 (1966).
217. H. Weber, *Deut. Med. Wochschr.*, *91*, 1927 (1966).
218. M. Knedel and R. Böttger, *Klin. Wochschr.*, *45*, 325 (1967).
219. M. Sakai, *Appl. Entomol. Zool.*, *2*, 111 (1967).
220. F. Bufardeci and V. Buosanto, *Boll. Soc. Ital. Biol. Sper.*, *43*, 1365 (1967).
221. M. A. Karahasanoğlu and P. T. Özand, *Turk. J. Pediat.*, *8*, 1 (1966); *J. Lab. Clin. Med.*, *70*, 343 (1967).
222. G. Voss, *J. Econ. Entomol.*, *59*, 1288 (1966).
223. G. Voss, and H. Geissbühler, *Medd. Rijksfacul. Landbouwwetensch.*, *32*, 877 (1967).
224. G. Voss, *Residue Rev.*, *23*, 71 (1968).
225. G. Voss, and J. Schuler, *Bull. Environ. Contamin. Toxicol.*, *2*, 357 (1967).
226. G. Voss, *Bull. Environ. Contamin. Toxicol.*, *3*, 339 (1968).
227. P. Juul, *Clin. Chim. Acta.*, *19*, 205 (1968).
228. G. I. Klingman, J. D. Klingman, and A. Poliszczuk, *J. Neurochem.*, *15*, 1121 (1968).
229. D. C. Leegwater and H. W. van Gend, *J. Sci. Food Agr.*, *19*, 513 (1968).
230. K. Wilhelm, *Archiv. Hig. Rada Toksikol.*, *19*, 199 (1968).
231. R. L. Metcalf, *J. Econ. Entomol.*, *44*, 883 (1951).
232. J. de la Huerga, C. Yesinick, and H. Popper, *Amer. J. Clin. Pathol.*, *22*, 1126 (1952).
233. S. Janah, B. R. Chatterjee, D. N. Bhowmik, B. C. Rudra, and S. N. Chaudhuri, *Ann. Biochem. Exp. Med.*, *16*, 17 (1956).
234. H. J. Wetstone and G. N. Bowers, Jr., *Standard Methods of Clinical Chemistry*, *4*, 47 (1963).
235. H. J. Wetstone, R. Tennant, and B. V. White, *Gastroenterology*, *3*, 41 (1957).
236. J. H. Fleisher and E. J. Pope, *Arch. Ind. Hyg. Occup. Med.*, *9*, 323 (1954).
237. J. H. Fleisher, E. J. Pope, and S. F. Spear, *Arch. Ind. Health.*, *11*, 332 (1955).
238. M. London, *Igienă*, *6*, 226 (1957); from *Chem. Abstr.*, *53*, 11491 (1959).
239. K. Irino, *Shikoku Igaku Zasshi*, *15*, 558 (1959); from *Chem. Abstr.*, *54*, 2543 (1960).
240. S. L. Bonting and R. M. Featherstone, *Arch. Biochem. Biophys.*, *61*, 89 (1956).
241. B. J. Kallman, *Chemist-Analyst*, *51*, 75 (1962).
242. D. Vincent, *Clin. Chim. Acta*, *3*, 104 (1958).
243. D. Vincent and G. Segonzac, *Ann. Biol. Clin.*, *16*, 227 (1958).
244. W. Pilz, *Klin. Wochschr.*, *36*, 1017 (1958).
245. W. Pilz and G. Kimmerle, *Z. Physiol. Chem.*, *327*, 280 (1962).
246. W. Pilz, I. Johann, and E. Stelzl, *Klin. Wochschr.*, *43*, 1227 (1965).
247. W. Pilz, *Z. Klin. Chem.*, *3*, 89 (1965).
248. W. Pilz, in *Methoden der enzymatischen Analyse*, 2. Aufl., Verlag Chemie, Weinheim/Bergstr., 1969.
249. K. Kuroda, M. Fujino, and K. Irino, *Tokushima J. Exp. Med.*, *6*, 73 (1959).

250. E. D. Nesheim and J. W. Cook, *J. Ass. Offic. Agr. Chem.*, *42*, 187 (1959).
251. D. F. McCaulley and J. W. Cook, *J. Ass. Offic. Agr. Chem.*, *42*, 200 (1959).
252. H. von Willgerodt, H. Theile, and K. Beyreiss, *Z. Klin. Chem. Klin. Biochem.*, *6*, 149 (1968).
253. W. Pilz, *Z. Anal. Chem.*, *162*, 81 (1958).
254. W. Pilz, *Z. Anal. Chem.*, *193*, 338 (1963).
255. W. Pilz, *Z. Anal. Chem.*, *166*, 189 (1959).
256. W. Pilz and A. Eben, *Arch. Toxikol.*, *23*, 17 (1967).
257. P. N. Meulendijk, *Pharm. Weekblad*, *94*, 623 (1959).
258. H. Guyot, *C. R. Soc. Biol.*, *152*, 1511 (1958); *155*, 1649 (1961).
259. D. Vincent, G. Segonzac, and G. Sesque, *Ann. Biol. Clin.*, *18*, 489 (1960).
260. Z. M. Muravéva, *Vopr. Med. Khim.*, *7*, 97 (1961); from *Chem. Abstr.*, *55*, 13827 (1961).
261. W. A. Groff, L. A. Mounter, and M. van Sim. *Technicon Symp.*, *1967*, 498 (1967).
262. J. W. Cook, *J. Ass. Offic. Agr. Chem.*, *37*, 561 (1954); *38*, 664 (1955).
263. N. R. Rosenthal, *J. Ass. Offic. Agr. Chem.*, *43*, 737 (1960).
264. H. O. Fallscheer and J. W. Cook, *J. Ass. Offic. Agr. Chem.*, *39*, 691 (1956).
265. K. Kojima and T. Ishizuka, *Botyu-Kagaku*, *25*, 30 (1960).
266. G. Yip and J. W. Cook, *J. Ass. Offic. Agr. Chem.*, *42*, 194 (1959).
267. R. Mestres and C. Chave, *Ann. Fals. Exp. Chim.*, *60*, 29 (1967).
268. E. Heilbronn, *Acta Chem. Scand.*, *10*, 337 (1956).
269. R. Fried, *Fed. Proc.*, *23*, 492 (1964).
270. A. N. Panlukov, *Vopr. Med. Khim.*, *12*, 88 (1966).
271. R. M. Lee and B. H. Livett, *Biochem. Pharmacol.*, *16*, 1757 (1967).
272. O. Fehér and E. Bokri, *Acta Physiol. Acad. Sci. Hung.*, *18*, 1 (1960).
273. H. Benger and E. Kaiser, *Sci. Pharm.*, *25*, 1 (1957).
274. G. Aksnes and K. Sandberg, *Acta Chem. Scand.*, *11*, 876 (1957).
275. J. S. Hanker, A. Gelberg, and B. Witten, *J. Amer. Pharm. Assoc.*, *47*, 728 (1958).
276. W. Kalow and H. A. Lindsay, *Can. J. Biochem. Physiol.*, *33*, 568 (1955).
277. W. Kalow, *Pharmacogenetics. Heredity and the Response to Drugs*, Saunders, Philadelphia, 1962, pp. 69–92.
278. W. Kalow and K. Genest, *Can. J. Biochem. Physiol.*, *35*, 339 (1957).
279. H. Harris and M. Whittaker, *Ann. Hum. Genet.*, *27*, 53 (1963).
280. H. Harris, *Recent Advan. Clin. Pathol.*, Ser. IV, S. C. Dyke, Ed., Churchill, London, 1964, pp. 83–105.
281. H. Lehman and J. Liddell, *Progr. Med. Genet.*, *3*, 75 (1964).
282. M. Whittaker, *Acta Genet.*, *18*, 325, 335 (1968).
283. A. Doenicke, T. Gürtner, G. Kreutzberg, I. Remes, W. Spiess, and K. Steinbereithner, *Acta Anaesthesiol. Scand.*, *7*, 59 (1963).
284. H. Vergnes and T. Hobbe, *Ann. Biol. Clin.*, *25*, 687 (1967).
285. E. Kovacs, I. Niedner, and G. Wagner, *Z. Naturforsch.*, *18b*, 869 (1963).
286. M. N. Linyuchev, *Vopr. Med. Khim.*, *6*, 427 (1960); from *Chem. Abstr.*, *55*, 8523 (1961).
287. F. P. W. Winteringham and R. W. Disney, *Nature*, *195*, 1303 (1962).
288. H. W. Goędde, K. R. Held, and K. Altland, *Mol. Pharmacol.*, *4*, 274 (1968).
289. D. Neubert, J. Schaefer, and H.-D. Belitz, *Arch. Exp. Pathol. Pharmacol.*, *239*, 492 (1960).
290. F. P. W. Winteringham and R. W. Disney, *Biochem. J.*, *91*, 506 (1964).
291. F. P. W. Winteringham and R. W. Disney, *Bull. World Health Organ.*, *30*, 119 (1964).

292. F. P. W. Winteringham and R. W. Disney, *Lab. Practice*, *13*, 739 (1964).
293. R. W. Disney, *Amer. J. Med. Electronics*, *4*, 70 (1965).
294. R. W. Disney, *Biochem. Pharmacol.*, *15*, 361 (1966).
295. F. P. W. Winteringham and K. S. Fowler, *Biochem. J.*, *99*, 6 P (1966).
296. F. P. W. Winteringham and K. S. Fowler, *Biochem. J.*, *101*, 127 (1966).
297. F. P. W. Winteringham, *Nature*, *212*, 1368 (1966).
298. F. P. W. Winteringham, *J. Insect Physiol.*, *12*, 909 (1966).
299. F. P. W. Winteringham, *Bull. World Health Organ.*, *35*, 452 (1966);
300. H. Marchart, *Mikrochim. Acta*, *1968*, 669.
301. D. J. Reed, K. Goto, and C. H. Wang. *Anal. Biochem.*, *16*, 59 (1966).
302. C. H. Frady and S. E. Knapp, *J. Parasitol.*, *53*, 298 (1967).
303. W. W. Berg and R. P. Maickel, *Life Sci.* (Oxford), *7*, 1197 (1968).
304. L. T. Potter, *J. Pharm. Exp. Ther.*, *156*, 500 (1967).
305. S. Gaballah, *Proc. Soc. Exp. Biol. Med.*, *129*, 376 (1968).
306. M. W. McCaman, L. R. Tomey, and R. E. McCaman, *Life Sci.* (Oxford), *7*, 233 (1968).
307. S. Koslow and E. Giacobini, *J. Neurochem.*, *16*, 1523 (1969).
308. Y. Takahashi, I. Aoyama, F. Ito, and Y. Yamamura, *Clin. Chim. Acta*, *18*, 21 (1967).
309. H. J. Raderecht, *Clin. Chim. Acta*, 8, 307 (1963).
310. R. L. Smith, H. Loewenthal, H. Lehmann, and E. Ryan, *Clin. Chim. Acta*, *4*, 384 (1959).
311. U. Rossi, D. Davies, and H. Lehmann, *St. Bartholomew's Hosp. J.*, Jan. 1963, Suppl.
312. S. Kimura, *Seikagaku*, *29*, 351 (1957).
313. K. Kuroda, M. Fujino, and K. Irino, *Tokushima J. Exp. Med.*, *4*, 83 (1957).
314. J. P. Liberti, *Anal. Biochem.*, *23*, 53 (1968).
315. A. R. Main, *Biochem. J.*, *79*, 246 (1961).
316. A. R. Main, K. E. Miles, and P. E. Braid, *Biochem. J.*, *78*, 769 (1961).
317. G. Szász, *Clin. Chem.*, *14*, 646 (1968).
318. R. B. McComb, R. V. LaMotta, and H. J. Wetstone, *Fed. Proc.*, *23*, 280 (1964).
319. R. B. McComb, R. V. LaMotta, and H. J. Wetstone, *Clin. Chem.*, *11*, 645 (1965).
320. M. Geldmacher-von Mallinckrodt and I. Kaiser, *Z. Klin. Chem. Biochem.*, *6*, 141 (1968).
321. J. Paul and P. Fottrell, *Biochem. J.*, *78*, 418 (1961).
322. D. J. Ecobicon and W. Kalow, *Biochem. Pharmacol.*, *11*, 573 (1962); *Can. J. Biochem.*, *42*, 277 (1964).
323. R. S. Holmes, C. J. Masters, and E. C. Webb, *Comp. Biochem. Physiol.*, *26*, 837 (1968).
324. R. S. Holmes and C. J. Masters, *Biochim. Biophys. Acta*, *151*, 147 (1968).
325. H. A. Ramsay, *Clin. Chem.*, *3*, 185 (1957).
326. T. R. F. Wright and K. Keck, *Anal. Biochem.*, *2*, 610 (1961).
327. H. S. Funnell and W. T. Oliver, *Can. J. Biochem.*, *44*, 953 (1966).
328. K.-B. Augustinsson, *Biochim. Biophys. Acta*, *159*, 197 (1968).
329. W. Pilz, *Microchim. Acta*, *1961*, 614.
330. G. Dávid, L. Gyarmati, and I. Fránczi, *Kiserl. Orvostud.*, *12*, 201 (1960).
331. G. G. Guilbault and D. N. Kramer, *Anal. Chem.*, *37*, 120 (1965).
332. G. G. Guilbault and D. N. Kramer, *Anal. Chem.*, *37*, 1675 (1965).
333. G. J. Siegel, G. M. Lehrer, and D. Silides, *J. Histochem. Cytochem.*, *14*, 473 (1966).

334. H. Harris and E. Robson, *Lancet*, *1963II*, 218.
335. N. E. Simpson and W. Kalow, *Amer. J. Human Genet.*, *17*, 156 (1965).
336. G. Radam, *Deut. Gesundheitsw.*, *21*, 1620 (1966).
337. G. Lee and J. C. Robinson, *J. Med. Genet.*, *4*, 19 (1967).
338. K. I. Beynon and G. Stoydin, *Nature*, *208*, 748 (1965).
339. U. Lippi and E. Pulido, *Boll. Soc. Ital. Biol. Sper.*, *42*, 1246 (1966).
340. A. C. Morrow and A. G. Motulsky, *J. Lab. Clin. Med.*, *71*, 350 (1968); *Science*, *159*, 202 (1968).
341. S. J. Holt, *General Cytochemical Methods*, Vol. 1, Academic Press, New York, 1958, pp. 375–398.
342. J. Günther and D. Ruff-Sondermeier, *Naturwissenschaften*, *52*, 540 (1965).
343. D. Ruff-Sondermeier, J. Günther, and M. Zobel-Bansi, *Deut. Gesundheitsw.*, *22*, 743 (1967).
344. H. F. Linskens, *J. Chromatogr.*, *1*, 202 (1958).
345. C. E. Mendoza, P. J. Wales, H. A. McLeod, and W. P. McKinley, *Analyst*, *93*, 34 (1968).
346. R. Schutzmann and W. F. Barthel, *J. Ass. Offic. Anal. Chem.*, *52*, 151 (1969).
347. P. J. Wales, H. A. McLeod, and W. P. McKinley, *J. Ass. Offic. Anal. Chem.*, *51*, 1239 (1968).
348. J. Matousek, J. Fisher, and J. Cerman, *Chem. Zvesti.*, *22*, 184 (1968).
349. D. N. Kramer and G. G. Guilbault, *Anal. Chem.*, *36*, 1662 (1964).
350. G. G. Guilbault, M. H. Sadar, R. Glazer, and C. Skou, *Anal. Lett.*, *1*, 365 (1968).
351. D. N. Kramer and R. M. Gamson, *Anal. Chem.*, *30*, 251 (1958).
352. T. E. Archer and G. Zweig, *J. Agr. Food Chem.*, *7*, 178 (1959).
353. R. W. P. Master *Biochim. Biophys. Acta*, *39*, 159 (1960.
354. T. H. Ridgway and H. B. Mark, Jr., *Anal. Biochem.*, *12*, 357 (1965).

Measurement of Choline Acetylase

J. Schuberth, *Psychiatric Research Center, Ulleråker Hospital, Uppsala, and B. SÖRBO, Division of Experimental Defence Medicine, Research Institute of National Defence, Sundbyberg, Sweden*

I. INTRODUCTION

Determinations of the enzyme choline acetylase (or choline acetyltransferase, acetyl-CoA: choline O-acetyltransferase EC 2.3.1.6) have obvious applications in neurophysiology and pharmacology. Thus the presence of acetylcholine, choline acetylase, and cholinesterase in a group of neurones should be required for the identification of the neurones as being cholinergic. Furthermore, the recent introduction of compounds active as inhibitors against choline acetylase (1,2) has put new tools in the hands of the pharmacologist (3), but their proper use obviously requires reliable methods for the determination of choline acetylase activity.

Choline acetylase was discovered and given its name by Nachmansohn and Machado (4). They demonstrated that homogenized rat brain could synthesize acetylcholine if supplied with choline, acetate, ATP, fluoride, and eserine (the latter two compounds required in order to inhibit ATP-ase and cholinesterase, respectively). It was postulated that the hydrolysis of ATP provided the energy necessary for the acetylation of choline. Further work by different workers demonstrated that acetylcholine synthesis could also be obtained with aqueous extracts of acetone-dried brain and furthermore that a coenzyme was necessary for the reaction. The latter could be identified as Lipman's coenzyme A. Choline acetylase activity was detected not only in vertebrate nervous tissues, richer sources of the enzyme were in fact found in invertebrate nervous tissue such as the squid head ganglion (5), in a nonnervous mammalian tissue, human placenta (6), and in the bacterium *Lactobacillus plantarum* (7). It then turned out that the original choline acetylase system actually consisted of two enzymes, one producing the acetylated form of coenzyme A in an ATP-requiring reaction and the other catalyzing the reaction between the acetylated coenzyme and choline. The name choline acetylase was reserved for the latter enzyme (8). During the later 1940s the importance of coenzyme A for different biochemical acetylation reactions had been demonstrated and a most important step was taken when Lynen and co-workers (9) isolated and chemically identified the acetylated coenzyme A as the thiolester of the coenzyme. In this fundamental paper reference is made to unpublished work on the acetylation of choline by acetyl-coenzyme A in the presence of choline acetylase from rabbit brain, but the formal demonstration of the enzyme-catalyzed reaction between acetyl-coenzyme A and choline, giving acetylcholine and coenzyme A, is due to Korkes et al. (10). The foundations were now laid for a better understanding of the enzyme and its properties. Improved assay procedures were worked out and highly purified (although not yet homogenous) choline acetylase preparations have been obtained from rat brain (11), human placenta (12,13), squid head ganglia (14), and housefly brain (15).

This review outlines, in more or less detail, available methods for the determination of choline acetylase from different sources using different analytical techniques. This is only a representative number of the possible methods, which vary widely in sensitivity, rapidity, reproducibility, and interference from adventitious substances. An account is first given of those properties of choline acetylase that are of particular importance for the determination of the enzyme. Different procedures for following the appearance of products or disappearance of substrates in the enzyme-catalyzed reaction are then presented. Special emphasis will be laid on methods where the formation of isotopically labeled acetylcholine from ^{14}C- or ^{3}H-labeled acetyl-coenzyme A is measured, as these methods are very precise, can be applied to the enzyme of any state of purity, and have not been described in previously published reviews on choline acetylase (16–22).

II. THE CHOLINE ACETYLASE REACTION

1. Substrate Specificity

Choline acetylase is not absolutely specific with respect to the acyl acceptor, as certain N-alkyl substituted choline analogs are also substrates for the enzyme, although inferior in that respect to choline (23–26). The obvious choice of acyl acceptor for determinations of enzyme activity is thus choline, mostly used as the chloride. The hygroscopic nature of this salt should be noted and the commercial product should be dried *in vacuo* and stored in a desiccator before use. Choline bromide and iodide are not hygroscopic, but the latter salt is unsuitable as substrate if the choline acetylase reaction is followed colorimetrically with the hydroxamic acid method (see below) as iodide interferes in this determination (25).

Also, with respect to the acyl donor, choline acetylase is not absolutely specific as propionyl-coenzyme A is a substrate, but less efficient than the acetyl analog (23,27). Acetyl-coenzyme A is therefore the substrate of choice in the assay of the enzyme. Earlier reports claimed (15,23,27–29) that the maximum enzyme activity which could be obtained with synthetic acetyl-coenzyme A as substrate was much inferior to that obtained when choline acetylase was coupled with another enzyme that provided a continuous supply of acetyl-coenzyme A. Recent investigations have shown, however, that if the enzyme activity is measured at optimum salt concentration and in the presence of a suitable activator, similar rates will be achieved with synthetic and enzymatically generated acetyl-coenzyme A (30,31). As the use of coupled enzyme systems has certain drawbacks (to be discussed below), we recommend that synthetic acetyl-coenzyme A be used as substrate. Both isotopically labeled and unlabeled preparations are commercially available but are expensive and not always reliable. It is therefore recommended that they be synthetized in the laboratory. Acetyl-coenzyme A is conveniently prepared from coenzyme A and acetic anhydride (labeled or unlabeled) according to the method of Simon and Shemin (32), which in essence consists of treating coenzyme A dissolved in 0.1 M KHCO$_3$ at 0°C with a slight excess of acetic anhydride. Various modifications of the original procedure have been described (11,14,33). Too large an excess of acetic anhydride should not be used in the acetylation reaction, as less active products are obtained in this way (34), but on the other hand the coenzyme A should be completely acetylated (as checked by a colorimetric test for free sulfhydryl groups) as coenzyme A inhibits the reaction (33). Labeled acetic anhydride is also commercially available as a benzena solution which may be used for acetylation of coenzyme A by following the directions of Bové et al. (35). Acetyl-coenzyme A may furthermore be prepared from labeled acetate by conversion of the latter to the mixed anhydride of acetic acid and ethyl hydrogen carbonate, which is then used to acetylate coenzyme. A. The procedure is described in detail by Lynen et al. (36). Alternatively, labeled acetate may be used for acetylation of coenzyme A with the aid of acetyl-CoA synthetase (37). Coenzyme A can be acetylated by thiolacetate (38), but a considerable excess of the reagent must be used, and labeled thiolacetic acid is not commercially available, which restricts this procedure to the preparation of unlabeled acetyl-coenzyme A. It should be noted that commercial coenzyme A preparations contain impurities which are not removed by conversion to acetyl-coenzyme A and may inhibit choline acetylase. Thus acetyl-coenzyme A purified by chromatography on DEAE-Sephadex (39) gave twice as high activity with bovine brain choline acetylase as the unpurified material (40). Acetyl-coenzyme A may also be purified by ion-exchange chromatography on DEAE-cellulose (41) and by paper electrophoresis (11,42). Purified acetyl-coenzyme A should be used in kinetic studies of the enzyme, but for routine assay the nonpurified material may also be used. Acetyl-coenzyme A preparations should be assayed for their acetylated coenzyme content by enzymic methods based either on the phosphotransacetylase reaction (43) or citrate synthetase coupled with malate dehydrogenase (44). Their thiolester content may also be assayed by the hydroxamic acid procedure (45), but this method is less specific.

The following description of preparation of acetyl-coenzyme A is based on the Simon and Shemin method, but the present method gives a salt-free product because a strong cation exchange resin in the hydrogen form is used to remove NaHCO$_3$ (40).

Synthesis of ^{14}C-labeled acetyl-coenzyme A. Forty milligrams of coenzyme A of 85% purity (41 μM) are dissolved in 10 ml 0.1 M NaHCO$_3$ and treated at 0°C with 70 μM (acetic-1-^{14}C) anhydride, dissolved in 0.5 ml ice-cold water. The specific activity of the anhydride should be somewhat higher than that desired in the final product. After 15 min at 0°C, an aliquot of the solution is tested for the presence of free sulfhydryl groups by the Ellman reaction (46) and if the latter is positive, 30 μM (3 μl) of unlabeled acetic anhydride are added. The sulfhydryl test should then become negative in a couple of minutes. The reaction mixture is now acidified to pH 3 (indicator paper) by addition of Dowex 50 W-X8 (100–200 mesh) in the hydrogen form, which requires about 10 g of the resin. After gas evolution has subsided the suspension is transferred to a small column tube and washed 4 times with 5 ml of water. The combined effluents are taken to dryness on a rotary evaporator and the residue dried *in vacuo* over solid KOH. The yield of acetyl-coenzyme A, as assayed enzymatically, is better than 90%, calculated from coenzyme A.

2. Activators and Inhibitors

The rate of the choline acetylase catalyzed reaction is greatly influenced by the presence of different salts. Thus, it has been shown that fairly high concentrations (0.1–0.3M) of salts such as sodium and potassium chloride stimulate the rate of the reaction catalyzed by rat brain choline acetylase (11,30), placenta enzyme (13), and Squid ganglion enzyme (14). Some evidence has been presented that the salt effect is due to the anion and that chlorides and bromides are more efficient than other anions tested (13). Potassium iodide also stimulates the rate of the choline acetylase reaction but only in concentrations up to 0.2M. Higher concentrations of the iodide fail to have a stimulatory effect (14).

Since choline acetylase probably contain thiol groups in the "active site" of the enzyme, it is inhibited by different SH reagents (34,47,48). Under certain conditions inactivation of the enzyme occurs even when it is stored as a precipitate in ammonium sulfate of analytical grade, probably due to oxidation of the thiol groups of the enzyme. The full activity can, however, be restored by treating the enzyme solution with cysteamine prior to desalting the enzyme preparation by Sephadex chromatography (13). Cysteine has often been added to the incubation medium to protect the thiol groups of the enzyme (12,25,26,28,49–52). The effect of cysteine and various other sulfhydryl compounds on the enzyme was studied by Morris (31), who found that cysteine at low concentrations had a small stimulatory effect but higher concentrations were inhibitory. This was attributed to the fact that cysteine is acetylated to N,S-diacetylcysteine by acetyl-coenzyme A (27,53) and may thus compete with the enzyme for the acetyl donor. Thioglycollate was more stimulatory than cysteine and did not inhibit at concentrations below 0.02M, and even better was cyanide, which gave good stimulation and no inhibition even at the highest concentration studied (0.04M). EDTA has been reported to activate the enzyme from Squid head ganglia (34), but this was not found with the enzyme from human placenta (54).

3. pH and Temperature Dependency

Choline acetylase from Squid head ganglion has a pH activity optimum of 6.9 (34). However, it has been shown that the pH activity curve of choline acetylase from human placenta is affected by sodium chloride, resulting in a shift of the pH maximum from pH 6.3 without sodium chloride to pH 7.0–8.0 in the presence of the salt (13). It was also found that the temperature optimum was lowered from 51°C without sodium chloride to 47°C in the presence of the salt (13). There are some data which indicate that the temperature optimum of choline acetylase from fish and squid is lower [30–32°C, squid ganglion (34); 25°C, fish brain (55)] than that of the enzyme from warm-blooded animals [42°C, rabbit brain (55)].

A $0.1–0.01M$ phosphate buffer of pH 6.8–7.0 is appropriate for routine choline acetylase assay. The incubation temperature should be selected with respect to the source of choline acetylase preparation used. For most purposes, however, an incubation temperature at 30°C is suitable.

4. Equilibrium and Kinetic Constants

The reversibility of the choline acetylase reaction was originally demonstrated by Schuberth (13) who used a partially purified enzyme preparation of human placenta. The equilibrium constant, as defined by the relationship $K = $ [acetylcholine][CoA]/[acetyl-CoA][choline], was estimated to be 145 in the presence of $0.3M$ sodium chloride and 5100 without the salt. Potter et al. (11), using a choline acetylase preparation from rat brain, found an equilibrium constant of 514 in the presence of $0.15M$ potassium chloride. Since the equilibrium of the choline acetylase reaction favors acetylcholine formation, it is evident that for routine analysis of the enzyme only acetyl-coenzyme A and choline are suitable as substrates.

The kinetic data of choline acetylase from human placenta show (13) that acetyl-coenzyme A and choline react with the enzyme according to a "ping-pong" mechanism, to use the nomenclature proposed by Cleland (56). In the presence of rat brain choline acetylase, on the other hand, both these substrates combine with the enzyme before any products are released (11). The apparent Michaelis constants of the substrates estimated with different enzyme preparations are shown in Table I. As seen, the values differ a great deal even with enzyme preparations

TABLE I

Substrate	K_m (mM)	Source of choline acetylase	Ref.
Choline	0.5	Squid ganglion	34
	0.4[a]	"	14
	1.9[b]	"	14
	1.6	Rat brain	25
	2.3	"	26
	0.5	"	30
	0.039	"	11
	3.5	Rabbit brain	26
	2.7	Goat sciatic nerve	26
	2.5	Human placenta	26
	0.4[a]	" "	13
	1.1[b]	" "	13
Acetyl-CoA	0.022	Rabbit brain	28
	0.1	Bacteria	33
	0.01	Mouse brain	33
	0.025	Rat brain	30
	0.019	" "	11
	0.08[a]	Squid ganglion	14
	0.1[b]	" "	14
	0.07[b]	Human placenta	13
	0.02[b]	" "	13
Acetylcholine	4[a]	" "	13
	2[b]	" "	13
CoA	0.02	" "	13

[a] Incubation at low ionic strength.
[b] Incubation at high ionic strength.

from the same species. This may at least in part be explained by differences in the experimental conditions during incubation and also by differences in the purity of the acetyl-coenzyme A used. Variations in the salt concentration during the incubation have been shown to influence the K_m values of placenta enzyme (13) and Squid ganglion enzyme (14). Since excess choline or acetyl-coenzyme A do not inhibit acetylcholine formation, the concentration of these substrates used for choline acetylase analysis should be reasonably high, i.e., $10^{-2}M$ choline and $5 \times 10^{-4}M$ acetyl-coenzyme A. For reasons of economy the concentration of acetyl-coenzyme A can be kept as low as $5 \times 10^{-5}M$. This concentration will not result in a lowering of more than 10–15% of the maximal enzyme activity when choline acetylase from rat brain (11,30) and from human placenta (13) is used. It should be kept in mind, however, that crude enzyme extracts usually contain acetyl-coenzyme A hydrolase (EC 3.1.2.1) which may decrease the available acetyl-coenzyme A and thus influence the choline acetylase measurements. If long-time incubations are necessary, it may be advisable to use acetyl-coenzyme A concentrations higher than $5 \times 10^{-5}M$ to maintain zero-order kinetics during the entire incubation period.

5. Inhibition of Interfering Enzymes

Acetylcholine formed through the action of choline acetylase is rapidly broken down by the cholinesterase which is present in many tissues where choline acetylase is found. Choline acetylase assays on tissue preparations and partly purified enzyme preparations must therefore be carried out in the presence of a cholinesterase inhibitor. Eserine (physostigmine) or prostigmine (neostigmine) are often used. The concentration in the assay should be $10^{-5}-10^{-4}M$. Organophosphorus anticholinesterases such as TEPP (8) or DFP (26) have also been used, but we prefer eserine because it is nonvolatile and much less toxic. When choline acetylase was assayed in test systems coupled to acetyl-coenzyme A synthetase, some workers have added fluoride of about $0.04M$ in order to inhibit ATP-ase (4,57). This may not always be necessary (8,27); e.g., if acetone-dried tissues are used as the source of choline acetylase, as acetone treatment destroys ATP-ase. It should be noted in this connection that fluoride does not inhibit choline acetylase (13,14).

6. Definition of the Choline Acetylase Unit

Unfortunately, great confusion exists in the literature in the ways used to express the activity of the enzyme. In order to facilitate comparisons between the results given by different authors, we strongly advocate that the recommendations of IUB should be followed (*Enzyme Nomenclature, Recommendations (1964) of the International Union of Biochemistry on the Nomenclature and Classification of Enzymes, together with their Units and the Symbols of Enzyme Kinetics*, Elsevier, Amsterdam, 1965). One unit of choline acetylase should then be defined as that amount of enzyme which catalyses the transformation of 1 μM of substrate per minute. The reaction should, if possible, be measured at optimum pH and substrate concentration. The temperature should be 30°C according to the recommendations; this is a favorable choice in the case of choline acetylase (*vide supra*) because higher temperatures may inactivate the enzyme from certain sources (34).

III. COUPLING OF CHOLINE ACETYLASE WITH ACETYL-COENZYME A FORMING ENZYMES FOR ASSAY PURPOSES

It has been claimed by different authors (28,29,31,34,57) that higher choline acetylase activity is obtained when acetyl-coenzyme A is continuously generated during the incubation than when stoichiometric amounts of acetyl-coenzyme A are used. This assumption has, however, not been proved by direct comparison of both systems used under optimal conditions on the same choline acetylase preparation. Although the acetyl-coenzyme A generating technique may offer some advantages, e.g., that zero-order reaction kinetics is maintained in long-time incubation experiments (57), it has certain drawbacks. When varying the experimental conditions, e.g., by changes in the incubation mixture (in studies of ionic effects or inhibitors) or by using enzyme extracts containing interfering substances, they should always be controlled so that the rate limiting step of the coupled reactions is determined by choline acetylase and not by the generation of acetyl-coenzyme A. Since the steady-state concentration of acetyl-coenzyme A is unknown when a coupled enzyme system is used, it is evident that this technique is of only limited value when the kinetics of the choline acetylase reaction are being studied. However, since acetyl-coenzyme A generating systems have frequently been used for the assay of choline acetylase, some of the enzymic systems employed will be described.

1. Acetyl-CoA Synthetase

Acetyl-CoA synthetase (acetate: CoA lyase) (EC 6.2.1.1) catalyzes the acetyl-coenzyme A formation from acetate, ATP, and coenzyme A. The enzyme can conveniently be prepared from pigeon liver (58). Since the enzyme requires magnesium, incubation should be performed in the absence of EDTA. The system has been used for the assay of choline acetylase by various authors (27,28,49,57,59–61).

2. ATP-Citrate Lyase

The citrate cleavage enzyme (ATP: citrate oxaloacetate-lyase) (EC 4.1.3.8) catalyzes the acetyl-coenzyme A formation from citrate, ATP, and coenzyme A. The crude liver enzyme (58) has been used for choline acetylase assay together with the acetyl-coenzyme A synthetase system (49).

3. Phosphotransacetylase

Phosphotransacetylase (acetyl-CoA: orthophosphate acetyl-transferase) (EC 2.3.1.8) catalyzes the acetyl-coenzyme A formation from acetyl-phosphate and coenzyme A. The enzyme is conveniently obtained from *Clostridium kluyverii*, which is commercially available. This acetyl-coenzyme A generating system has been used for the choline acetylase determination by various workers (12,23,26,29,51,52,62). If the acetylcholine formed is determined by the colorimetric method according to Hestrin, excess acetylphosphate has to be destroyed by boiling for 4 min at pH 4.5 before assay (23).

IV. PREPARATION OF TISSUE SAMPLES FOR CHOLINE ACETYLASE ASSAY

Hebb and Smallman (49) noticed that isotonic sucrose homogenates of rabbit brain showed much less choline acetylase activity than extracts of acetone powder made from an equivalent weight of brain. The activity of the homogenates could, however, be increased by treatment with ether, acetone, or chloroform. As the choline acetylase activity was found to be associated with subcellular particles, originally thought to be mitochondria, the explanation was put forward that the enzyme inside these particles was partially prevented from access to the substrate by surrounding membranes. The activating effect of organic solvents was attributed to a disruption of the membranes. Later work by Whittaker et al. (63,64) and DeRobertis et al. (65) demonstrated that choline acetylase activity was not localized to the mitochondria but to detached nerve endings. The nerve endings, later called "synaptosomes," sedimented as mitochondria in the differential-centrifugation method used in the earlier work. The question of the localization of choline acetylase within the synaptosome has been much debated; according to Whittaker et al. (64) the enzyme is found in the soluble cytoplasm of the synaptosome whereas DeRobertis et al. (65) claimed that it was bound to small particles, the synaptic vesicles, which are present in the synaptosomes and contain most of their acetylcholine. More recent work convincingly establishes that choline acetylase is a soluble enzyme but may be artificially adsorbed to smaller fragments during isolation of subfractions of the synaptosomes (66,67).

Returning to the problem of obtaining maximum enzyme activity when assaying tissue samples, Hebb advocated (21) that extracts should be made from acetone powder of the tissue or that tissue homogenates should be treated with ether. These procedures cannot be used on tissue samples smaller than 50 mg, but Bull et al. (68) devised alternative methods for activating choline acetylase in small samples of tissue. One method consisted of pipetting small volume (10–15 μl) of tissue homogenates onto filter paper and drying the sample in a stream of cold air. The paper was then cut into small pieces which were added directly to the enzyme assay system. The second method, suitable only for soft tissues, consisted of squeezing a small tissue sample (0.5–2 mg) between cellophane and adding the cellophane with the tissue attached to the incubation system. In the third method, fresh-frozen sections of tissue were placed on a small piece of cellophane which was added to the assay system. These three methods gave comparable results when tested on the same tissue. Fonnum (57) found that addition of certain detergents, Triton X-100 or Nonex 501, to a final concentration of 0.5% in homogenates was superior to ether treatment for activating choline acetylase. The addition of n-butanol (final concentration 1%) to the assay system has also been suggested (11) and its activating effect on choline acetylase in homogenates of bovine *N. caudatus* has been confirmed (40). On the other hand McCaman et al. (30,69) claimed that brain homogenates made in distilled water or isolated subcellular brain fractions treated by suspension in water gave higher choline acetylase activities than samples treated with ether according to Hebb and Smallman (49). This was not confirmed by Tucek (70) and in a later modification of the choline acetylase assay method of McCaman and Hunt (71,72) the addition of detergents was found to activate the enzyme in sympathetic or spinal ganglia. However, no activating effects of detergents was noticed when choline acetylase was measured in the abdominal ganglion from the mollusk Aplysia (73). The necessity of special treatments for activation of choline acetylase may thus depend upon the tissue being studied. We strongly recommend that each investigator verify if his tissue samples give maximum activity by simple homogenization in water or buffer or if any of the special treatments described above are necessary. In this connection it should be noted that detergents have an unfavorable effect in the biological assay of acetylcholine and cannot be used to activate choline acetylase if the enzyme is determined by bioassay.

V. CHOLINE ACETYLASE DETERMINATIONS BASED ON BIOASSAY OF ACETYLCHOLINE

In the first method reported for the demonstration of choline acetylase (4), the acetylcholine formed by the enzyme was determined by bioassay on the eserinized frog rectus abdominis muscle. A considerable number of methods for the determination of choline acetylase based on this principle have later been published, see earlier reviews (16,19) and, e.g., (52,74). Choline acetylase determinations have occasionally been made by bioassay of acetylcholine on the dorsal muscle of the leech (26,

31,52,75) and also by using the effect of acetylcholine on the carotid blood pressure in the neostigmine-treated dog (76). Excellent reviews are available on the bioassay of acetylcholine (77–79) and should be consulted by anybody attempting to determine choline acetylase by bioassay. These procedures are very sensitive (although the radiometric or fluorimetric methods may compete in this respect) but are not very precise and are time consuming. They are also affected by tissue constituents and compounds added to the incubation system, such as ATP, choline, KCl, and trichloroacetate (16). Errors due to the sensitizing or inhibiting effects of such compounds in the bioassay may be overcome by the precautions described by Feldberg (16) although the presence of ATP may still cause trouble (80). In summary, we feel that choline acetylase determinations by bioassay should only be attempted if considerable experience of these bioassay procedures are already available in the laboratory or if radiometric methods cannot be used due to lack of isotope counting equipment.

VI. ASSAY METHODS BASED ON COLORIMETRIC OR FLUORIMETRIC DETERMINATIONS

The choline acetylase reaction involves the disappearance of two substrates, choline and acetyl-coenzyme A, and the appearance of two products, acetylcholine and coenzyme A. Assay methods for choline acetylase have been described which are based on colorimetric or fluorimetric determinations of either acetyl-coenzyme A, acetylcholine, or coenzyme A, but not for choline, as simple and sensitive methods for the determination of this compound are not available. Some of the assay methods for the enzyme to be described offer certain advantages in comparison with those based on isotope techniques. Thus colorimetric assay is often the method of choice for routine determinations of purified enzyme preparations due to the simplicity of the procedure. The colorimetric methods are, however, less sensitive than the isotopic method. Since most spectrophotometers may be equipped for measurements on 1 ml samples, the sensitivity of the methods described below is expressed as μM of the compound measured in 1 ml sample.

1. Acetyl-Coenzyme A

Burgen et al. (24) developed a method for the assay of choline acetylase which was used for studying the specificity of the enzyme with respect to its acetyl acceptors. Synthetic acetyl-coenzyme A was used as acetyl donor and the remaining acetyl-coenzyme A in the assay system was determined colorimetrically by a modification of the hydroxamic acid method of Lipman and Tuttle (45). Acetylcholine does not react with hydroxylamine at the neutral pH used in this method. A disadvantage of this choline acetylase assay is its low precision, presumably due to the fact that it measures the remaining substrate, and the enzyme activity is then obtained from the difference between two absorbancy measurements. The method is also fairly insensitive, requiring a disappearance of about 0.1 μM acetyl-coenzyme A in the test system. Furthermore, the presence of acetyl-coenzyme A deacylase in the choline acetylase preparation interferes. This error can be compensated by running a sample from which choline is omitted as a blank. Acetyl-coenzyme A may also be measured spectrophotometrically at 230 mμ, where the acyl thioester bond has a fairly high absorbancy (81). This method was applied by Berry and Whittaker (27) to choline acetylase assay. A disappearance of 0.003 μM of acetyl-coenzyme A may be measured (81). The method is simple and especially suited for kinetic experiments, as it allows continuous registration of enzyme activity but is obviously limited to measurements on purified choline acetylase preparations.

2. Acetylcholine

A. COLORIMETRIC PROCEDURE

Acetylcholine may be determined colorimetrically by conversion to acethydroxamic acid with hydroxlamine at an alkaline pH (82). Acethydroxamic acid is then measured as the ferric-ion complex at 540 mμ. The method has been applied to choline acetylase assay by a number of workers (25,34,61,83,84). Since acetyl-coenzyme A also reacts with hydroxylamine under the conditions used, choline acetylase can only be measured in an assay system where acetyl-coenzyme A is generated in a coupled enzyme system (see Section III). The sensitivity corresponds to about 0.1 μM of acetylcholine formed in the system.

B. FLUORIMETRIC PROCEDURE

A method for fluorimetric assay of acetylcholine was recently devised and adapted to the determination of choline acetylase (85). It is based on the fluorimetric determination of the acetylhydrazyl salicylhydrazone formed after hydrazinolysis of acetylcholine to acetylhydrazide. The method involves the precipitation of acetylcholine with iodine and liberation of acetylcholine from the precipitate by extraction of iodine with ether. Acetylcholine is then adsorbed on a cation-exchange resin column and hydrazinolysis is performed on the adsorbed acetylcholine by addition of hydrazine to the column. The acetylhydrazide formed is eluted and converted to a fluorescent compound by reaction with salicylaldehyde. Spectrophotofluorimetric assay of the latter is made at 370 mμ/475 mμ. The sensitivity of the method (the value corresponding to three times the blank fluorescence) is in the order of $7 \times 10^{-4}\,\mu$M acetylcholine, but the method is fairly complicated and even less precise than bioassay procedures.

3. Coenzyme A

The other product formed in the choline acetylase reaction, coenzyme A, may easily be measured by colorimetric methods with thiol reagents. Berman et al. (23) and Berman (34) used the nitroprusside reagent for determining the appearance of free SH groups of coenzyme A. The sensitivity of this method is about 0.02 μM/ml. The color developed by nitroprusside is, however, unstable, but DTNB [5,5'-dithiobis(2-dinitrobenzoic acid)] (46) is a much more favorable reagent in this respect. It has successfully been used for the assay of choline acetylase activity by different authors (11,13,14). The sensitivity of the DTNB method as devised by Schuberth (13) is 0.005 μM/ml of formed CoA-SH. The DTNB method fulfills the requirements of simplicity, accuracy, and reasonable sensitivity, and is the method of choice when a simple method is wanted for assay of partially purified choline acetylase preparations. Crude enzyme preparations, on the other hand, which contain interfering substances like thiols and acetyl-coenzyme A hydrolase, give a high blank absorbancy and are therefore not suitable for assay by the DTNB method (or by any other colorimetric procedure). It should also be noted that DTNB cannot be used for continuous registration of choline acetylase activity since it is an efficient inhibitor of the enzyme (11,48). A detailed description of the DTNB method, which was developed for the assay of partially purified choline acetylase from human placenta (13), follows.

Procedure. 0.75 ml 0.01M phosphate buffer (pH 7.4) containing 0.5 mM EDTA, 0.3M sodium chloride, $2 \times 10^{-4}M$ acetyl-coenzyme A, and 10 mM choline chloride (no choline in the controls) is prewarmed for 5 min at 30° and 5–20 μl choline acetylase extract is then added. The reaction is terminated by the addition of 0.25 ml trichloroacetic acid (20% w/v). The sample is centrifuged and the pH of 0.5 ml of the clear supernatant is adjusted to pH 7.5 by adding 0.20 ml 1.0M tris/0.05M EDTA. The amounts of thiols are measured spectrophotometrically at 412 mμ 2 min after adding 5 μl of 0.01M DTNB (46). The difference between the absorbancy of the sample and blank (incubated without choline) is taken to represent the coenzyme A formed in the choline acetyltransferase catalyzed reaction. When using this procedure an absorbancy difference of 0.100 corresponds to $5.2 \times 10^{-3}\,\mu$M coenzyme A formed in the reaction.   A linear relationship between enzyme activity and absorbancy is obtained up to 25×10^{-3} μM coenzyme A formed.

VII. ASSAY METHODS BASED ON THE USE OF ISOTOPICALLY LABELED ACETYL-COENZYME A AND SEPARATION OF LABELED ACETYLCHOLINE

The use of isotopic techniques for choline acetylase assay offers undisputable advantages to other methods with respect to sensitivity and accuracy.   Sensitive methods are often required when the topical distribution of the enzyme in the nervous system is studied because determinations have to be carried out on small amounts of tissue.   To give some indication of the sensitivity of the reported isotopic methods, the blank values given by the different authors have been calculated in this review as μM of acetylcholine formed in the enzymatic reaction and the volume of the assay system has also been stated.   (The blank values are the radioactivity found in incubated samples from which choline or the enzyme have been omitted and include the "background" counts.) All isotopic methods reported use ^{14}C-labeled acetylcoenzyme A (or ^{14}C- or ^{3}H-labeled acetate in a coupled system) as substrate and measure the labeled acetylcholine formed.   This is due to the fact that acetylcholine can easily be separated from acetyl-coenzyme A or acetate. Choline acetylase assay by separation of labeled acetylcholine with labeled choline as substrate is not feasible, because the separation involves several steps (86).

1. Acetylcholine as a Nonvolatile Product

The first method based on an isotope technique for the assay of choline acetylase was described by Schuberth (61).   It took advantage of the fact that acetylcholine is a nonvolatile compound and can therefore easily be separated from the labeled precursor, ^{14}C-acetate, by evaporation of the acidified sample.   The ^{14}C-acetate is transformed to ^{14}C-acetyl-coenzyme A in the assay by the action of added acetylcoenzyme A synthetase (see Section III).   The blank value (choline omitted) corresponds to about 5×10^{-3} μM of acetylcholine with crude extracts of brain acetone powder as the source of the choline acetylase.   The volume of the assay system is 100 μl.

2. Precipitation of Acetylcholine

The precipitation of acetylcholine and other quaternary ammonium compounds as the reineckate is a well-known procedure.   As acetyl-coenzyme A is not precipitated by ammonium reineckate, choline acetylase may be determined by this principle as first described by McCaman and Hunt (30). ^{14}C-Labeled acetylcoenzyme A was used as substrate, and the acetylcholine reineckate obtained (as a coprecipate with choline reineckate) was dissolved in acetone and its radioactivity determined by liquid scintillation counting. Various modifications of the original method have been described (87–90), including the use of ^{14}C-acetate in the presence of acetylcoenzyme A synthetase as a source of acetyl-coenzyme A (87,89). The method can be made very sensitive by working on a microscale (88). The reineckate precipitation methods have, however, certain disadvantages. Ammonium reineckate is not a specific reagent for precipitation of acetylcholine, which may explain why the blank values found with these methods are fairly high, often corresponding to 1–2% of the total radioactivity present in the enzymatic assay (30). Furthermore, the reineckate ion is colored and gives a strong color quench effect in the liquid scintillation counting step unless very small amounts of the reineckate precipitate are used. The color quench effect is obviated if the radioactivity is measured with a Geiger-counter (87) but this is a minor improvement because self-adsorption effects appear. The blank value (enzyme omitted) of the original method (30) corresponds to $7 \times 10^{-6} \mu$M of acetylcholine formed in a volume of 11 μl. Buckley et al. (88) modified this method to the assay of choline acetylase in single cells from the sympathetic ganglion of the cat and claimed that the method would detect $1 \times 10^{-7} \mu$M of acetylcholine formed in an assay volume of 1 μl. However, the blank values reported by these authors correspond to $1 \times 10^{-6} \mu$M of acetylcholine formed.

Fonnum (57) described a method for choline acetylase determination based on the precipitation of acetylcholine as a tetraphenylborate complex. Sodium tetraphenylborate (Kalignost) was used as the precipitating agent and the precipitate was dissolved in acetonitrile-benzyl alcohol, which was added to a toluene-based liquid scintillation system. Very little quenching was obtained as the tetraphenylborate ion is colorless. However, acetyl-coenzyme A is apparently also carried down with the precipitate to a certain extent and special precautions were necessary in order to obtain satisfactory blank values. ^{14}C-Acetate in the presence of acetylcoenzyme A synthetase was thus used as a source of labeled acetylcoenzyme A, and the incubated assay samples were treated with hydroxylamine prior to the precipitation step in order to deacylate the labeled acetylcoenzyme A present in the samples.

It should be noted that acetyl-coenzyme A synthetase has a fairly low affinity for acetate (91), which necessitates the use of fairly high concentrations of acetate in the assay. This obviously limits the sensitivity of the method. The blank values obtained corresponds to $2 \times 10^{-2} \mu$M of acetylcholine formed in an assay volume of 2.5 ml (57).

Acetylcholine may also be precipitated as the periodide, a method which was recently applied by Goldberg et al. (90) to a modification of the choline acetylase assay of McCaman and Hunt (30). The precipitate was dissolved in ethanol-acetone and added to a scintillator solution containing Hyamine. The latter decomposes the brown-colored periodide ion to give a colorless solution and thus obviates the color quenching found in the original method. This makes the modified method also adaptable to a larger scale analysis. Furthermore, the modified method was found to give lower blank values. Those found by Goldberg et al. (90), corresponded to about $2 \times 10^{-6} \mu M$ of acetylcholine formed in an assay volume of 11 μl.

From the foregoing it appears that choline acetylase assay by isolation of labeled acetylcholine with precipitating reagents has become quite popular in recent years. The reason for this may be that these methods at first appear rather simple to perform, but we are of the opinion that the proper handling of small amounts of precipitate is fairly laborious and demands a high degree of experience.

3. Extraction of Acetylcholine

Fonnum recently reported (92,93) that the tetraphenylborate complex of acetylcholine is soluble in certain organic solvents and can be extracted from a water solution with ethyl butyl ketone. Acetylcoenzyme A is not extracted into the ketone phase. This procedure was applied to choline acetylase assay using ^{14}C-acetyl-coenzyme A as substrate. This method for choline acetylase determination may offer an attractive alternative to the precipitation methods described earlier.

4. Ion-Exchange Chromatography

Acetylcholine is cationic, whereas acetyl-coenzyme A is anionic. If labeled acetyl-coenzyme A is used as substrate, the labeled product obtained in the enzymatic reaction can conveniently be separated from the remaining substrate (and any labeled acetate obtained by hydrolysis of acetyl-coenzyme A) by ion-exchange chromatography. Schrier and Schuster (33) have described a method where the substrate is [1-^{14}C] acetyl-coenzyme A, which is removed at the end of the reaction on a column of an anion-exchange resin, and the labeled acetylcholine obtained in the effluent is determined by liquid scintillation counting. The method was successfully applied to the determination of choline acetylase from mouse brain, *Lactobacillus plantarum*, and chick spinal cord (94). An approach by Diamond and Kennedy (95) used ^3H-acetate and coenzyme A in the presence of acetyl-coenzyme A synthetase as the source of acetyl groups, and adsorbed the acetylcholine formed on a column of a cation-exchange resin. Acetylcholine was then eluted with hydrochloric acid and determined by liquid scintillation counting. A disadvantage of this method is that because acetylcholine is eluted in a fairly large volume of a strong salt solution, only an aliquot of the effluent can be used in the counting step.

The following is a description of a modification of the Schrier and Schuster method which has been developed and successfully used for the assay of choline acetylase from bovine *Nucleus caudatus* (40).

Reagents. *Potassium phosphate buffer*, 0.5*M* (pH 7.0) (giving a final pH in the system of 6.8).

[*1-¹⁴C*] *Acetyl-coenzyme* A, 3 mM (about 0.5 mCi/mM).

Choline chloride, 0.05*M*.

NaCl, 3*M*.

Eserine salicylate, 0.01*M*.

n-*Butanol*, 5% in water.

Dowex 1-X2 (200–400 mesh), chloride form. The commercial product is washed with 1*M* NaOH (until the washings are free from chloride ions) then with water, 1*M* HCl, and water.

"*Scintillation cocktail:*" 2 volumes of toluene, containing 4.0 g of PPO and 0.1 g of POPOP/1000 ml, is mixed with 1 volume of Triton X-100.

Procedure. The following are placed in a semimicro centrifuge tube: 40 μl of phosphate buffer, 5 μl of eserine, 10 μl of butanol solution, 20 μl of choline, 10 μl of acetyl-coenzyme A, and water to give a final volume of 200 μl. The tube is preincubated for 5 min at 30°C and the reaction is then started by the addition of the enzyme (5–40 μl). After 20 min incubation at 30°C, the contents of the reaction tube are rapidly transferred to a column 0.5 × 4 cm of Dowex 1 (chloride) made up in a 23-cm Pasteur pipet, which contains a small plug of glass wool in its constricted end. The incubation tube is immediately washed with 0.5 ml of water, which is run into the column. The reaction tube is then washed again with 1.0 ml of water, which is also applied to the column. The effluent from the latter is directly collected into a scintillation vial, containing 10 ml of "scintillation cocktail." Liquid scintillation counting is then carried out by conventional methods.

Remarks. This modified procedure has been changed from the original one in the following respects. The reaction is carried out at pH 6.8, which is closer to the pH optimum of the enzyme than the pH 6.0 used by Schrier and Schuster. The choline concentration has been reduced 10-fold and the acetyl-coenzyme A 2-fold from the original ones, whereas the NaCl concentration is twice that of the original procedure (cf. 11 and 94). Furthermore, butanol has been included in the assay because it was found to activate the enzyme from *Nucleus caudatus* and eserine is used instead of neostigmine for inhibition of cholinesterase. The reaction is carried out at 30°C, following the recommendations of IUB, and the reaction time is only $\frac{1}{3}$ that used by Schrier and Schuster. Zero-order kinetics is obtained for this period of time, provided that less than 15% of the acetyl-coenzyme A is utilized. Scintillation counting is carried out in the Triton X-100 system (96) which is more convenient to use than the dioxane-anisole-dimetoxyethane system of the original procedure. The blanks (enzyme omitted) contain less than 0.1% of the radioactivity added to the system. The blank values, including background counts, correspond to $1 \times 10^{-4}\,\mu M$ of acetylcholine. According to our experience, 20 samples may conveniently be run consecutively with a time interval of 1 min. The time required for application of a sample to the column is usually negligible, but if precise timing of the incubation is required, the enzymic reaction can be stopped by heating the reaction tubes in a boiling water bath for 2 min (33).

5. Electrophoresis

As previously mentioned, acetylcholine is cationic in contrast to acetyl-coenzyme A and can also be separated from the latter by paper electrophoresis. Assay procedures for choline acetylase based on this principle have recently been described (11,73) and applied to the rat brain enzyme and to the enzyme found in the abdominal ganglion of the marine mollusk Aplysia. ^{14}C-Acetyl coenzyme A is used as substrate and the assay system is directly applied to the electropherogram. After electrophoresis in $1N$ acetic acid (11) or buffer of pH 4.7 (73) the area containing acetylcholine is localized with iodine vapor and cut out. Its radioactivity is then determined by liquid scintillation counting. The yield of acetylcholine is reported to be quantitative, and enzyme blanks contain less than 0.1% of the radioactivity added (11). The blank value, including background counts, correspond to about $1 \times 10^{-5}\,\mu M$ of acetylcholine with an assay volume of 40 μl (11). These methods appear to be precise and not laborious but only small volumes of the assay system (10–20 μl) can be applied as a spot to the paper. Unless incubations are carried out on a microscale, a large fraction of the assay system is thus wasted.

References

1. C. J. Smith, C. J. Cavallito, and F. F. Foldes, *Biochem. Pharmacol.*, *16*, 2438 (1967).
2. B. O. Persson, L. Larsson, J. Schuberth, and B. Sörbo, *Acta Chem. Scand.*, *21*, 2283 (1967).
3. M. E. Goldberg and V. B. Ciofalo, *Psychopharmacol.*, *14*, 142 (1969).
4. D. Nachmansohn and A. L. Machado, *J. Neurophysiol.*, *6*, 397 (1943).
5. D. Nachmansohn and M. S. Weiss, *J. Biol. Chem.*, *172*, 677 (1948).
6. R. S. Comline, *J. Physiol.* (London), *105*, 6P (1946).
7. M. Stephenson, E. Rowatt, and K. Harrison, *J. Gen. Microbiol.*, *1*, 279 (1947).
8. S. R. Korey, B. de Braganza, and D. Nachmansohn, *J. Biol. Chem.*, *189*, 705 (1951).
9. F. Lynen, E. Reichert, and L. Rueff, *Ann. Chem.*, *574*, 1 (1951).
10. S. Korkes, A. del Campillo, S. Korey, J. Stern, D. Nachmansohn, and S. Ochoa, *J. Biol. Chem.*, *198*, 215 (1952).
11. L. T. Potter, V. A. S. Glover, and J. K. Saelens, *J. Biol. Chem. 243*, 3864 (1968).
12. D. Morris, *Biochem. J.*, *98*, 754 (1966).
13. J. Schuberth, *Biochim. Biophys. Acta*, *122*, 470 (1966).
14. A. K. Prince, *Proc. Nat. Acad. Sci. U. S.*, *57*, 1117 (1967).
15. K. N. Mehrotra, *J. Insect Physiol.*, *6*, 215 (1961).
16. W. Feldberg, *Methods Med. Res.*, *3*, 95 (1950).
17. D. Nachmansohn and I. B. Wilson, *Methods Enzymol.*, *1*, 619 (1955).
18. C. O. Hebb, *Physiol. Rev.*, *37*, 196 (1957).
19. B. N. Smallman, *Methods Med. Res.*, *9*, 203 (1961).
20. D. Nachmansohn, in *Handbuch der Experimentellen Pharmakologie, Ergänzungswerk XV*, G. B. Koelle, Ed., Springer Verlag, Berlin, 1963, p. 40.
21. C. O. Hebb, in *Handbuch der Experimentellen Pharmakologie, Ergänzungswerk XV*, G. B. Koelle, Ed., Springer Verlag, Berlin, 1963, p. 55.
22. W. Hardegg, in *Hoppe-Seyler/Thierfelder Handbuch der physiologisch- und pathologischchemischen Analyse*, Vol. VIB, K. Lang and E. Lehnartz, Eds., Springer Verlag, Berlin, 1966, p. 861.
23. R. Berman, I. B. Wilson, and D. Nachmansohn, *Biochim. Biophys. Acta*, *12*, 315 (1953).
24. A. S. V. Burgen, G. Burke, and M. L. Desbarats-Schonbaum, *Brit. J. Pharmacol.*, *11*, 308 (1956).
25. W. C. Dauterman and K. W. Mehrotra, *J. Neurochem.*, *10*, 113 (1963).
26. B. H. Hemsworth and D. Morris, *J. Neurochem.*, *11*, 793 (1964).
27. J. F. Berry and V. P. Whittaker, *Biochem. J.*, *73*, 447 (1959).
28. B. N. Smallman, *J. Neurochem.*, *2*, 119 (1958).
29. D. Morris and S. Tucek, *J. Neurochem.*, *13*, 333 (1966).
30. R. E. McCaman and J. M. Hunt, *J. Neurochem.*, *12*, 253 (1965).
31. D. Morris, *J. Neurochem.*, *14*, 19 (1967).
32. E. J. Simon and D. Shemin, *J. Amer. Chem. Soc.*, *75*, 2520 (1953).
33. B. K. Schrier and L. Shuster, *J. Neurochem.*, *14*, 977 (1967).
34. R. Berman-Reisberg, *Yale J. Biol. Med.*, *29*, 403 (1957).
35. J. Bové, R. O. Martin, L. L. Ingraham, and P. K. Stumpf, *J. Biol. Chem.*, *234*, 999 (1959).
36. F. Lynen, I. Hopper-Kessel, and H. Eggerer, *Biochem. Z.*, *340*, 95 (1964).
37. C. F. Fox and E. P. Kennedy, *Anal. Biochem.*, *18*, 286 (1967).

38. I. B. Wilson, *J. Amer. Chem. Soc.*, *74*, 3205 (1952).
39. B. Eriksson, *Acta Chem. Scand.*, *20*, 1178 (1966).
40. B. Mannervik and B. Sörbo, Biochem. Pharmacol., *19*, 2509 (1970).
41. D. H. Jones and W. L. Nelson, *Anal. Biochem.*, *26*, 350 (1968).
42. L. T. Potter and W. Murphy, *Biochem. Pharmacol.*, *16*, 1386 (1967).
43. E. R. Stadtman, *Methods Enzymol.*, *3*, 935 (1957).
44. D. J. Pearson, *Biochem. J.*, *95*, 23C (1965).
45. F. Lipman and L. L. Tuttle, *J. Biol. Chem.*, *159*, 21 (1945).
46. G. L. Ellman, Arch. *Biochem. Biophys.*, *82*, 70 (1959).
47. S. E. Severin and V. Artenie, *Biochimija*, *32*, 125 (1967).
48. B. Mannervik and B. Sörbo, *FEBS VI Meeting, Madrid* (1969).
49. C. O. Hebb and B. N. Smallman, *J. Physiol.* (London), *134*, 385 (1956).
50. C. O. Hebb and B. N. Smallman, *J. Physiol.* (London), *134*, 718, (1956).
51. C. O. Hebb and D. Ratkovic, *J. Physiol.* (London), *163*, 307 (1962).
52. C. O. Hebb, K. Krnjevic, and A. Silver, *J. Physiol.* (London), *171*, 504 (1964).
53. E. R. Stadtman, *J. Biol. Chem.*, *196*, 535 (1952).
54. J. Schuberth, Unpublished observations.
55. A. S. Milton, *J. Physiol.* (London), *142*, 25P (1958).
56. W. W. Cleland, *Biochim. Biophys. Acta*, *67*, 104 (1963).
57. F. Fonnum, *Biochem. J.*, *100*, 479 (1966).
58. N. O. Kaplan and F. Lipman, *J. Biol. Chem.*, *147*, 37 (1948).
59. C. O. Hebb, *Quart. J. Exp. Physiol.*, *40*, 176 (1955).
60. C. O. Hebb and A. Silver, *J. Physiol.* (London), *134*, 718 (1956).
61. J. Schuberth, *Acta Chem. Scand.*, *17*, S233 (1963).
62. M. Cohen, *Arch. Biochem.*, *60*, 284 (1956).
63. C. O. Hebb and V. P. Whittaker, *J. Physiol.* (London), *142*, 187 (1958).
64. V. P. Whittaker, I. A. Michaelson, and R. J. Kirkland, *Biochem. Pharmacol.*, *12*, 300 (1963).
65. E. DeRobertis, G. Rodriguez, L. Salganicoff, A. Pellegrino DeIraldi, and L. M. Zicher, *J. Neurochem.*, 10, 225 (1963).
66. F. Fonnum, *Biochem. J.*, *103*, 262 (1967).
67. F. Fonnum, *Biochem. J.*, *109*, 389 (1968).
68. G. Bull, C. Hebb, and D. Ratkovic, *Biochim. Biophys. Acta*, *67*, 138 (1963).
69. R. E. McCaman, G. Rodriguez de Lores Arnaiz, and E. DeRobertis, *J. Neurochem.*, *12*, 927 (1965).
70. S. Tucek, *Biochem. Pharmacol.*, *16*, 109 (1967).
71. G. Buckley, S. Consolo, E. Giacobini, and R. McCaman, *Acta Physiol. Scand.*, *71*, 341 (1967).
72. P. C. Marchisio and S. Consolo, *J. Neurochem.*, *15*, 759 (1968).
73. E. Giller, Jr., and J. H. Schwartz, *Science*, *161*, 908 (1968).
74. I. Nordenfelt, *Quart. J. Exp. Physiol.* *48*, 67 (1963).
75. O. Dann, H. Roselieb, and H. Sucker, *Ann. Chem.*, *710*, 161 (1967).
76. R. W. Morris, *Biochim. Biophys. Acta*, *73*, 511 (1963).
77. F. C. MacIntosh and W. L. M. Perry, *Methods Med. Res.*, *3*, 78 (1950).
78. J. Crossland, *Methods Med. Res.*, *9*, 125 (1961).
79. V. P. Whittaker, in *Handbuch der Experimentellen Pharmakologie, Ergänzungswerk XV*, G. B. Koelle, Ed., Springer Verlag, Berlin, 1963, p. 2.
80. I. H. Stockley, *J. Pharm. Pharmacol.*, *21*, 302 (1969).
81. F. Lynen, *Fed. Proc.*, *12*, 683 (1952).
82. S. Hestrin, *J. Biol. Chem.*, *180*, 249 (1949).

83. D. Nachmansohn, S. Hestrin, and H. Voripaieff, *J. Biol. Chem.*, *180*, 875 (1949).
84. G. Holan, *Nature*, *206*, 311 (1965).
85. J. H. Fellman, *J. Neurochem.*, *16*, 135 (1969).
86. J. Schuberth, B. Sparf, and A. Sundwall, *J. Neurochem.*, *16*, 695 (1969).
87. A. Alpert, R. L. Kisliuk, and L. Shuster, *Biochem. Pharmacol.*, *15*, 465 (1966).
88. G. Buckley, S. Consolo, E. Giacobini, and R. McCaman, *Acta Physiol. Scand.*, *71*, 341 (1967).
89. G. J. Maletta and P. S. Timiras, *J. Neurochem.*, *15*, 787 (1968).
90. A. M. Goldberg, A. A. Kaita, and R. E. McCaman, *J. Neurochem.*, *161*, 823 (1969).
91. L. T. Webster and F. Campagnari, *J. Biol. Chem.*, *237*, 1050 (1962).
92. F. Fonnum, *Biochem. Pharmacol.*, *17*, 2503 (1968).
93. F. Fonnum, *Biochem. J.*, *113*, 291 (1969).
94. A. M. Burt, *J. Exp. Zool.*, *169*, 107 (1968).
95. I. Diamond and E. P. Kennedy, *Anal. Biochem.*, *24*, 90 (1968).
96. M. S. Pattersson and R. C. Greene, *Anal. Chem.*, *37*, 854 (1965).

Author Index

Numbers in parentheses are reference numbers and show that an author's work is referred to although his name is not mentioned in the text. Numbers in *italics* indicate the pages on which the full references appear.

A

Aarsen, P. N., 69, *86*
Aarseth, P., *270*
Abbs, E. T., 169(230), *181*
Abdel-Rahman, M. A., 184(8), *212*
Abood, L. G., 50, *85*, 173(248), *182*
Abraham, D., 143(84), *151*
Acerboni, F., 238(157), *268*
Ackermann, D., 190(78), *214*
Ackermann, E., 157(98), *178*
Ackermann, H., 157(61), *176*
Adamstone, F. B., 62, *86*
Adie, P. A., 235(138), 237, *268*
Agioutantis, G., 223(29), *265*
Ahmed, A., 198(119), *215*
Aksnes, G., 254, *271*
Alberici, M., 169(204), 170(204), 173(204), *180*
Albers, R. W., 169(218), *181*, 246(214), 248(214), *270*
Albers, W., 169(209), 170(209), 173(209), *181*
Alcaraz, A. F., 126(40), 127(40), *150*
Alkon, D. L., 189(66), 196(66), *214*
Allen, N., 120(1), *149*
Alpert, A., 290(87), *296*
Altland, K., 218(5), 254(5), 255(288), 262(5), *264*, *271*
Alton, H., 120(8), *149*
Alton, J., 120(9), *149*
Anagnoste, B., 155(31), 160(150), *176, 179*
Anden, N. E., 14(1), *32*, 147, 148, *152*, 157(67,89), *177*
Anderson, P. J., 169(192), 170(192,239), *180, 181*
Andres, V., Jr., 244–246(209), 249(209), *269*
Andreyev, V. S., 230(86), *266*
Ansell, G. B., 122(19), 123, *150*
Anton, A. H., 123, 125, 130, 132, 134(61), 136(33), *150, 151*

Aoyama, I., 259(308), *272*
Aprison, M. H., 185, 188(34,152), 189(62), 197(112), 198(34,112), 199(34,62,112), *213–216*
Ara, R., 184(12,19), *213*
Archer, T. E., 264(352), *273*
Armstrong, J. C., 26(24), *33*
Armstrong, M. D., 144(89), *151*
Arqueros, L., 155(16), 160(16,122,140), 163(16,122,140), 165(122,140), 174(122), *175, 178, 179*
Artenie, V., 279(47), *295*
Arvidsson, U. B., 74, *87*
Ashcroft, G. W., 22(2), 23(2), *32*
Assicot, M., 169(213,238), 170(213), 173(213), *181*
Auditore, J. V., 223(35), *265*
Augstein, J., 160(152), *179*
Augustinsson, K.-B., 190(76), 191(83), 192(76), 200(76), *214*, 217, 218(1,3), 222(8–10,13), 223(13,18,23), 224(13), 228–230(13), 232(13), 234(13), 235(23), 238(13), 249(13), 250(13), 258(13), 259(3,8,10), 261(13,328), 263(13), 264(13), *264, 265, 272*
Aures, D., 60, 61, *86*, 101, 105–107, *117*, 155(42), 157(58), 158(110), *176, 178*
Austin, L., 160(145), 165(145), *179*
Awapara, J., 157(59,60,77), 158(60,77), *176, 177*
Axelrod, J., 2(4), 8(71), 14(71), 25(3,86), 31, *32, 34*, 55, 56, *86*, 95(10,11), *97*, 138(74), 140(80), *151*, 155(14), 157(84), 160(73,139), 163(139), 166(169–172,178,180,183–185), 167(170,178,188), 168, 169(190, 191,194–196,200,201,205,209,212, 218,224,225,227,241), 170(209, 225,241), 172, 173(209,212,224, 241,259), 174(265), *175, 177, 179–182*
Ayukawa, S., 155(29), 160(153), *176, 179*

297

G

Gaballah, S., 257(305), *272*

Gaddum, J. H., 188(109), 197(109), 199, *215*

Gage, J. C., 223(19), *265*

Gal, E. M., 26(24), *33*, 242(194,195), 243 (195), *269*

Gal, G., 157(101), *178*

Gale, E. F., 100(9), *117*

Gale, P. H., 169(203), *180*

Galehr, O., 188(153), *216*

Gallo, M. A., 233(119), *267*

Gamson, R. M., 264(351), *273*

Ganelin, R. S., 230(81), *266*

Gang, H., 155(31), *176*

Ganrot, P. O., 100(4), *116*

Garattini, S., 2(25), 12(25), *33*

Gardier, R. W., 169(237), *181*

Gardiner, J. E., 188(81), 190(81), *214*

Garry, P. J., 244, 245(212), 246, *270*

Garvey, T. Q., III, 160(134), *179*

Gaudin, D., 155(46), *176*

Gavin, G., 37(10), 65, *84*

Gazzaniga, P., 239, 241(186), *269*

Geffen, L. B., 161(158), *179*

Geissbühler, H., 244(223), 247, *270*

Gelberg, A., 254(275), *271*

Geldmacher-von Mallinckrodt, M., 260 (320), *272*

Gend, H. W. van, 247, 248(229), *270*

Genest, K., 254(278), *271*

Gerarde, H. W., 232(114), *267*

Gerber, C., 200(138), *216*

Gerez, C., 78, 79(110), *87*

Gerst, E. C., 128(49), 137, *150*

Gey, K. F., 38(30), *85*, 155(17), 157(100), *175, 177, 178*

Giacobini, E., 225, 226, 228(67), 257, *265, 266, 272*, 285(71), 290(88), *295, 296*

Giarman, N. J., 173(243), *181*, 184(22), 191(22), 192(22), 201(22), *213*

Gibb, J. W., 155(28), 156(28), 161(157), *176, 179*

Gibson, S., 154(12), 155(12,27), *175, 176*

Giglio, M., 238(157), *268*

Giller, E., Jr., 285(73), 293(73), *295*

Gillson, R. E., 18, *32*

Giltrow, J., 15(16), *32*

Ginsberg, B., 26(24), *33*

Ginsburg, S., 239, 241(178), *269*

Giordano, C., 38, 39(31), *85, 177*

Gitter, S., 242(188), *269*

Gjessing, L. R., 143(86,88), 144(88), 147, *151, 152*

Gladish, Y. C., 159(112), *178*

Glaesser, A., 157(54), *176*

Glamkowski, E. J., 157(101), *178*

Glazer, R., 264(350), *273*

Glenner, G. G., 61, 62, *86*

Glick, D., 14(58), *33*, 68, *86*, 91(4), 93(4), *97*, 114, 115, *117*, 189(61), *214*, 225, 226, *265*

Glover, V. A. S., 276(11), 278–282(11), 285(11), 288(11), 293(11), *294*

Glowinski, J., 160(73), *177*

Goa, J., 48, *85*

Goedde, H. W., 218, 254, 255, 262(5), *264, 271*

Goldberg, A., 191(86), *214*

Goldberg, A. M., 189(69), *214*, 290(90), 291, *296*

Goldberg, M. E., 276(3), 281(3), *294*

Goldenberg, M., 136(67), *151*

Golder, R. H., 66, *86*

Goldstein, M., 61, *86*, 154(11), 155(31), 160(119,121,127,133–135,138,146–148,150,151), 161(119,127,138), 162(121), 163(138), *175, 176, 178, 179*

Golz, H. H., 223(17), 232(114), *265, 267*

Gomes, B., 40, *85*, 173(244), *182*

Goodall, McC., 120(7–9), 126, 127(38), 139(7), *149, 150*

Goodson, L. H., 243(203), 246(203), *269*

Goodwin, W. J., 230(76), *266*

Gordon, C., 234(142), 235(142), 237(142), *268*

Gordon, E. K., 120(6), *149*

Gordon, R., 155(19), *175*

Gorkin, V. Z., 173(251), *182*

Goryachenkova, E. V., 65, *86*

Goshev, A. J., 238(163), *268*

Gosselin, L., 169(226), *181*

Goto, K., 256(301), *272*

Graham, R. C., 63, *86*

Grahame-Smith, D. G., 26(26), *33*

Grahn, B., 115, *117*

Subject Index

Methods of Biochemical Analysis

CUMULATIVE INDEX, VOLUMES 1–19 AND SUPPLEMENT

Author Index

CUMULATIVE INDEX, VOLUMES 1–19 AND SUPPLEMENT

Subject Index